it

45° N

RUSSIA

idō

40° N

Japan Trench

35° N

30° N

25° N

145° E

The Japanese archipelago

[adapted from NOAA and Google Earth imagery] NB latitude and longitude lines are approximate

Offshore isla	
1	Rebun-tō
2	Rishiri-tō
3	Yagishiri-tō
4	Teuri-tō
5	Okushiri-tō
6	Ōshima (Oshima-Ōshima)
7	Kojima (Oshima-kojima)
8	Tobi-shima
9	Awa-shima
10	Sado-shima (Sadoga-shima)
11	Oki islands: Dōgo
12	Oki Islands: Dōzen (Chiburi-jima, Nakano-shima, Nishino-shima)
13	Mi-shima
14	Tsushima
15	Iki (Ikino-shima)
16	Gotō-rettō
17	Danjo-guntō
18	Koshikishima-rettō

Nansei Shotō	
19	Kuro-shima
20	Iwō-jima, Take-shima
21	Tanega-shima
22	Yaku-shima
23	Kuchino-shima & Nakano-shima
24	Suwanose-jima & Akuseki-jima
25	Amami Ōshima
26	Kikai-jima
27	Tokuno-shima
28	Okinoerabu-jima
29	Yoron-jima
30	Iheya-jima, Izena-jima
31	Aguni-jima
32	Tonaki-jima

37	Tarama-jima
38	Ishigaki-jima
39	Iriomote-jima
40	Yonaguni-jima
41	Hateruma-jima

42	Minamidaitō-jima
43	Kitadaitō-jima

Izu Islands	
44	Ōshima
45	To-shima
46	Nii-jima
47	Kōzu-shima
48	Miyake-jima
49	Mikura-jima
50	Hachijō-jima
51	Aoga-shima

52	Tori-shima

Ogasawara Islands	
53	Nishino-shima
54	Muko-jima
55	Chichi-jima
56	Haha-jima

Iwō Islands	
57	Kita Iwō-jima
58	Iwō-jima
59	Minami Iwō-jima

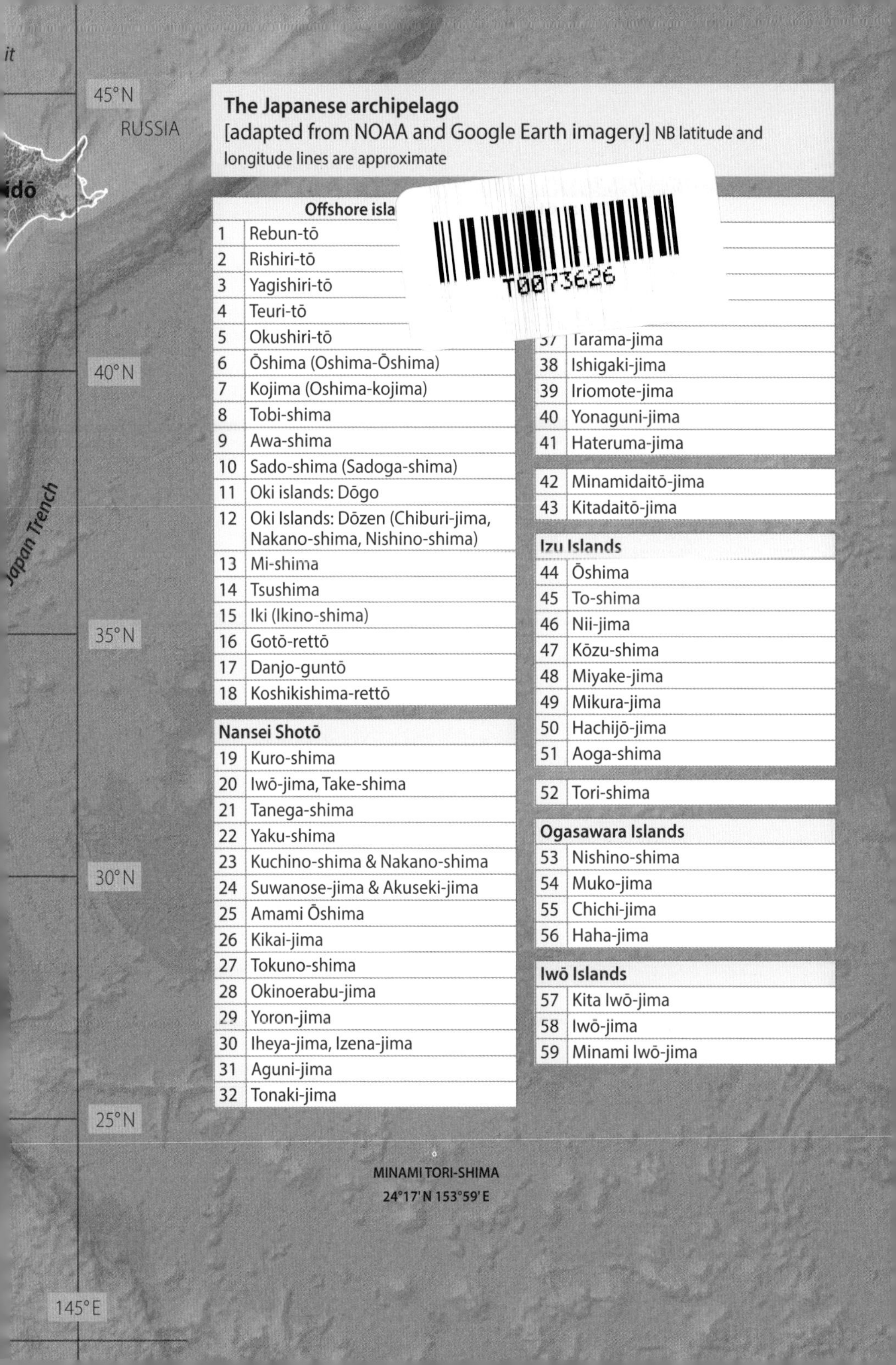

MINAMI TORI-SHIMA
24°17' N 153°59' E

JAPAN

THE NATURAL HISTORY OF AN ASIAN ARCHIPELAGO

MARK BRAZIL

WILD*Guides*

PRINCETON
press.princeton.edu

Published by Princeton University Press,
41 William Street, Princeton, New Jersey 08540
6 Oxford Street, Woodstock, Oxfordshire OX20 1TR
press.princeton.edu

First published 2022

British Library Cataloging-in-Publication Data is available

Library of Congress Control Number 2021944387
ISBN 978-0-691-17506-5
ISBN (ebook) 978-0-691-23097-9

Production and layout by **WILD***Guides* Ltd., Old Basing, Hampshire UK.
Printed in Italy

10 9 8 7 6 5 4 3 2 1

Front cover Japanese Macaque [HT]
Frontispiece Mt Fuji at dawn [CC]

Contents

Dedicated to the memories of:

Engelbert Kämpfer (1651–1716)
Carl Peter Thunberg (1743–1828)
Philipp Franz Balthasar von Siebold (1796–1866) and
Thomas Wright Blakiston (1832–1891)

each of whom did so much to teach the world about the
fascinating natural history of Japan

Facing page The endemic Japanese Serow [HT]

Preface

I feel tremendously privileged to have first been able to visit the islands of Japan while I was a post-graduate student, now four decades ago. I have been fascinated by Japan's natural heritage and natural history ever since. I found the landscapes and seasons of Japan to be attractive and appealing, and its biodiversity and geodiversity extraordinary. At the same time, I found the approaches of the Japanese people to the natural world around them to be inspiring, while also paradoxical and bewildering. My fascination with the country and its wildlife grew as I explored the archipelago, from the high mountains to the coasts, from northernmost Hokkaidō to the southernmost of the Nansei Shotō Islands and the Ogasawara Islands, and encountered more and more of the species that are at home there. A major allure that Japan holds for me is the considerable number of endemic species and subspecies found here, a consequence of the long geographical isolation that the archipelago has experienced as a whole and, more specifically, that of each subsidiary island group. The extent and length of the isolation experienced by life on these islands have been so great as to render Japan a veritable Galápagos of East Asia.

My own early explorations were hampered by an almost complete lack of guidebooks in English indicating places suitable for watching wildlife and helping with the identification of the species occurring there. There were few resources that could help the overseas visitor unable to read Japanese to understand the natural heritage, natural history and ecology of this special group of islands. That lack of information inspired me to write site guides and field guides – the very kind of books which I had wished were available during my first explorations of Japan. This new book is the culmination of my final dream, which is to make Japan's natural history interesting and accessible to even the most casual of visitors and travellers to these islands. It is exactly the kind of book I wished I could have read when I first visited. I hope that the personal and non-technical style will make the subject matter interesting, approachable and informative to the novice naturalist and specialist alike, and that it will allow another generation of Japan-obsessed naturalists to delve much further and deeper into Japan's fascinating natural history.

Japan is widely perceived as an intensively urbanized and highly industrialized techno-marvel, complete with environmentally destructive policies, yet the country hosts a delightful array of wild creatures to excite any visiting naturalist (as I have previously described in *The Nature of Japan*). As a temperate island archipelago Japan cannot match large continental or tropical areas for their natural diversity, and Japan's range of animals and plants does face threats from the overwhelming exploitation of both the marine and the terrestrial environments, yet, as you delve into this book, you will discover that it is home to many interesting, intriguing and even iconic species. Japan can relate some major conservation success stories, as exemplified by the recovering populations of endangered species such as Short-tailed Albatross [VU][1], Red-crowned Crane [VU], Oriental Stork [CR] and Crested Ibis [EW]. At the same time, it battles with growing conflict between people and certain common wildlife species in rural farming areas. As human populations in rural areas decline, mammals such as the Japanese Deer[2], Japanese Macaque and Wild Boar become emboldened, expand their ranges and put pressure on farmland where there are now too few people to deter them.

In this book, I introduce the natural heritage, natural history and wildlife of Japan, and reveal where Japan's relationship with nature shines and where it is tarnished. The structure of the book offers a progression of chapters describing the underlying form and functioning of the country, the elements that have shaped and continue to shape its natural history, and which concern the identity

1 The Ministry of the Environment's (2002) categories for populations of rare species in Japan include Extinct in the Wild [EW], Critically Endangered [CR], Endangered [EN], Vulnerable [VU] and Near Threatened [NT]. These categories are indicated in this book when a species is first mentioned.

2 The widespread English name Sika (pronounced *see-ka*) for this species of deer is derived from a confusing mis-transliteration of the Japanese name, which is pronounced *sh-ka* or *she-ka*. Given that the Japanese word *sh-ka* means deer, Sika Deer is tautological; hence throughout this book I have referred to it as the Japanese Deer.

and fascinating characteristics of individual species. This book is designed also for dipping into. Short stories elucidate aspects of the country's natural history and illuminate the ecology and behaviour of some highlighted species.

My approach in this book is centred largely on terrestrial wildlife, as birds and mammals have always been my own primary interest and provide the main extent of my experience. Given the size constraints of a volume of this type, I have made no attempt to turn this book into a complete field guide. For those looking to understand Japan's avifauna, my field guide *Birds of Japan* fills that niche. Those wanting a guide to the mammalian fauna need look no further than *A Guide to the Mammals of Japan* and *The Wild Mammals of Japan*, both of which are in English.

The country is now well provided with field guides in Japanese for most groups of organisms, including plants, freshwater invertebrates, terrestrial insects (especially butterflies and dragonflies), marine fish, reptiles and amphibians. I have therefore recommended a number of these references in the bibliography for those who wish to explore beyond the pages of this book.

As many of the species mentioned here are likely to be unfamiliar to readers, I have included their scientific names in the index, after their English names. I have made no attempt to reference every fact, but have provided an extensive bibliography and list of recommended reading.

I hope that this book will provide you with insights into the natural history of a fascinating archipelago and will encourage you to explore these islands with fresh and open eyes.

Subtropical forest can be found on the islands of the southwestern archipelago [TsM].

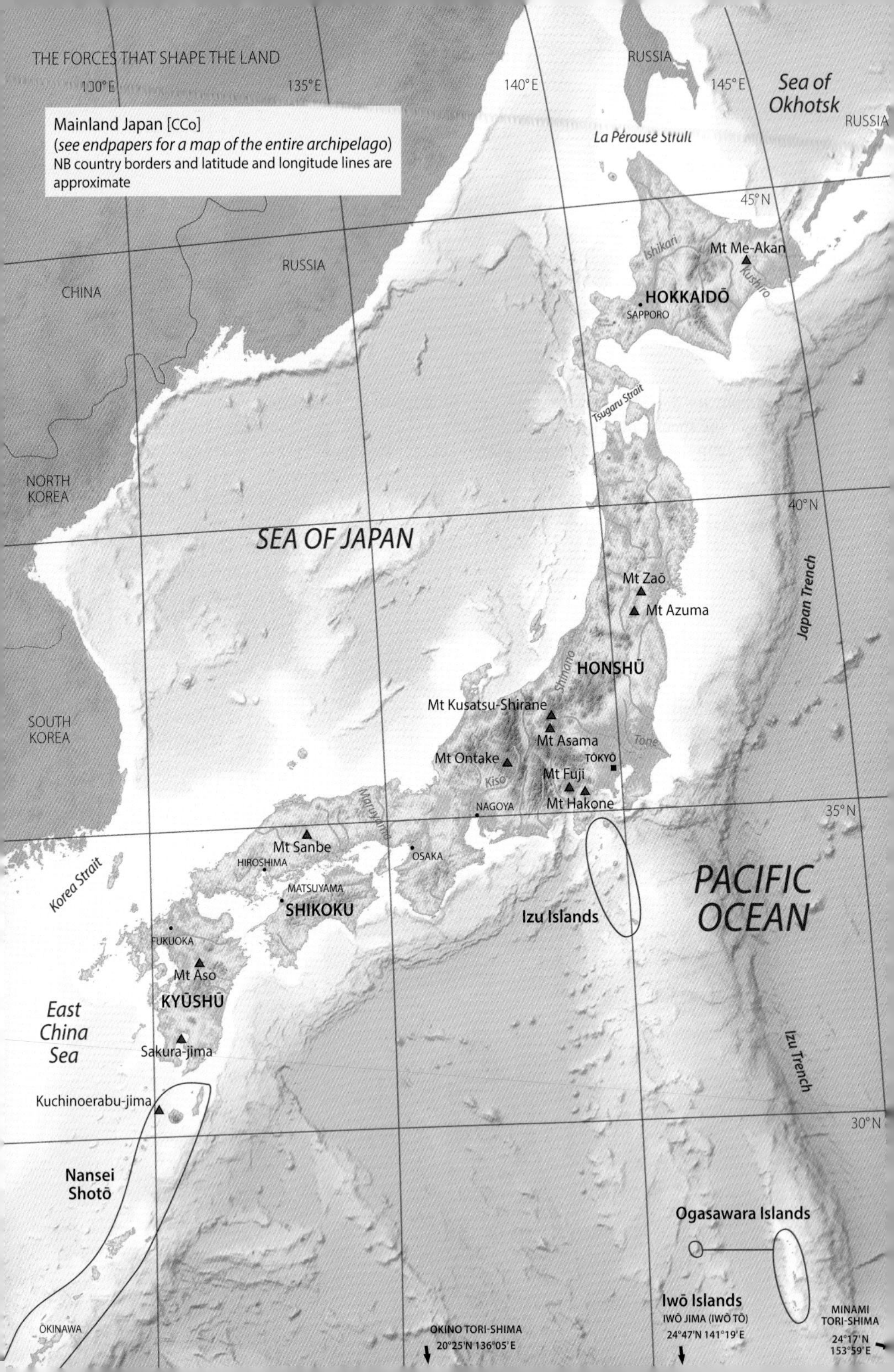

Mainland Japan [CCo]
(see endpapers for a map of the entire archipelago)
NB country borders and latitude and longitude lines are approximate
130°E
135°E
140°E
145°E
45°N
40°N
35°N
30°N
RUSSIA
CHINA
NORTH KOREA
SOUTH KOREA
Sea of Okhotsk
La Pérouse Strait
SEA OF JAPAN
PACIFIC OCEAN
East China Sea
Korea Strait
Tsugaru Strait
Japan Trench
Izu Trench
HOKKAIDŌ
SAPPORO
Ishikari
Kushiro
Mt Me-Akan
HONSHŪ
Mt Zaō
Mt Azuma
Shinano
Tone
Mt Kusatsu-Shirane
Mt Asama
Mt Ontake
Kiso
TŌKYŌ
Mt Fuji
Mt Hakone
NAGOYA
Maruyama
OSAKA
Mt Sanbe
HIROSHIMA
MATSUYAMA
SHIKOKU
Izu Islands
FUKUOKA
Mt Aso
KYŪSHŪ
Sakura-jima
Kuchinoerabu-jima
Nansei Shotō
OKINAWA
OKINO TORI-SHIMA
20°25'N 136°05'E
Ogasawara Islands
Iwō Islands
IWŌ JIMA (IWŌ TŌ)
24°47'N 141°19'E
MINAMI TORI-SHIMA
24°17'N
153°59'E

THE FORCES THAT SHAPE THE LAND

Wild Japan in Context

On a global scale, the Japanese archipelago may seem a small and inconsequential group of islands situated in the northwest Pacific when compared with the immense scale of the Eurasian continent to its west. Islands, however, are invariably fascinating. The Japanese archipelago spans 3,000 km from northeast to southwest, encompasses more than 6,000 islands, and contains such a wide range of habitats, from subarctic and alpine to subtropical as well as from high mountain tops to great ocean depths, that its biodiversity[1] is astonishing, as is its underlying geodiversity[2]. The long isolation of many of the islands in the archipelago has resulted in the evolution of local forms and species of plants and animals. This, combined with the extraordinary contrasts in climate, fauna and flora found between the almost subarctic north and the subtropical south, results in the fascinating natural diversity of the Japanese archipelago.

Images of Japan as a tiny, crowded island nation, although largely promulgated by the Japanese themselves, are clichéd and misleading. In fact, Japan is larger than either Britain or New Zealand and is larger than most countries in Europe except Spain, and almost exactly equal in size to Germany.

To grasp the scope of Japan's wildlife diversity requires an understanding of the scale of the country itself, of the fabric of the land, its underlying geology, its geography, and its location and climate. The early sections of this book deal with these topics, while the main chapters focus on each region and the special wildlife to be found there.

Common images of Japan involve shiny cars, the latest electronic goods, and regimented ranks of commuters in business suits. Japan is, after all, an intensely developed country with a population of 126·71 million (2017)[3]. The Japanese people's belief in their racial and social homogeneity is an interesting one, not supported by observation. Furthermore, their self-image as 'loving nature' is a surprising one when one travels around an archipelago showing few clear signs of that 'love'. Yet, behind the façade of development, outside the bounds of the metropolitan areas, and beyond the supposed homogeneity, there is another, wilder Japan.

Japan's numerous cities are densely developed and house more than half of the population [UNSP].

Japan consists of thousands of islands, such as tiny Higashi-jima, in the Pacific Ogasawara Islands [MOEN].

1 The diversity of life forms living in a particular region or ecosystem.
2 The diversity of geological features in a particular region.
3 Declining from a peak of 127·09 million in 2015.

The range of habitats includes subtropical forest, such as in northern Okinawa [BOTH KuM].

Wild Japan is astonishingly diverse, and the reasons, although apparent, are little known among those who travel the standard tourist circuit of the country. Perhaps Japan's most significant feature is that it is not just an island country, but a country of islands. Furthermore, it is situated where the Oriental and the Temperate regions meet, meaning that Japan is situated at a significant natural-history crossroads.

The northern island of Hokkaidō[1] has a subarctic feel, while in the far south the Nansei Shotō, the islands that stretch between Kyūshū and Taiwan, lie within the subtropical zone. In between, there exists just about every habitat imaginable, including brackish coastal wetlands and high alpine meadows, dune forests and peat swamps, sea-ice-battered northern coasts and subtropical coral reefs. In the sand dunes of Tottori Prefecture there is even a hint of a desert landscape.

Following the geological process known as 'back-arc spreading' that separated what are now the Japanese islands from the edge of the Asian continent about 20–15 Mya, there have been numerous glacial episodes during which sea levels have fallen, as water was locked away in ice during glacial maxima, and risen, as that ice melted during interglacial periods. During past periods of lower sea levels the Japanese islands were connected to the Asian continent in three areas, via Sakhalin to the north, the Korean Peninsula to the west, and Taiwan to the south. These land bridges allowed land mammals to colonize what later became far more restricted islands as the sea levels rose once more. As a result, some of Japan's 170 or so mammal species, the Asiatic Black Bear and Eurasian Red Squirrel for example, are widespread across Asia, while others, including the Japanese Dormouse and Japanese Serow, are specific to these islands. Similar patterns are to be found here among most terrestrial groups of organisms.

Islands isolated for long periods are crucibles for the evolution of new species, and Japan's thousands of islands therefore support a fascinating mix of widespread Eurasian species and many more Asian species, along with numerous local endemic plants and animals. Many species, such as the evergreen tree Itaji Chinkapin, survived the last major glacial period (the Last Glacial Maximum, or 'LGM', was 21,000–18,000 years ago) in various refugia here, leaving Japan today with a fauna and flora that has both an ancient history and a modern diversity, reflecting a series of colonizations between ice ages.

Ancient Japanese people referred to their homeland as *Akitsu Shima*, a name that means 'Islands of the Dragonflies'. Quite rightly so, because not only is Japan recognized as an important centre of diversification of Odonata but, moreover, dragonflies and their cousins, the damselflies, are so common all summer and autumn long as to be among the most familiar and popular insects in the country; they are even commemorated in popular song and are regarded as deeply symbolic of autumn.

1 Previously Ezo (a word that appears in many Japanese species names), but named Hokkaidō, meaning the North Sea Road, in 1869 and formally adopted in 1910.

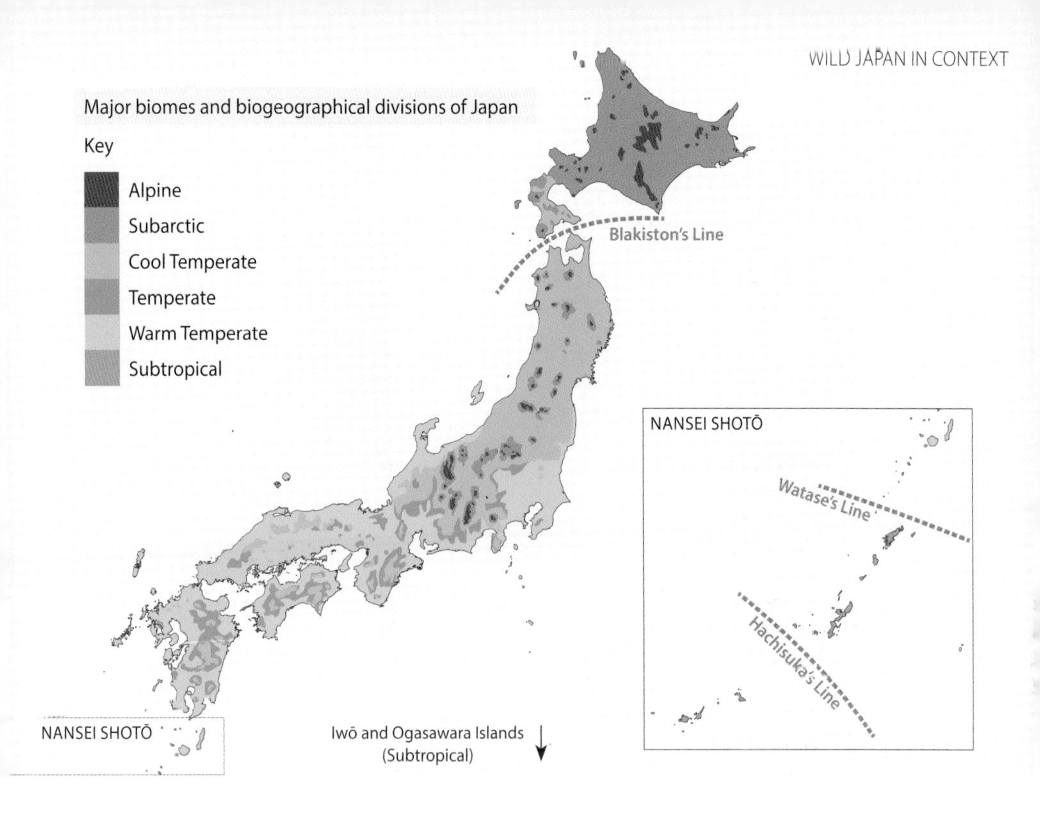

Dragonflies, such as this Foot-tipped Darter, are abundant in Japan in autumn [TaM].

Kämpfer, Thunberg, Siebold and Blakiston – Famous Names in Japanese Natural History

Wander through the pages of any Japanese natural-history volume and you will notice that the majority of names, whether English, Japanese or scientific, refer to the unique distinguishing characteristics of a species, perhaps its colour, voice, morphology, habitat or behaviour, or something else that is striking. This is the fascinating (to naturalists, at least) realm of taxonomic etymology.

In some cases the names of people appear in species' names. Some such names are given to honour individuals, perhaps referencing a sponsor, a mentor or a famous figure. Such is the case with the gecko *Homonota darwinii*, named after the great British naturalist Charles Darwin; the rare butterfly *Euptychia attenboroughi*, named in 2017 after Sir David Attenborough; and the particularly golden-haired horsefly *Scaptia beyonceae*, recently named after the singer and dancer Beyoncé. Bob Marley, J. R. R. Tolkien, President Barack Obama, Frank Zappa and many, many other public figures have been recognized in this way by taxonomic biologists. Being commemorated in such naming conveys a kind of inescapable immortality. Taxonomists are human, and some of the names they have chosen are not so much in honour as in contempt of someone with whom they have crossed scientific swords. The father of modern taxonomy, the Swede Carolus Linnaeus, even stooped to damning unfortunate colleagues with negative names. For example, Linnaeus named a small, unattractive and unpleasantly sticky plant with tiny flowers *Siegesbeckia orientalis*, because Prussian botanist Johann Siegesbeck had criticized Linnaeus's revolutionary binomial naming system. Much more recently, a Canadian entomologist named a new species of North American moth *Neopalpa donaldtrumpi*, because its yellowish-white head scales were reminiscent of Donald Trump's golden pompadour, although whether in honour or as insult is not clear.

While our human lives are brief, the echoes of some resonate through the centuries as cultural memes that pass from generation to generation, perhaps never to be forgotten. The natural history of Japan is replete with examples, but four names are outstanding: Kämpfer, Thunberg, Siebold and Blakiston. These four names and their histories span four centuries, but in natural history they are immortal.

During the Tokugawa Shogunate, Japan passed through centuries of *Sakoku*, from the early 1600s to the mid-1850s. The country was essentially closed to the outside world as the shogunate controlled and regulated trade and other relationships between Japan and the outside world. Foreign nationals were barred from entering Japan, and opportunities for the exchange of knowledge between Japan and the wider world were extremely limited. Like a chink in a solid door, the tiny island of Dejima, in the bay of Nagasaki, acted as the narrowest of conduits through which knowledge and trade with the West could pass. This permitted, but highly restricted, Dutch trading enclave was confined on a tiny fan-shaped island of just 1·9 ha. Those who worked there, and their overseers, nevertheless required their own physicians. The physicians of Dejima proved to be enterprising men who imparted a vast store of

The tiny island of Dejima, Nagasaki, was a crucial trading enclave [CNMW].

The German physician Engelbert Kämpfer [*left* CCo] and the Japanese Larch named after him [*right* MAB].

knowledge about Japan to the Western world and, conversely, introduced many aspects of Western life to Japan, not least among these being elements of Western medicine.

The first who garnered international attention was Engelbert Kämpfer (1651–1716), a German who had trained as a physician. Like so many physicians and clergymen of centuries past, he was also an ardent naturalist. His travels as a physician allowed him to collect specimens of previously undescribed species from then poorly known parts of the world, including Japan.

In September 1690, when Kämpfer arrived at Dejima, it was the only Japanese trading port open to westerners. Kämpfer was to remain in Japan for a little over two years (1690–1692) but was permitted on occasions to leave the tiny island of Dejima. It was during one of those brief excursions, to a temple in Nagasaki, that he noted and described the then poorly known and now widely planted living fossil the Ginkgo or Maidenhair Tree. At the time it was known in Europe only from fossils and was presumed long extinct. Among the many species named after him, one in particular – the Japanese Larch *Larix kaempferi* – will be familiar to anyone travelling in central or northern Honshū or in Hokkaidō. This colourful and deciduous conifer is native to subalpine slopes of Honshū, but it is a successful pioneer species thriving in open areas with dry soil and plentiful sunlight. It has been widely planted in Japan and, because of its rapid growth, it is an important forestry species. It can be seen especially in east Hokkaidō, where it is a common windbreak tree. It was named in honour of Kämpfer.

A half-century later, Dejima was once more home to a foreign naturalist who has left a significant mark. Swedish naturalist Carl Peter Thunberg (1743–1828) has been honoured as the 'Japanese Linnaeus', so great was his contribution to Japan. A student under Linnaeus at Uppsala University, Thunberg studied medicine and natural philosophy, and was therefore invited to collect botanical specimens from Dutch colonies, including the tiny Dutch enclave in Dejima.

Swedish naturalist Carl Peter Thunberg [GUUM]

First, as a ship's surgeon, Thunberg devoted himself to learning the Dutch language so that he could pass as a Dutch merchant (as they were the only Westerners allowed into Japan at that time). He arrived at Dejima in August 1775 and became its chief surgeon, remaining there, as had Kämpfer, for just two years. Trading his knowledge of Western medicine for permission to visit Nagasaki itself and even to travel outside the city to collect local botanical specimens, he began his exploration of Japan's flora.

Thunberg was extraordinarily fortunate to have been allowed to travel from Nagasaki to Edo (Tōkyō) in 1776 as part of the Dutch representative's annual visit to the Shogun's court, and that overland journey enabled him to collect and preserve numerous plant specimens. His studies and his collections allowed him to undertake the first description of the Japanese flora, which was ultimately published as *Flora Japonica* in 1784, after his return to Sweden. His meandering return journey from Japan, from where he set out in November 1776, took him three years. On his way back to Sweden he visited London, where he met the great British botanist Sir Joseph Banks. Banks is himself commemorated in innumerable plant names, in particular the generic name *Banksia* for the many species commonly known as Australian honeysuckles. In London, Thunberg was able to examine the collection of Engelbert Kämpfer, his predecessor at Dejima, and we can imagine how thrilling an experience that must have been for him.

In the measured footsteps of Kämpfer and Thunberg came Philipp Franz Balthasar von Siebold (1796–1866), another German physician, who was destined to leave behind an astonishing legacy. Building on the works of his predecessors, Siebold's contributions to our knowledge of the Japanese flora and fauna were enormous. As with Thunberg, Siebold became a ship's doctor, learned Dutch and, like both Kämpfer and Thunberg before him, was posted to Dejima, arriving there in June 1823.

Trading his scientific and medical knowledge for information on Japan's customs, culture and natural history, at a time when, it seems, Japanese people were increasingly hungry for such knowledge, Siebold was eventually granted more extended access beyond the close confines of Dejima, even 'marrying' a local woman and fathering a daughter. He honoured his wife by naming a hydrangea, Taki's Hydrangea, after his pet name for her, and honoured his daughter with sufficient training and skills for her to become a highly regarded physician in her own right. Siebold's own collections rapidly expanded as his grateful patients 'paid' him in kind with objects ranging from ethnographic artefacts and artworks to natural-history specimens. He also sent Japanese specimen-collectors into the countryside. As was typical in that period, the specimens which he collected, including the first specimen of the huge Japanese Giant Salamander (now [NT]), were sent in shipments to various European museums.

Siebold's enquiring mind and acquisitive collecting eventually landed him in trouble, as during a journey to Edo he was caught with several maps in his possession (a treasonous act at that time). He was seized and placed under house arrest. Then, ultimately in October 1829, his six-year sojourn in Japan

Philipp Franz Balthasar von Siebold [CNMW]

Taki's Hydrangea, named for Siebold's wife [SABB].

Japanese Giant Salamander was one of many significant specimens collected by Siebold [NBCL].

Southern Japanese Hemlock is one of many plant species named after Siebold [NZUE].

came to an end when he was expelled, having been accused of spying on behalf of Russia. Settling in the Dutch city of Leiden, he established a small private museum and wrote a number of books on Japanese ethnography, geography and natural history, as well as collections of literature and a dictionary.

I first came across Siebold's illustrious name in the context of the multi-volume *Fauna Japonica*, published from 1833 to 1850, which relied heavily on his specimen collection. It is renowned as having made the Japanese fauna the best-described non-European fauna of those times. Further serving to establish Siebold's name as one to live for ever was his work *Flora Japonica* (1835–1847), co-authored with German botanist Joseph Gerhard Zuccarini, much of which was published posthumously. Siebold's legacy lives on not only in his published works and the species named after him, but also in the species that he introduced to Europe and which have subsequently spread around the world. These include a range of species from azaleas and butterburs to the larch tree, named after his predecessor, Kämpfer. His knowledge of Japan, then closed and isolated, made him an invaluable expert and adviser, and even the American Commodore Matthew Perry, renowned for 'opening' Japan, is said to have consulted Siebold.

Little known in Europe or North America today outside the limited spheres of botany and horticulture, Siebold remains well known in Japan (as *Shiborudo*) and among naturalists, both as an important historical figure and as someone amply and eponymously commemorated in species' scientific names. These include a bird, White-bellied Green Pigeon, along with various trees and shrubs, among them Siebold's Magnolia, Siebold's Viburnum and Southern Japanese Hemlock, and a fern, Siebold's Wood Fern.

Whereas in earlier centuries the fragment of Nagasaki known as Dejima was the only conduit for knowledge exchange between Japan and the West, the second half of the 19th century, especially following the Meiji Restoration in 1868, brought the relaxation of certain restrictions. A number of other ports were opened to overseas trade, including Kōbe, Yokohama and especially Hakodate. It was to Hakodate, which had opened fully as a treaty port in 1859, that the British merchant and naturalist Thomas Wright Blakiston[1] (1832–1891) moved, initially to run his trading enterprise the West Pacific Company.

With far fewer restrictions placed on the movements of foreigners in Japan at that time, Blakiston was able, during his long, although not entirely continuous, residence in Hakodate from 1861 to 1884, to collect natural-history specimens himself, exchange specimens with others and buy specimens on offer. In collaboration with the entomologist Henry Pryer (1850–1888), who lived in Yokohama, Blakiston published a *Catalogue of the Birds of Japan*, the first work of its kind for Japan. This groundbreaking ornithological work provided the basis for a later compendious volume, *Birds of the Japanese Empire*, by Henry Seebohm, and stimulated, a century later, my own *Birds of Japan*.

1 Previously Captain Blakiston, but he resigned his army commission in 1862.

Blakiston was the first to recognize the fundamental difference between the fauna of Hokkaidō, which was more closely related to that of northern Asia, and that of Honshū, which was more closely related to the southern Asian fauna. In recognition of this insight, he is commemorated not just in the names of species, but also in the name of a globally significant biogeographical boundary.

The fragmented crescent of islands that form the Japanese archipelago is divided by seawater channels, some deep, some with strong currents (see *p. 18–31* for a geological history of the formation of Japan). Some of those channels are so ancient as to dictate very strongly the distribution of wildlife in Japan today. One of the most important of these channels is Tsugaru Kaikyo, the strait that separates the Japanese main island of Honshū from Hokkaidō to the north.

The monument in Hakodate to British merchant and naturalist Thomas Wright Blakiston [MAB].

So significant are the depth and age of the Tsugaru Kaikyo that whole suites of species are divided by it, and major distinctions between the fauna and flora north and south of it are recognized. Such biogeographical borders are given special names. Perhaps the most famous is Wallace's Line in Indonesia, named after one of the two original proponents of evolution by natural selection, Alfred Russel Wallace. In Japan, too, we have such lines. The one running through the Tsugaru Strait that separates Hokkaidō from Honshū has, since Blakiston's time, been recognized as a significant zoogeographical boundary and is now known as Blakiston's Line[1]; identified by Blakiston, it was immediately named after him. Blakiston's Line continues to be recognized today.

North of the biogeographical divide of Blakiston's Line occur such notable species as Blakiston's Fish Owl [CR] and Brown Bear. South of Blakiston's Line we find Japanese Serow, Japanese Giant Salamander, Asiatic Black Bear and Japanese Macaque. Some species, such as Red Fox, occur naturally on both sides of the 'line', indicating either their much more ancient distributions or that they have been able to colonize the northern and central parts of Japan successfully and separately more than once from the continent of Asia. A small selection of species occurs both north and south of Blakiston's Line, but not through natural causes. The Japanese Marten, a predatory denizen of the main islands south of the Tsugaru Strait, has been introduced into Hokkaidō and currently thrives there, mainly in the southwest. Conversely, the vegetarian rodent Siberian Chipmunk, considered appealing by tourists to Hokkaidō, has been introduced into various parts of Honshū, Shikoku and Kyūshū.

One of the specimens collected by Blakiston and shipped to Britain was of the enormous owl that now bears his name – Blakiston's Fish Owl. In 2021 it is both rare and endangered. I have had the good fortune of watching it frequently in the wild in Hokkaidō and observing it in the collection of specimens housed in Tring, in the English countryside, including the original specimen taken by Blakiston. Today, the Tring-based collection, which began as the private collection of Lord Lionel Walter, Baron Rothschild (1868–1937), houses almost 750,000 specimens of more than 95 percent of the world's bird

1 Named after the businessman and naturalist Thomas Wright Blakiston (TWB), who lived in Hakodate. The term was first used by John Milne in the questions session after TWB's presentation of the evidence to the Asiatic Society of Japan in Tōkyō.

species. Among them are 8,000 type specimens (the representative specimens from which species are first described), including Blakiston's eponymous owl. The enormous diversity of priceless material there was certainly overwhelming, but, in the company of fish-owl researcher Mr Yamamoto Sumio, only one specimen captured and held our attention – the very specimen of the owl sent by Blakiston. Having stood together on cold winter nights beside frozen rivers in east Hokkaidō while watching for and listening to this owl, on seeing that rare specimen in Tring we felt as if we had been transported back in time. We could imagine what a thrill it must have been to be an ornithologist in Japan in the late 1800s.

Numerous Japanese naturalists have each left their mark on the natural history of the country, but few have achieved the great significance and international fame accrued by Kämpfer, Thunberg, Siebold and Blakiston, whose pioneering work allowed the West to learn so much about this fascinating country.

Blakiston's Fish Owl occurs only north of Blakiston's Line, which separates Hokkaidō from Honshū [MAB].

Along the Arc of Fire

Plate tectonics and its significance in Japan

Earth's rigid outermost shell, known as the lithosphere and consisting of the crust and upper mantle, is broken into seven or eight major tectonic plates along with many minor plates. The first model describing the pattern of our continents and their movement was developed during the early 20th century. Building on this theory of continental drift, a new concept, that of plate tectonics, was developed when spreading of the ocean floor was confirmed during the 1950s and 1960s. These tectonic processes began approximately 3·5 billion years ago and continue today at speeds of up to 100 mm a year. The motion of the plates and the boundaries between them are what cause mountain-building, earthquakes, volcanic activity, and the formation of deep ocean trenches.

Japan's extraordinary tectonic and volcanic past and its ongoing geological processes create hazards and difficulties for its human population. They also, however, provide a wide range of benefits, including the rich volcanic soils that underpin its agriculture, geothermal energy providing hot springs, and an inspirational landscape of mountains, volcanoes, rivers, forests, rugged coasts, and islands large and small. The natural heritage, comprising the biotic and underlying geology, hydrology and climate (abiotic) environments, and cultural life of Japan are defined by geological processes combined with patterns of atmospheric and oceanic circulation.

The ongoing shaping of Japan is a consequence of three simultaneous forces at work: (1) horizontal and vertical movements of the earth's crust which include the formation of mountains; (2) volcanic eruptions and the subsequent deposition of rocks and ash; and (3) rapid erosion and weathering of those deposits by water and by wind, leading to such events as slope failure and mountain collapse.

Iconic Mt Fuji, Japan's highest peak and largest stratovolcano, is a testament to Japan's geologically violent history [PP].

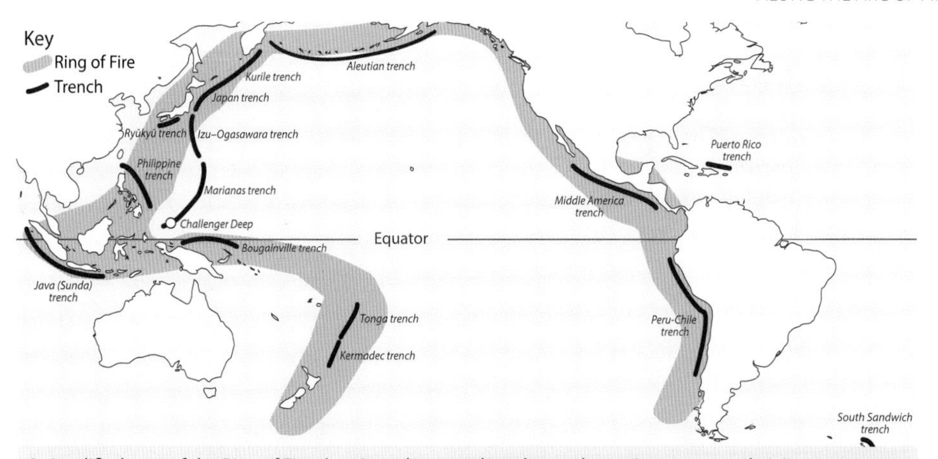

A simplified map of the Ring of Fire showing where earthquakes and eruptions occur and giving context to Japan's location.

A simplified geological map of Japan

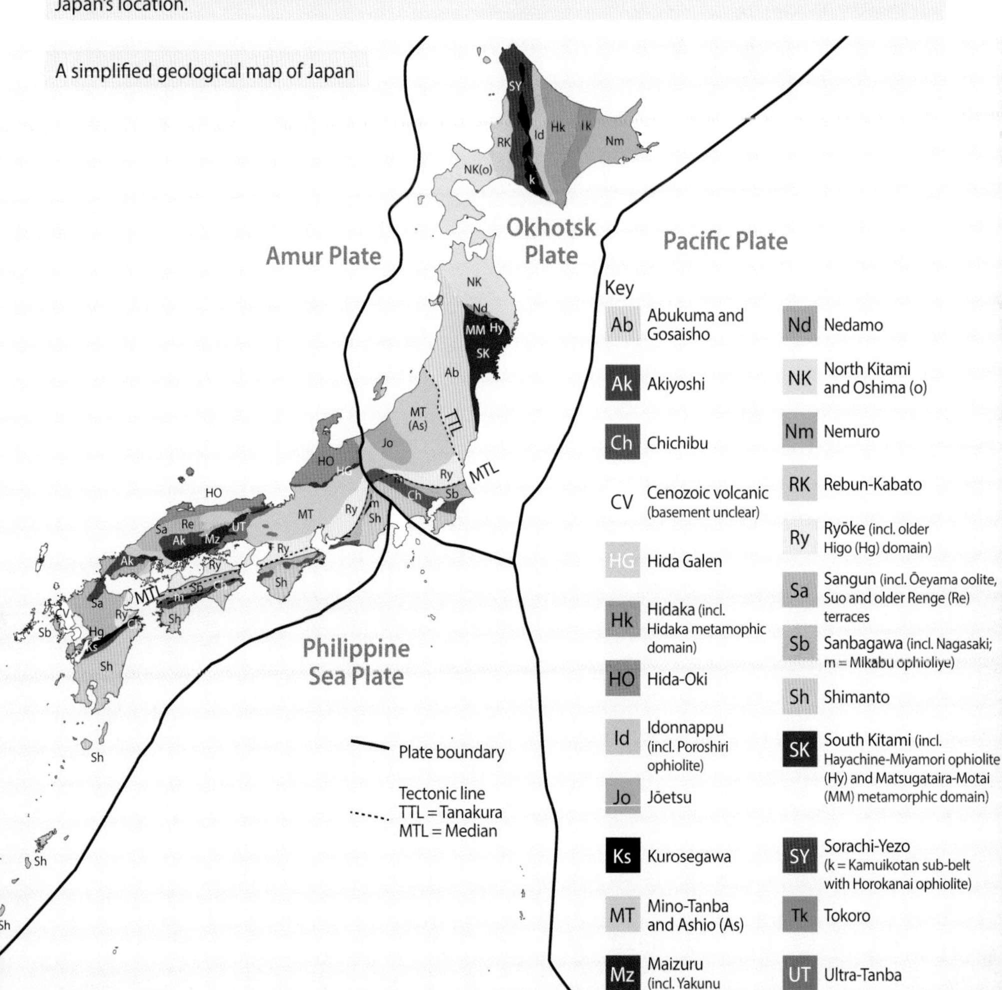

The majority of the country consists of steep-sided mountains interspersed with narrow valleys [SHUT].

Today, Japan is separated from the eastern coast of the Asian continent by marine channels and seas, but that was not always the case. The land that is now Japan is thought to have begun some 750–700 Mya[1] with the breakup of the Rodinia Supercontinent[2], the rocks of its eastern margin going on to form the Japanese islands.

Japan's geohistory as a supercontinental fragment can be traced through five transformative phases: first at the continental margin; second as an island arc; and third as an 'accretionary complex', a mix of materials some terrestrial, some basalts scraped from the ocean floor, and some marine sediments (all three of these occurred during the Palaeozoic Era[3] between 541 and 252 Mya); fourth as an accretionary complex during the age of the dinosaurs, the Mesozoic Era, which lasted from approximately 252 to 66 Mya; and fifth as an island arc again, during the Cenozoic Era, the age of the birds and mammals, which has lasted from about 66 Mya to the present. A key recent element in this transformation was the opening of the Sea of Japan[4] some 25–15 Mya through a process known as back-arc spreading, which was caused by the sliding of an oceanic plate beneath the continental margin in a process known as subduction. The separation of Japan from the East Asian continental margin, and the formation of the present Japanese island arc through these five phases, ultimately led to the distinctive geological and geographical features of the Japanese archipelago that is familiar to us today.

Japan's oldest known basement rocks[5] are ancient granites and gneisses dating back to the Precambrian–Palaeozoic eras, some 2,000–1,500 Mya, in the Kamisao formation in central Honshū (displayed in the Hichiso Precambrian Museum in Gifu Prefecture). These are overlain now by younger, highly deformed sedimentary and metamorphic rocks[6]. For more than 500 My, the archipelago has experienced the

1 My = million years; Mya = million years ago.

2 This supercontinent existed between 1·1 billion years ago and 700 Mya.

3 See Cohen *et al.* (2013) and International Commission on Stratigraphy www.stratigraphy.org v 2019/05 for the most recent dates of Eras and Periods.

4 To the north of Japan, the Sea of Okhotsk (which will feature later in this book) opened during the same period and through similar processes.

5 Basement rocks are the older, 'foundation' rocks of a continent lying below a younger sequence of sedimentary rocks.

6 Sedimentary rocks form under water as grains and fragments of pre-existing rocks (sediments) become glued together. Metamorphic rocks have been changed under pressure and heat.

continual sliding of an oceanic plate beneath the Asian continental margin leading to frequent earthquakes and continual intermittent volcanic activity.

The geological agitation beneath Japan is both ancient and ongoing. The landscape is scarred, dotted, and pockmarked with features that tell stories of upheaval and eruption. Visitors to Japan will see volcanoes, craters and calderas, and witness geothermal activity in the form of hot springs and steam and sulphur vents. Longer-term visitors are very likely to experience the physical evidence of turmoil in the form of earthquakes.

A simplified diagram showing how mountain-building and earthquakes are associated with plate tectonics and subduction.

Situated on the northwestern portion of the Pacific Ring of Fire, which wraps 40,000 km around the Pacific Ocean, Japan sits astride several floating crustal plates. Some of these are being driven down (subducted) beneath others that ride over them in the process of plate tectonic movement.

Active subduction of oceanic plates beneath Japan is known to have been taking place for around 145 My since the beginning of the Cretaceous Period. In particular, the Pacific Plate to the east and the Philippine Sea Plate to the south both began sliding beneath the Amur Plate to the west and the Okhotsk Plate to the north during the Neogene period, approximately 23–2·6 Mya. This process shaped the archipelago and led to the generation of Japan's four distinct geological domains. From north to south these are the western Kuril Islands and eastern Hokkaidō Arc, the western Hokkaidō–Honshū Arc, the Ryūkyū Arc, and the Izu–Ogasawara Arc.

The first of the four, the Kuril and eastern Hokkaidō Arc, stretches more than 1,000 km between the Kamchatka Peninsula and central Hokkaidō. Oblique subduction of the Pacific Plate beneath the Kuril Arc resulted in a collision zone in central Hokkaidō that gave rise to the Hidaka Mountains. One of the unique characteristics of Hokkaidō is that the eastern third of the island belongs to the Kuril and

Kussharo Caldera is the largest of many calderas in Japan [MiH].

eastern Hokkaidō Arc, while the western two-thirds belong to the Hokkaidō–Honshū Arc.

The second, the western Hokkaidō–Honshū Arc, is aligned north–south and stretches another 1,000 km south to the Izu Collision Zone, with two prominent parallel arrays of volcanoes. To the west of that collision zone, the southwest Honshū part of this arc extends about 600 km in a more east–west direction and includes the island of Shikoku; it is paralleled offshore to the south by the Nankai Trough. The 11 March 2011 megaquake and megatsunami were generated by the relatively rapid subduction of the Pacific Plate in the Japan Trench off the northeastern part of this arc. As these southwestern and northeastern parts of the arc have converged and collided with the Izu Collision Zone, central Honshū's distinctive highly rugged, mountainous topography has been generated.

The third, the Kyūshū–Ryūkyū Arc, extends over 1,000 km from Kyūshū to Taiwan, the

Mt Shinmoe erupting in February 2011 [MAB].

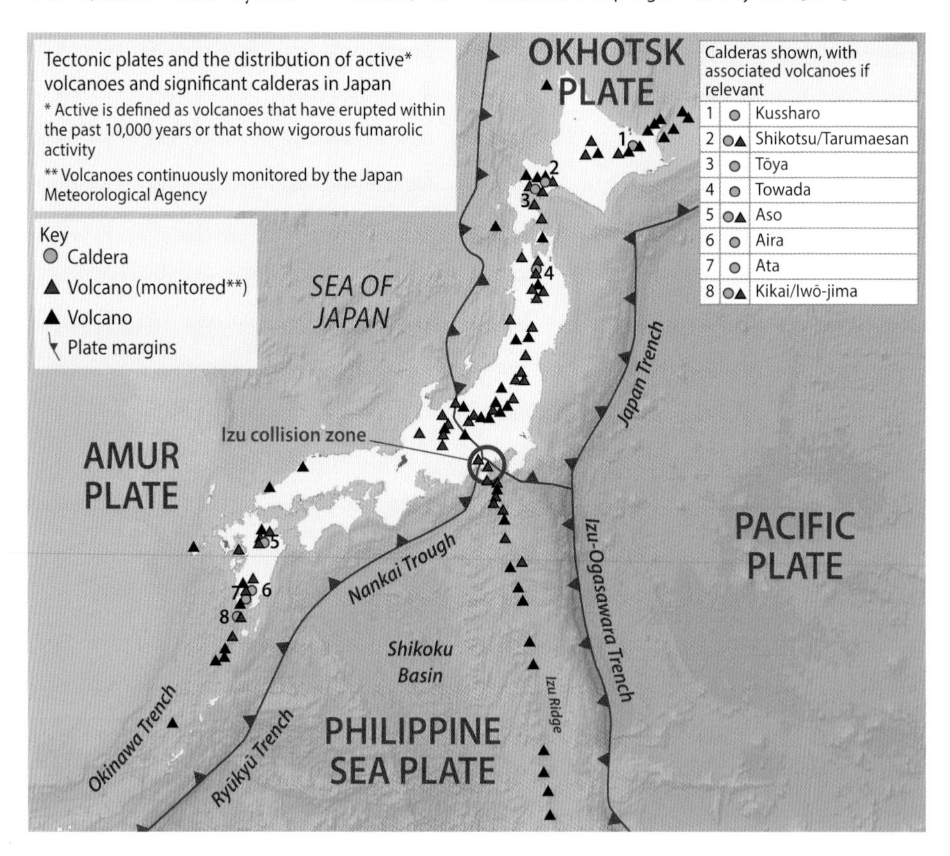

northern portion sharing affinities with Kyūshū and the southern portion sharing affinities with Taiwan. The Ryūkyū Trench, parallel to and east of this arc, is a manifestation of the subduction of the Philippine Plate beneath the Amur Plate. Parallel to this arc, and between it and the continental shelf, is the Okinawa Trough.

Finally, the Izu–Ogasawara–Mariana Arc consists of an arc of largely submerged oceanic islands stretching from the Izu Collision Zone in the north (where this arc has been colliding with the Hokkaidō–Honshū Arc contemporaneously with the opening of the Sea of Japan) to the Mariana Islands in the south.

Vulcanism

The distribution of volcanoes in Japan is not random, but it reveals the locations of plate activity. In northeastern Japan, volcanoes are found along a front known as the East Japan Volcanic Belt. This belt is situated essentially parallel to the largely north–south Japan Trench lying just offshore, where the Pacific Plate subducts beneath northern Japan. In southwestern Japan, volcanoes are distributed along the West Japan Volcanic Belt, nearly parallel to the roughly southwest–northeast Nankai Trough. In each volcanic belt the density of volcanoes is greatest closest to the volcanic front. Large-scale eruptions in Japan have generated calderas, and these can be found in Hokkaidō, northern Tōhoku, and central and southern Kyūshū. The largest eruption in Japan in the last 150,000 years occurred in Kyūshū when Mt Aso erupted 90,000 years ago. That eruption, which was ranked as Volcanic Explosivity Index 7 (of a maximum of 8), produced an astonishing volume of ejecta amounting to more than 600 km^3 and created the second largest caldera in Japan, the Aso Caldera. The ash fallout from that eruption spread over the entire length of Japan and deposited up to 15 cm of ash as far north as northern Hokkaidō, approximately 2,000 km away.

Rocks basal to the Japanese islands, dating back to the Precambrian Era, prior to 541 Mya, are exposed here and there in Japan's four major islands; at the same time completely new rocks are created during each volcanic eruption. Furthermore, Japan's current mountain ranges are the results of several mountain-building

Volcanic activity is widespread. Mt Iwō, in the Akan–Mashū National Park, continuously vents superheated steam and sulphur [MAB].

Bubbling mud pots, known as bokke, occur at many sites around Japan [MAB].

Although geysers are rare in Japan, they occur from Hokkaidō to Kyūshū [MAB].

From Kyūshū to Hokkaidō the country's scenery is dominated by alpine mountains and volcanoes [ALL MAB].

periods, so that mountain ranges with different origins and of different ages overlap and overlay one another, making Japan's topography extremely complex. The complexity of Japan's geological and geomorphological phenomena may seem baffling. To make these subjects more accessible and understandable to the Japanese public, and to visitors, special educational facilities have been developed at sites of particular geological interest, and these are now known as Geoparks (see *p. 129*).

Vulcanism Today

Japan, for all its societal stability and security, is one of the geologically most active places on earth. Its geological background serves to explain and illustrate the shaping of the modern Japanese archipelago. The current activity of its volcanoes (on the Ring of Fire), and its earthquakes, along with its numerous hot springs, fumaroles and mud springs, define a geology that is in constant turmoil.

Geology is inescapable in Japan. The setting of the archipelago amid extreme tectonic and volcanic activity means that, for Japanese people, the unpredictable and devastating nature of their geological surroundings is ever-present in their mindset, in their culture and in their society.

This turmoil originates from the very formation of the Japanese archipelago, or, more specifically, from the formation of the Japan island-arc system. It cannot be divorced from a hugely significant event, namely the opening of the Sea of Japan. A typical back-arc basin, the Sea of Japan lies behind both the northeastern and the southwestern Japan arcs. The combined rotation of the southwest Japan arc clockwise and of the northeast Japan arc anticlockwise has produced the gracefully curved shape of the Japanese archipelago that we see today.

To the average visitor, the landscape of Japan is dominated by two contrasting types of mountain features: abrupt ridges of alpine peaks, and isolated or clustered volcanic peaks. Associated with them are highly visible craters and calderas, and less visible glacial and periglacial features. The lakes and river systems of Japan are easy to understand. The close proximity of the sea to every part of such a narrow archipelago ensures that all rivers are short. The extreme altitudinal variation between the many peaks of the Japan Alps (rising to more than 3,000 m) and sea level, combined with proximity to the ocean, ensures that most rivers are both short and fast-flowing, and in their upper reaches filled with cold, oxygen-rich water. The same upland features mean that the impacts of seasonal rainfall, the rainy season, and the typhoon season, are all dramatic and leave much of lowland Japan prone to flooding while intermediate elevations are prone to erosion, landslides and slope failure. In fact, Japan is one of the more landslide-susceptible countries in the world, thanks to its complex geology, its high rate of tectonic activity and the unpredictability of its monsoon climate. In the early summer of 2018, in western Japan, record-breaking rainfalls led to many landslides, widespread destruction, and more than 200 deaths. Landslides in Japan may be induced by rainstorms, earthquakes and combinations of the two. The frequency of both heavy and intense rainfall events has increased in line with climate change. Landslides in June and July are associated with the *Baiu* (or rainy season) front, whereas those in August and September are associated with typhoons, resulting from heating of the water in the northwest Pacific to form a strong low-pressure cell.

To the ordinary person it is Japan's surface features that dominate – the plains and the mountains, the lakes, and the streams and rivers. To the geologist, what lies below the landscape, and what led to its very generation, is more meaningful than the exposed surface features.

Volcanoes

Japan, a country the size of Germany, has 110 active volcanoes, and in most years there is at least some form of eruption. Their immense destructive power may be what first springs to mind when you hear the word 'volcano'. While this aspect of them is a reality, they are also massive forces of creation. Although large-scale caldera-forming events are generally infrequent, ever since the opening of the Sea of Japan many such events have occurred in the archipelago, with dramatic, relatively young examples in Hokkaidō, northern Honshū, and central and southern parts of Kyūshū.

Volcanoes are as much landscape-builders as destroyers of landscape, and from Hokkaidō to Kyūshū a considerable amount of Japan's dramatic scenery owes its existence to volcanic forces. Japan is dotted with the results of immense geological forces working through volcanoes. These include the enormous

Central Kyūshū is dominated by the enormous and spectacular Mt Aso Caldera [AGPC].

Volcanic islands, such as the Izu Islands, serve as natural biological laboratories, allowing the study of life in isolation [OMTA].

Aira Caldera (formed about 22,000 years ago), in southern Kyūshū, the hugely inspiring Mt Aso Caldera (formed approximately 33,000 years ago), in the centre of Kyūshū, Japan's largest caldera at Lake Kussharo, in Hokkaidō (formed in a series of eruptions between 400,000 and 30,000 years ago), and the extraordinarily beautiful Lake Mashū, a much younger caldera lake (about 7,000 years old) in east Hokkaidō.

Ōnuma Quasi National Park, Hokkaidō, includes volcanic Komagatake [MAB].

Independent of other mountain-building forces, volcanoes can spring from the landscape, as did the lava dome of Showa Shinzan, in west Hokkaidō, between 28 December 1943 and September 1945. Volcanoes are natural islands, sometimes completely isolated on land surrounded by quite different landscapes and habitats, as is Mt Fuji, for example. Other volcanoes have arisen from the seabed, making them as isolated as land can be, as exemplified by the Izu Islands to the south of Tōkyō.

Volcanic islands, on land or in the sea, provide opportunities to observe the processes of nature. We can see, in precisely aged locations, how plants colonize new ground. We can measure how different species appear and disappear, and record how new communities stabilize in new habitats. Much older volcanoes, such as the Izu Islands, allow us to see the results of long-term isolation, and to identify the endemic species that have evolved *in situ*.

Like so many of Japan's volcanoes, Shinmoe-dake, in the Kirishima range of Kyūshū, erupts with ash [MAB].

Many of Japan's volcanoes are within national parks or other protected areas, although nothing we humans can do can either conserve or significantly damage such enormous symbols of power. What is protected, however, is the natural history of these majestic volcanic peaks and the forests that flank them, making them excellent destinations for the naturalist.

Active volcanoes may produce gases, ash or lava. Mt Fuji is a stratovolcano or composite volcano, a conical volcano built up by many layers of hardened lava, tephra, pumice and ash. The iconic and sacred Mt Fuji, at 3,776 m, is the tallest peak in the country. Outranking all others in size, it is undeniably the largest stratovolcano in Japan, having an estimated volume of 400–500 km3. It first erupted into life about 100,000 years ago (referred to as Ko-Fuji), then about 10,000 years ago (Shin-Fuji), although there is evidence of even older volcanoes in the same location. Mt Fuji has erupted frequently during historic times, but the summit crater last erupted about 2,000 years ago, while all subsequent eruptions have issued from its flanks. It has languished, almost entirely quiescent, for more than 300 years since the Hoei Eruption from 16 December 1707 to 1 January 1708 formed a crater on the southeast flank, disturbing the perfect symmetry of the sacred mountain. That is not the case for a number of Japan's other volcanoes. Since 1970, various eruptions have disrupted life in Japan: in 1977–8 and 2000 Sakura-jima, in Kyūshū, and Usu, in Hokkaidō; in 1983 and 2000 Miyake-jima, in the Izu Islands; in 1986–9 Izu Ōshima, in the Izu Islands; in 1990–5 Unzen, in Kyūshū; in 2011, 2017 and 2018 Shinmoe-dake, in Kyūshū; in 2014 Mt Ontake and in 2015 Mt Asama, in Honshū; and in 2017 Sakura-jima, in Kyūshū.

Volcanic Tori-shima, on the Izu–Bonin–Mariana arc (*left*) is a precarious home for Short-tailed Albatross (*right*) [BOTH TR].

Significantly longer ago, Japan experienced a number of catastrophic volcanic events or large-scale caldera-forming eruptions. Such events are measured on the Volcanic Explosivity Index (VEI[1]), and a number of Japan's events are ranked at seven on a scale of eight. During the last 1,000,000 to 500,000 years these have included the formation of the Kussharo, Akan and Tōya calderas of east and southwest Hokkaidō; of Towada of northernmost Honshū; and of the Aso, Kakuto-Kobayashi, Aira, Ata and Kikai calderas of Kyūshū. Each one of these events would have had a devastating impact on the natural environment. If any of them were to erupt in a similar fashion today, they would likely render entire islands uninhabitable so great would be the impacts on human life, health and agriculture. For example, the ash falls generated by Kyūshū's Mt Aso eruption event not only reached Hokkaidō 2,000 km away at the opposite end of the country, but also deposited there an ash layer of more than 10 cm. The Kussharo eruption, approximately 400,000 years ago, generated Japan's largest caldera, measuring 26 km by 20 km, and ejected an unimaginably large, but unknown, amount of material. The smaller Mashū Caldera, *only* 7·5 by 5·5 km, formed on the eastern rim of the Kussharo Caldera about 7,700 years ago and is estimated to have ejected approximately 18·6 km^3 of tephra. This, however, is nothing compared with the 600 km^3 of tephra ejected during Kyūshū's Aso-4 eruption about 90,000 years ago.

Japan is indeed fortunate that recent eruptions have not exceeded the VEI 5 of Honshū's Mt Fuji in 1707 and Hokkaidō's Mt Tarumae in 1739. Since then, the largest have been those of Sakura-jima, in Kyūshū, in 1914 and Hokkaidō-Komagatake in 1929, both VEI 4.

Some of Japan's volcanoes are remarkable for the size of the calderas which they have created and the volume of the ejecta which they have produced, others for the frequency of their eruptions. Among the most frequent is Mt Usu, on the southern rim of the Tōya Caldera in southwest Hokkaidō. The volcano itself dates back about 110,000 years, while its most recent period of activity began in 1663. Five eruptions occurred in the late 17th century, followed by eruptions in 1769, 1822, 1853, 1880–9, 1910, 1943–5, 1977–8 and 2000.

Among Japan's most active volcanoes is that on Miyake-jima, 180 km south of Tōkyō. The island is renowned among birdwatchers as home to all of the Izu Island endemic avian species. There, eruptions have been repeated every 20 years or so for three centuries, although the earliest documented eruption was in 1085 CE.

Shinmoe-dake is an active volcano in the Kirishima range in Kyūshū. The formation of the Kirishima range, a cluster of more than 25 volcanic cones, began some 600,000 years ago and continues to this day. Shinmoe-dake is just one of the currently active volcanoes there, and I was fortunate to witness its eruption in 2011. Seeing it belch a column of ash high into the cold winter atmosphere in a matter of moments was a very sobering experience.

---

1 The Volcanic Explosivity Index (VEI) is an open-ended logarithmic scale estimating the volume of tephra on a scale from zero to eight, with the largest eruptions ever given magnitude eight. Qualitative observations range from gentle to megacolossal and record cloud column heights up to 20 km.

Ceaseless tectonic grinding has produced the Kyūshū–Ryūkyū Arc and Izu–Bonin–Mariana arc systems (see *p. 23*). Journeying along this latter arc, 2,800 km from the Izu Peninsula southwards all the way to Guam, takes the traveller past a series of six visible island volcanoes and nine hidden submarine volcanoes. The islands, aligned slightly east of due south, begin with Ōshima, then Miyake, Hachijō, Aoga-shima, Sumisu and Tori-shima. The first three are populated and are popular tourist destinations, while Miyake and Hachijō are well known among naturalists for their endemic species. Once populated by humans, but now abandoned to its birds, Tori-shima is famously home to the bulk of the world's population of Short-tailed Albatross. South of the Izu Islands, following the same line, the first island encountered is Sōfugan, then Nishino-shima, followed by the Ogasawara Island group of Yome-jima, Chichi-jima, Haha-jima and Muko-jima, and finally Kita Iwō, Iwō-to and Minami Iwō.

Hokkaidō, Japan's northernmost island, belongs to the Okhotsk Plate. Hokkaidō represents the collision zone between the Kuril and Honshū arc-trench systems, and thus has geological affinities with each of its adjacent landmasses: Honshū to the southwest, Sakhalin to the north, Sikhote-Alin to the west and the Kuril Islands to the east. Furthermore, Hokkaidō is washed by the waters of three ocean basins: the Sea of Okhotsk to the northeast, the Sea of Japan to the west and the Pacific to the south. The underlying rocks date to long before the Sea of Japan and Sea of Okhotsk basins opened, and are related to the subduction of oceanic plates, with some interesting characteristics. Take, for example, eastern Hokkaidō. Almost one third of Hokkaidō belongs to the Kuril Arc rather than the Honshū Arc, with affinities to Sakhalin and the Kuril Islands. The remainder of the island, central and western Hokkaidō, is presumed to have been attached to Sikhote-Alin and part of the continental margin before the opening of the Sea of Japan.

Earthquakes and Tsunami

All the vulcanism described above is a consequence of Japan's geology, in one of the most active tectonic zones on earth (see *p. 21*). Japan is in a region where four, and perhaps even six, tectonic plates converge, and where the Pacific and Philippine oceanic plates are subducting beneath the continental

Earthquakes are a daily occurrence, but only rarely are they devastating [BOTH KOCI].

Megathrust earthquakes off the Pacific coast may trigger tsunami [*top* FLIC; *bottom* SHUT].

plates on which most of Japan rests. Frequent volcanic activity, earthquakes and tsunami are all consequences of its position. Seismic activity is a daily occurrence in Japan, where about 10 percent of all of the planet's earthquakes occur; most of these are subliminal, but others are of immense power.

Hot springs are the best appreciated aspects of volcanic activity. Indoor and outdoor hot pools can be found throughout the country [TOOI].

Earthquakes have been reported throughout Japan's recorded history. The oldest historical account was documented in the year 416 CE, in Japan's second oldest volume of classical history – the *Nihonshiki*. The most powerful earthquake in the country's modern history was recent, occurring on 11 March 2011 in northeastern Japan, in the region of Tōhoku. This megaquake registered magnitude 9·0 on the Richter scale, and it and the subsequent tsunami devastated northeastern Honshū and led to terrible loss of life (about 20,000 dead or missing). The damage included the meltdown of three reactors at the Fukushima Daiichi nuclear power station complex. This concatenation of events impacted the economy of the entire country in what was a quadruple national disaster (earthquake, tsunami, nuclear disaster, collapse of tourism).

Megathrust quakes offshore have triggered enormous tsunami through Japan's long recorded history, most recently on 11 March 2011. Before that, the most powerful had been the Hoei event (M 8·6)[1] of 1707, which triggered a tsunami up to 26 m high from Kantō to Kyūshū. Although not so powerful, the Taishō Kantō earthquake and tsunami of 1923 (M 7·9) struck at the heart of Japan, and caused phenomenal loss of life when large areas of Tōkyō, Yokohama, Yokosuka and Chiba Prefecture were devastated and more than 105,000 people died. It is large, shallow earthquakes near the ocean floor that trigger such deadly tsunami.

Technically, the Japanese archipelago experiences three different kinds of large earthquake: the subduction megathrust; the intraslab or intraplate earthquake; and the shallow inland crustal earthquake. The first of these, the subduction megathrust, occurs where one tectonic plate is forced downwards and beneath another, essentially at a plate boundary, as is the case along the eastern seaboard of Japan. The second, the intraslab earthquake, takes place within a tectonic plate, and the third, the shallow inland crustal earthquake, eventuates below shallow coastal regions and inland and has points of origin, or hypocentres, shallower than 20 km. To those killed, injured or displaced, nothing matters more than the power of the earthquake causing the destruction and damage, irrespective of the earthquake's type.

The high frequency of earthquakes in Japan is well documented throughout the country's long history and is something that residents take in their stride with astonishing stoicism and fortitude. Japan even has its own unique way of measuring how earthquakes *feel* at each given location, using the Shindo Scale, a seismic-intensity scale developed in 1884 by the Japan Meteorological Agency. This differs from the scientifically accurate but harder-to-grasp Richter Magnitude Scale (developed in 1935 by Charles Richter), which rates the energy output of earthquakes on a logarithmic scale.

Earthquake-resistant architecture is steadily improving and is remarkably successful, but almost nothing can stand in the way of a megathrust quake and associated megatsunami, as was seen in the vivid footage of the 11 March 2011 disaster as it unfolded. Tsunami occur fairly often, perhaps even annually, although their scale varies enormously, and very few, perhaps only one a century, reach the monumental proportions of the 21-m wave generated on 11 March 2011, which towered an astonishing 39 m high when

---

1 M stands for magnitude on the Richter scale.

it washed ashore on the east coast of the Tōhoku region of northern Honshū and flowed inland for up to 10 km!

Hot Springs

Associated with volcanic activity are spectacular steam vents, brilliant sulphurous fumaroles (as at Iwō-zan, in Akan–Mashū National Park), bubbling mud pools (as at Beppu, in Kyūshū), and geothermally heated soil, beaches, and many warm springs at various locations around the country. Some of these springs appear on land, forming hot pools and streams. Others arise underwater, keeping lakes ice-free in winter. Wherever they are, it is likely that human or other animal life will be taking advantage of them.

Outdoor hot springs, known as *rotenburo*, make wonderful bases from which to observe wildlife. At one location in central Honshū, it is possible not only to watch monkeys from an outdoor bath but even to share the bath with them. Many of these baths are set amid trees, or in remote valleys in the mountains, and taking an early-morning dip is therefore ideal for watching local wildlife. On the Shiretoko Peninsula sea eagles can be watched from the bath, while in Nagano Prefecture there is a *rotenburo* from which sharp-eyed observers can see Japanese Serow.

Mammals and birds are often found around such pools, but there is life in the pools, too. Particularly in the hottest of all, at temperatures beyond those bearable by humans, there are highly specialized organisms, thermophilic bacteria that survive only at extreme temperatures. They form multicoloured filamentous streamers and microbial mats where they have colonized the bottom and sides of the spring. A little farther away from the intense heat one may find warm-water insects and fish active throughout the winter. Because the air at such places is warmer than that in the surroundings, insects are often able to remain active for longer there than elsewhere. From one *rotenburo* at the south end of Lake Kussharo, in Hokkaidō, it is possible to hear tiny (6–12 mm) Band-legged Ground Crickets calling throughout the depths of winter, when all else around is snow-covered and frozen. This is possible simply because these crickets live in the microclimate where geothermal activity warms the surface soil and their microhabitat.

Warm microclimates associated with geothermal activity provide niches suitable for some insects, such as Band-legged Ground Cricket, to remain active year-round [KaS].

Japanese Red Pine is typical of black-soil areas [MAB].

Soils

In relation to the underlying geology and vulcanology, combined with the wide variety of geographical and climatic conditions experienced throughout Japan, there are as many as ten soil types to be found in these islands. We can reduce this complexity to four broad and easily recognized soil types: black soil, red soil, podsol, and brown forest soil.

Typical black soils, derived largely from volcanic-ash deposits, are very compact and wet, and occur widely throughout Japan. In their natural state, they support native grasslands and Japanese Red Pine forest, or plantations of Japanese Larch, Japanese Cedar or Hinoki Cypress (also known as Japanese Cypress).

Red soils (lateritic soils) range in colour from orange-yellow to dark red and owe their colour to iron oxidation. They are rich in iron and aluminium and are produced in areas with hot, humid conditions where intense rock-weathering occurs. These are largely fossil soils dating back to the Pleistocene, and they occur mostly in lowland areas with low erosion. They naturally support Japanese Red Pine and are sometimes cultivated for bamboo or oranges.

Podsols are typical of subarctic or subalpine coniferous forests and alpine meadows. They are rather acidic and may support Japanese Stone Pine, Hinoki Cypress, Thujopsis, Northern Japanese Hemlock and Southern Japanese Hemlock, Veitch's Silver Fir or Maries' Fir[1].

Brown forest soil is typical of cool-temperate areas with moderate humidity, and supports mostly broadleaf deciduous trees. These soils range from dry to moist and the surface humus becomes well distributed in deeper layers. In northern areas they support Siebold's (or Japanese) Beech, Thujopsis and Japanese Red Pine, and in central areas of Japan Nikko Fir and Momi Fir.

The links between climate, soils and vegetation are complex, and there is ongoing debate over the relationship between the natural processes of soil formation and the human influences affecting them.

1 Named after the English plant-collector Charles Maries (1851–1902), who introduced the species to Britain in 1879.

Siebold's Beech favours areas of brown forest soil with a cool-temperate climate [AK].

Above Japan consists of archipelagos within archipelagos, and numerous isolated islands [OCVB].
Below The islands of Japan's subtropical south are ringed with coral reefs [ALL SM].

The Story Begins with the Landscape

The Japanese archipelago is situated off the east coast of the Eurasian continent and occupies the eastern border of the Asian continental shelf. As many as 6,800 islands stretch more than 3,000 km along the northeast coast of the Asian continent and cover an area of approximately 380,000 km^2 in the form of a long crescent, its outer curve facing the Pacific Ocean. From the farthest tip of Hokkaidō in the northeast to remote Yonaguni Island in the Nansei Shotō in the southwest, Japan has a coastline of more than 27,000 km. It spans almost 25° of latitude and 30° of longitude, at 20–45° north and approximately 123–154° east.

Japan's four main islands are, in decreasing order of size: Honshū, Hokkaidō, Kyūshū and Shikoku. Honshū means literally the 'mainland'. Together these four islands comprise 90 percent of Japan's land area. From a naturalist's perspective we may view these islands in two parts, as there is a great, natural disjunction between Hokkaidō in the north and the three large islands to the south, namely Honshū, Shikoku and Kyūshū, which from here onwards will be referred to as Central Japan[1]. Hokkaidō very noticeably shares its faunal and floral affinities with northern East Asia, whereas Central Japan shares its affinities with southern East Asia.

While there are many islands and rocks scattered off the coasts of each of Japan's four main islands, two subsidiary archipelagos form vital elements of the whole country. The first lies to the south of the central part of Honshū. This chain of islands is itself broken into a series of isolated volcanic islands: the Izu Islands in the north, the Ogasawara Islands, and even farther south the Iwō Islands, which are among the southernmost islands of Japan's territory. The second of these subsidiary archipelagos stretches between Kyūshū and Taiwan. Known variably as the Nansei Shotō (all of the islands) or the Ryūkyū Islands (strictly speaking, only from Okinawa southwards), they can be described as 'a string of pearls', as 'Japan's finger in the subtropical pie', and as an 'Asian Galápagos'.

Japan's altitudinal diversity is enormous, from 8,046 m below sea level to 3,776 m above sea level. The highest point *above* sea level is the summit of Mt Fuji. That peak stands alone as an isolated, active stratovolcano about 100 km to the southwest of Tōkyō, just beyond the western extremity of the Kantō Plain. Elsewhere are the extensive ranges of the Japanese Alps, where a number of peaks rise to more than 3,000 m. In contrast, Japan's lowest point lies to the east of Honshū, extending submerged 8,046 m *below* sea level in the Japan Trench. This deep oceanic trench continues to be created by the subduction process as the oceanic Pacific Plate advances beneath the continental Okhotsk Plate, in a process that continues to unleash earthquakes and tsunami in northern Japan.

The remote Ogasawara Islands lie approximately 1,000 km south of Tōkyō [MAB].

1 For convenience, and as a form of shorthand, I have adopted the term Central Japan to distinguish Honshū, Shikoku and Kyūshū as a group separate from Hokkaidō, the Nansei Shotō, the Izu, Ogasawara and Iwō island chain, and any other subsidiary island groups.

More than 70 percent of Japan is mountainous, consisting of mountain ridges uplifted by the shifting tectonic plates. More than 18,000 peaks have been named. One hundred and eighty or more of these mountains are volcanoes, 110 of which are considered active[1], and 50 of these are monitored by the Japan Meteorological Agency. The two most active are less than 50 km apart in Kagoshima Prefecture (see map *p. 22*)[2]: Sakura-jima[3], in Kagoshima Bay, and Shinmoe-dake, in the Kirishima Mountains, the first of these being at alert level three (the highest level).

Japan is prone to natural disasters of almost every kind, including earthquakes, volcanic eruptions, tsunami, typhoons, floods, landslides and avalanches. Given the number and extent of Japan's mountains, there are sensible religious, cultural and practical taboos against building on steep and unstable mountain slopes. As a consequence, the majority of the human population and the entire agricultural and industrial economy of Japan are concentrated in less than 30 percent of the country, in areas that consist of coastal plains and inter-mountain basins. It is these crowded basins, plains and coastal lowlands that have helped to create an image of Japan as small, cramped and crowded.

Japan has few plains, and they consist largely of highly fertile volcanic soils washed out from the mountains into alluvial fans, lowlands and deltas, making them ideal for agriculture. Ironically, much of this rich farmland has been consumed by expanding urbanization and industrialization. The largest of Japan's plateaux, the Kantō Plain of coastal central Honshū, has an area of 17,000 km^2. It was once covered with forest interspersed with rivers, swamps and low hills, but now much of it consists of intensively farmed land and vast conurbations, including Tōkyō, the country's modern capital. In addition, the coast around Tōkyō Bay has been drastically modified by land reclamation.

In the far southwest lie the Yaeyama Islands, including Ishigaki and Iriomote [MC].

Steep terrain is prone to rapid erosion during the rainy season and typhoon season [BOTH MAB].

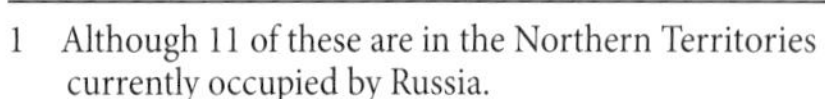

1 Although 11 of these are in the Northern Territories currently occupied by Russia.

2 Japan, since 1890, has had 47 administrative districts, known as prefectures.

3 In Japanese, the suffix shima (sometimes -jima), means 'island', hence Sakura-jima translates as Sakura Island, and Kirishima means Kiri Island.

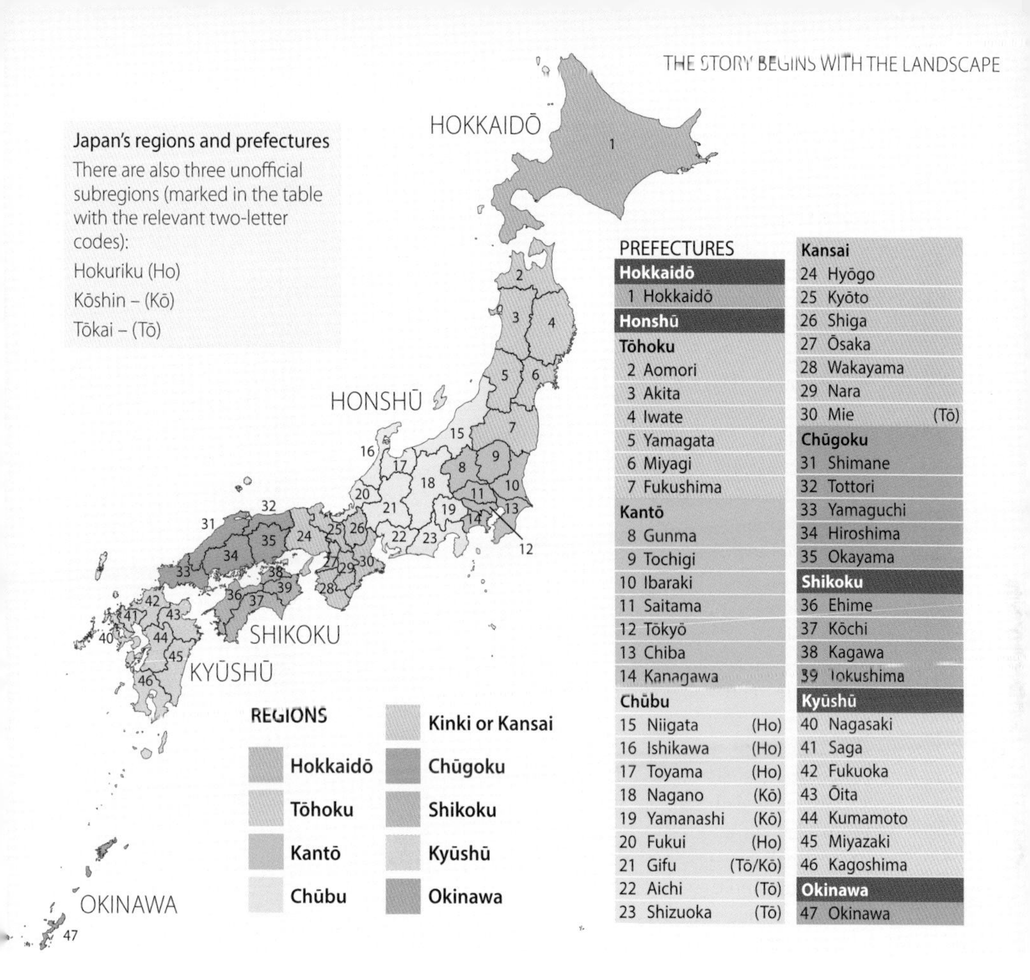

Japan's regions and prefectures

There are also three unofficial subregions (marked in the table with the relevant two-letter codes):

Hokuriku (Ho)

Kōshin – (Kō)

Tōkai – (Tō)

PREFECTURES

No.	Prefecture	Subregion
Hokkaidō		
1	Hokkaidō	
Honshū		
Tōhoku		
2	Aomori	
3	Akita	
4	Iwate	
5	Yamagata	
6	Miyagi	
7	Fukushima	
Kantō		
8	Gunma	
9	Tochigi	
10	Ibaraki	
11	Saitama	
12	Tōkyō	
13	Chiba	
14	Kanagawa	
Chūbu		
15	Niigata	(Ho)
16	Ishikawa	(Ho)
17	Toyama	(Ho)
18	Nagano	(Kō)
19	Yamanashi	(Kō)
20	Fukui	(Ho)
21	Gifu	(Tō/Kō)
22	Aichi	(Tō)
23	Shizuoka	(Tō)
Kansai		
24	Hyōgo	
25	Kyōto	
26	Shiga	
27	Ōsaka	
28	Wakayama	
29	Nara	
30	Mie	(Tō)
Chūgoku		
31	Shimane	
32	Tottori	
33	Yamaguchi	
34	Hiroshima	
35	Okayama	
Shikoku		
36	Ehime	
37	Kōchi	
38	Kagawa	
39	Tokushima	
Kyūshū		
40	Nagasaki	
41	Saga	
42	Fukuoka	
43	Ōita	
44	Kumamoto	
45	Miyazaki	
46	Kagoshima	
Okinawa		
47	Okinawa	

Extensive plains are a rare feature in mostly mountainous Japan [SHUT].

Lake Biwa, Japan's largest and oldest lake, is one of the oldest surviving lakes in the world. It was formed by tectonic activity about 4 Mya [ChM].

Other plains are much smaller than Kantō. They include the Ishikari Plain (3,800 km^2), in western Hokkaidō; the Tokachi Plain (3,600 km^2), in south-central Hokkaidō; the Nōbi Plain (1,800 km^2) of Gifu and Aichi Prefectures; and the Abeno or Ōsaka Plain (1,600 km^2), home to Ōsaka, another of Japan's major conurbations.

The mountainous topography of Japan, combined with its small size, means that the country's hundreds of streams and rivers are short, steep and thus fast-flowing. Their upper reaches flow through freshly carved V-shaped valleys, while their lowest reaches form alluvial plains. The longest and widest of Japan's rivers, the Shinano, runs just 367 km from its source in the Japanese Alps below Mt Kobushi (draining most of Nagano and Niigata Prefectures) to the Sea of Japan.

While montane ponds, inland pools and coastal lagoons are common, Japan has few major inland freshwater lakes. The largest and oldest is Lake Biwa (670·3 km^2), in the Kansai Region, while the second largest, Kasumigaura (167·6 km^2), is situated on the Kantō Plain. Hokkaidō has three of Japan's ten largest lakes and each of them is a caldera lake: Lake Kussharo (79·3 km^2), Lake Shikotsu (78·4 km^2) formed about 32,000 years ago and Lake Tōya (70·7 km^2) formed about 25,000 years ago.

Streams and rivers are typically short, steep and fast-flowing [*above* TsM; *below* AK].

Oceans, Seas and Currents

A storm strikes Japan's Pacific coast [TsM].

Nowhere in Japan is far from the sea. While not always in sight, the sea nevertheless exerts a powerful influence on the islands, the climate, the fauna and flora, and the people of Japan.

Sometimes placid, sometimes violent, the North Pacific Ocean washes the entire eastern seaboard of Japan. This coast has few continental islands, but many volcanic islands arise south of central Honshū (the Izu, Ogasawara and Iwō island groups) and northeast of Hokkaidō (the Kurile island chain). To the north of Hokkaidō lies the Sea of Okhotsk, which, although cold even in summer, is frigid during the winter months when covered with sea-ice. To the west of Honshū and separating it from both Russia and the Korean Peninsula is the Sea of Japan, which harbours a scattering of islands, the largest of which are Sado and Tsushima. To the west of Kyūshū and the island chain of the Nansei Shotō is the very much warmer East China Sea. Extending between Kyūshū, Honshū and Shikoku is the Inland Sea or *Setonaikai*, with numerous tiny islands. This sea has a surprisingly Mediterranean climate.

Several ocean currents meet and mix at the mid-latitude location of the Japanese archipelago. Surface ocean currents that flow past the archipelago have a strong influence on the climate of Japan. They also affect the distribution of marine planktonic organisms, the organisms that feed on them, and so, too, the fish and birds that form the marine food web. Two of these currents deserve special mention. The warm, subtropical *Kuroshio* current flows northwards past the southern coasts of Kyūshū, Shikoku and Honshū. The cold *Oyashio* current carries cold subarctic water southwards from the Bering Sea, the Kuril Islands and the Sea of Okhotsk past the east coast of Hokkaidō and the northeast coast of Honshū. The strong *Kuroshio* exerts a powerful warming influence on the oceanographic conditions and biodiversity of the Nansei Shotō and western Japan. Flowing up the Pacific Ocean and passing Taiwan, it penetrates the Yonaguni Depression between Taiwan and Japan's Yaeyama Islands, and passes up the East China Sea before flowing back into the Pacific through the Tokara Strait near the northern end of the Nansei Shotō, between Amami Ōshima and Tanega-shima. Thereafter, the *Kuroshio* flows up the east coast of Kyūshū and northeast off the southern coast of Shikoku, passing the Kii, Izu and Bōsō peninsulas before flowing eastwards out into the Pacific.

The *Oyashio* follows a much simpler route from its origins in the Arctic Ocean. It flows via the Bering Sea down the Pacific side of the Kuril Islands, passing Hokkaidō and northeast Honshū before it, too, flows out eastwards into the Pacific. The precise location where the two currents meet and mix varies with season, and where they meet the result is a significant increase in the abundance and diversity of marine life. The cold surface waters of the *Oyashio* carry a high nutrient load and, when they mix with the warm, more saline *Kuroshio*, the resulting eddies cause upwelling of nutrient-rich deeper water to the surface. The abundant nutrients greatly increase primary productivity which supports a large and diverse food chain above it, and this in turn sustains prolific marine life.

Three further currents strongly influence Japan. The warm Tsushima Current, a branch of the *Kuroshio*, flows from the East China Sea northeastwards between the Korean Peninsula and Japan and along the Sea of Japan coast to northern Honshū. The Tsugaru Current flows eastwards from the Sea of Japan into the Pacific through the Tsugaru Strait (just 19·5 km across at its narrowest point) separating Honshū from Hokkaidō. The warm Sōya Current flows from the Sea of Japan into the Sea of Okhotsk via the La Pérouse Strait between Hokkaidō and Sakhalin.

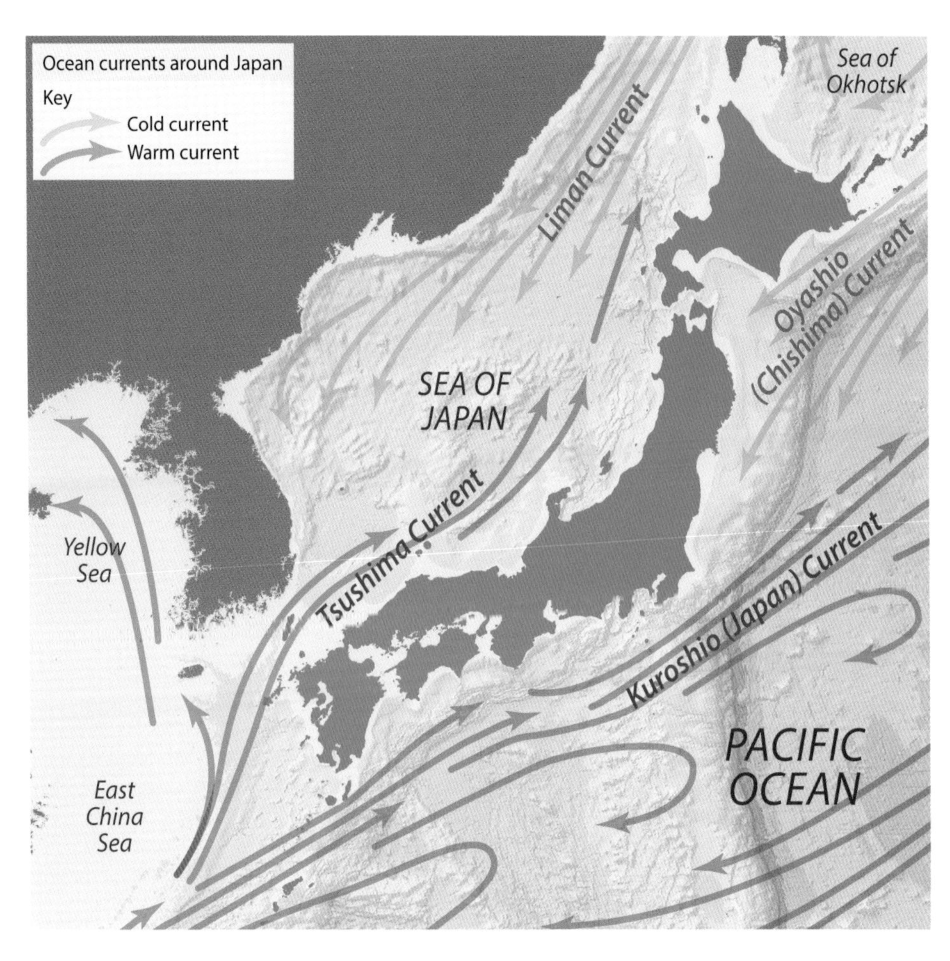

Drift-ice reaches the Sea of Okhotsk coast of Hokkaidō each winter; it flows round the Shiretoko Peninsula into the Nemuro Strait [MAB].

The southern islands bask in subtropical warmth, and have coral-sand beaches and reefs offshore [*left* MAB; *right* TsM].

Cold currents, warm currents and their mixing, along with the various countercurrents and gyres that occur around Japan, contribute to the great diversity of marine life around the country. In the warmer, southern regions coral reefs and tropical fish occur, as do various species of turtle and shark. In colder, northern parts kelp beds and northern fish abound. The whole region also supports many marine mammals, including Humpback Whale, which breeds in the south, and Sea Otter, Steller's Sea Lion and Northern Fur Seal, which occur only in the far north. Furthermore, the marine and coastal currents also exert an enormous influence on the climate, and thus on the flora and fauna onshore.

Japan is home to considerable marine diversity, including migratory Humpback Whales [ZWWA].

Climate

Japan and the concept of four clear seasons seem to be culturally inseparable, yet if we count them as Japanese people describe them – winter, spring, rainy season (the early-summer monsoon period), summer, typhoon season, and autumn – the country in fact experiences six notable seasons, each with very different characteristics. Not surprisingly then, Japan's climate is diverse, both seasonally and regionally, with seasonal temperatures ranging locally from below −30°C up to above +40°C. Each of Japan's seasons impacts on the environment and has a profound influence on the flora and fauna of the country.

Japan is frequently described as having a generally temperate climate that ranges from subarctic in the north to subtropical in the south, with differing conditions between the Sea of Japan side and the Pacific side of the country, and

Contrasting seasons not only have a profound influence on the country's biodiversity but also bring drama to the scenery, especially in winter [WM].

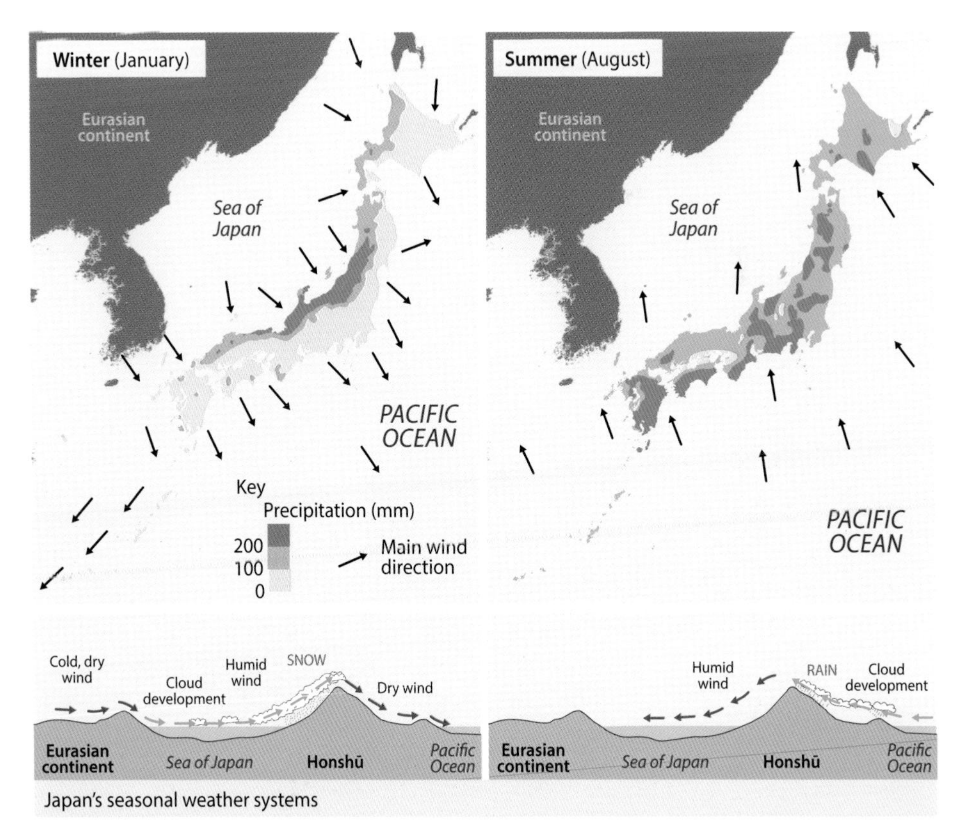

Japan's seasonal weather systems

While flowering cherries are an iconic symbol of Japan in spring (here at Hirosaki Castle, Aomori Prefecture), many other species of flowering tree occur here too [FT].

As spring fades, the early-summer rainy-season front reaches southern Japan [HT].

Strong sunshine and high humidity mark summer, here on Amami Ōshima [TsM].

high precipitation throughout the year. This is a great simplification of complex conditions that arise around the country.

The length and orientation of the Japanese archipelago ensures that it spans a wide range of climatic zones[1] and experiences great variation in time and space. Northern regions, such as Hokkaidō and Tōhoku, enjoy mild summers and have very cold winters. Northern central (or eastern[2]) Japan, particularly the Hokuriku, Kantō/Kōshin and Tōkai regions, and the Kantō Plain and the Japanese Alps of central Honshū, has hot, humid summers and cold winters, with heavy winter precipitation. Southern central (or western) Japan, in the Kinki, Chūgoku, Shikoku and north Kyūshū regions, has very hot, humid summers and only moderately cold winters with little snow. To the south, in south Kyūshū (Amami Ōshima) and Okinawa, the climate is distinctly subtropical and oceanic, with hot, humid summers and mild winters.

During winter (December, January and February), high-pressure (Siberian High) systems develop over the Eurasian continent northwest of Japan while low-pressure (Aleutian Low) systems develop over the northern North Pacific Ocean. The prevailing winds during winter are from the northwest, carrying cold air and heavy clouds to northern Japan and bringing heavy snowfalls to the Sea of Japan coast and the mountainous spine of the northern part of the archipelago. Meanwhile, the Pacific side enjoys cold sunny weather whereas Amami Ōshima and Okinawa, in the farther subtropical south, have mild winters.

During spring (March, April and May), cyclonic (low-pressure) and anticyclonic (high-pressure) systems move eastwards across Japan, bringing steadily warming temperatures from the south. Towards the end of spring, in May, Okinawa and then Amami Ōshima begin to experience the *Baiu* or rainy season.

Summer (June, July and August) begins with the rainy season, an early-summer monsoon dominating much of southern Japan east to Tōkyō. The heavy precipitation at this time of year is a consequence of the stationary *Baiu* front forming where the warm tropical maritime air mass of the south meets the cool polar maritime air mass of the north. Heavy rains frequently bring local flooding and cause numerous landslides. This explains the extent of concrete poured in Japan – its purpose being to shore up hillsides and to canalize rivers, hastening the rainwater to the sea.

1 Six zones are commonly defined: Hokkaidō, Sea of Japan side, Pacific Ocean side, Inland Region, Inland Sea (Setouchi), and Southwestern Islands (Nansei Shotō) . These are clearly strong factors driving the distribution of Japan's flora and fauna.

2 The shape of the Japanese archipelago is such that it can be divided into eastern and western regions, everywhere west of Tōkyō being western. Equally, it can be seen that everywhere 'east' of Tōkyō is in reality to the north!

Eventually this frontal system disappears, giving way as the North Pacific High extends across Japan from the southeast. With it come the hot, sunny conditions of true summer. Southern and central regions are affected annually by a rapidly advancing strong, warm spring wind known as *Haru Ichiban*, the very first experience of which each year is normally between mid-February and mid-March in southern and western Japan. This is followed by a spring or early-summer monsoon or rainy season, with typically hot summers and a powerful late-summer or autumn typhoon season, which typically lasts from July to October, although this period is steadily becoming longer.

Meanwhile, in the south, in Okinawa and Amami Ōshima, the first battering impact of seasonal typhoons[1] is felt usually during late summer. Like hurricanes, typhoons are tropical cyclones that form over warm ocean water near the equator. Warm moist air serves as fuel to these mighty engines of wind and rain, driving them in a spiral fashion on relatively predictable tracks, although often with unpredictable strengths (strengths likely to increase as the oceans warm). As the global climate has warmed in recent decades seasonal boundaries have become blurred, typhoons sometimes overlapping with the rainy season (May) and even occurring late into autumn (November); typhoons become increasingly powerful as the season progresses.

Autumn (September, October and November) sees temperatures falling steadily combined with high precipitation, especially in September, associated with an active autumnal rain front and numerous tropical depressions, tropical storms and typhoons. Typhoons regularly batter southern Japan, and more rarely northern Japan, even on occasion reaching Hokkaidō, with powerful winds exceeding 120 kph, and sometimes exceeding 190 kph in violent typhoons. Anticyclonic systems bring cool sunny conditions to Japan in October, by which time the first snows usually fall in the northern mountains. During November, cold northwesterly air flows increasingly across Japan, temperatures fall further and precipitation increases.

Battering winds and lashing rains mark the annual typhoon season [TsM].

1 Tropical cyclones in south and east Asia and the western Pacific are known as typhoons, a name derived from Chinese tai fung, meaning big wind. Those that form over the Atlantic Ocean or eastern Pacific Ocean are known as hurricanes, a name derived via Spanish, probably derived from Taino hurakán, meaning 'god of the storm'.

Autumn in central and northern Japan is rich and colourful [*above left* AK; *above right* MAB; *below* TK].

Ice and Sea-ice

Japan's great length allows for tremendous contrasts in climate and habitats [MAB].

There are interesting choices to make in Japan, this being due to the broad climatic variation experienced by the archipelago. In February, for example, there is a choice of reclining on a white coral-sand beach while layering on sunblock at a resort in Okinawa or, alternatively, layering on clothes and visiting northern Hokkaidō. There, the volcanic-sand beaches are intrinsically grey or black rather than coral-white, although in February this is often obscured by snow and ice. This is the season of sea-ice, when great blocks and packs of ice drift in on the wind and currents from farther north in the Sea of Okhotsk.

Asymmetric heating of land and sea and seasonal changes in atmospheric circulation lead to the precipitation associated with a monsoon. In addition to its summer monsoon, Japan experiences the East Asian winter monsoon. This typically brings cold dry northwesterlies that blow from the continent across the Sea of Japan to Japan's western shores. The winter monsoon's cooling effect lowers the surface-water temperatures of the Sea of Okhotsk, thereby facilitating sea-ice formation, which is further seeded by freshwater ice flowing out of the mouth of the Amur River in the northwest Sea of Okhotsk. This freezing has reinforcing effects in intermediate and deeper waters in both the Sea of Okhotsk and the Sea of Japan. As seawater begins to freeze and forms ice crystals, typically at temperatures below −1·9°C, salt is excluded in a process known as 'brine rejection'. As a consequence, the surface seawater surrounding the ice becomes both saltier and denser and sinks beneath the surface waters, carrying its load of oxygen with it and replenishing the waters at middle and deep depths.

As the winter progresses, the northern part of the Sea of Okhotsk freezes over and ice drifts

Currents and northerly winds drive drift-ice south to the Sea of Okhotsk shore of Hokkaidō [ABTA].

southwards, eventually fetching up against the Hokkaidō coast between Wakkanai and the Shiretoko Peninsula. In some years ice also drifts around into the channel separating Hokkaidō from Kunashiri[1], and sea-ice may then be found off Cape Nosappu near Nemuro, and sometimes even reaches Kushiro.

The startling spectacle of an ice-filled horizon, instead of the usual sea, is fascinating for naturalists. The cold water associated with the drifting and melting ice is home to numerous planktonic creatures and various kinds of fish, and hence to fish-eating mammals such as sea lions and seals, and of course birds. Seabirds, particularly diving sea-ducks, haunt the breaks and leads within the ice, while two species of fish-eating eagle (the enormous Steller's Eagle [VU], boldly patterned with huge white shoulders, tail and thighs, and a massive yellow beak, and the slightly smaller White-tailed Eagle [NT]) use upraised ice-blocks as lookout posts and resting

1 Etorofu (Iturup), Kunashiri, Shikotan and the Habomai Archipelago, known as the Northern Territories in Japan, were taken by the Soviet Union in 1945 and have remained Russian-occupied ever since.

The drifting ice represents a mobile wildlife habitat [ABTA].

A range of species is associated with the ice, from massive Steller's Sea Lions to delicate Sea Angels [ABTA].

Steller's Eagles are among several species that use ice as a floating platform [MAB].

sites in their relentless search for fish prey. Having caught food, the eagles bring it back to the ice, which then serves as their dining table.

Eagles are not the only fishers here. Seals, too, follow the ice. Steller's Sea Lions migrate to the waters off Hokkaidō from their breeding grounds in the northern Kuril Islands, and they can be seen tearing through the water around the sea-ice or occasionally even hauled out and resting on the ice itself.

Drift-ice at sea is by no means the only form of ice to be found in winter in Japan. Bizarre ridges of ice form where enormous plates of lake ice meet and crush together, uplifting ridges that snake for kilometres across frozen lakes. Ice flowers blossom on smooth lake ice and prolonged spells of freezing weather force needles of ice up from the soil or beneath leaf litter, forming strange bundles of elongated crystals. Meanwhile, in caves, Moomin-like figures appear to gather in groups as stalagmites of ice form beneath dripping ice stalactites. The wind then sculpts and flutes them into dramatic blades and points.

Other icy phenomena revealed as winter begins to wane are the hummocks and mounds of sorted gravels piled on beaches where drift-ice has been driven ashore, washed back out and ultimately overturned. When the ice chunks are flipped over, the gravels that first become frozen to the underside of the ice are exposed to the warmth of the sun, and now they steadily melt their way down and drain through, sorted by size as they go. Onshore, spring warmth strikes tree trunks, which then radiate back that heat, melting out inverted cones around their bases and thereby forming tree wells deep enough sometimes to swallow, temporarily, spring skiers.

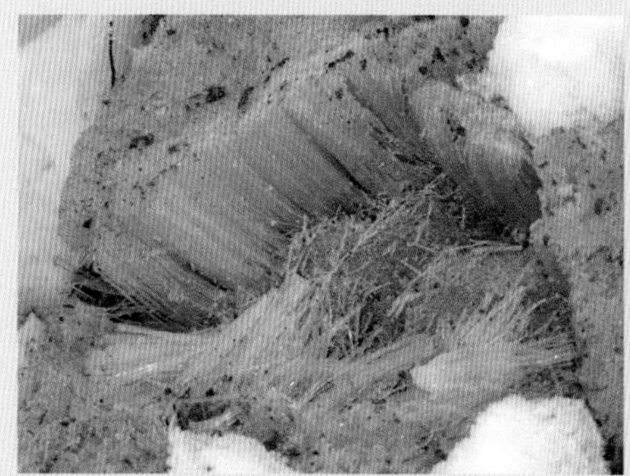

Closely-packed, elongated columns of ice emerge from the soil and are known as ice needles or needle ice [MAB].

Pressure ridges of ice build and snake across several frozen lakes in northern Japan. These are known locally as the paths of the gods, or goblin ways [MAB].

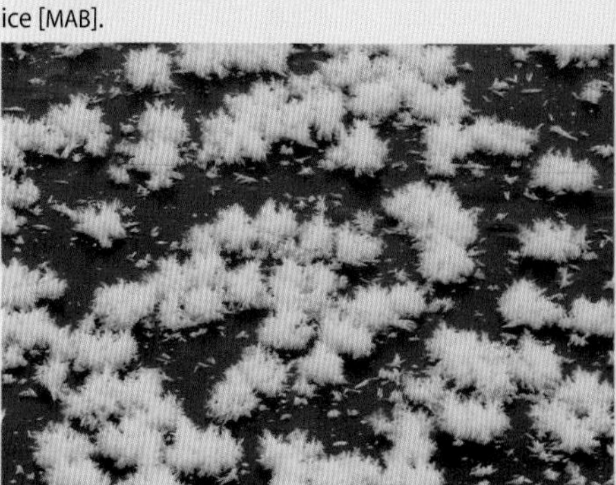

Frost flowers on Lake Kussharo, Hokkaidō [MAB]

Cave ice forms in fantastic shapes [MAB].

Above left Blocks of ice, jumbled ashore by storm surges and high tides, become stranded. Beach sands, gravels and pebbles beneath them freeze to the underside of the ice; further tide surges and currents may drag that ice back out to sea, only for it to be jumbled and flipped as it is driven back ashore. What was once the underside of the ice, with its load of sand, gravel and pebbles, becomes the top surface exposed to the sun. The shoreline debris is warmed and, with help from gravity, steadily melts its way down through the ice and is roughly sorted by size as it goes. Eventually, the material will melt its way completely through the ice to drain out on to the beach below. Once all of the beached sea-ice has melted away in spring, strange deposits of sorted and graded sands, gravels and pebbles can be found on the beach, as if some landscape artist has been at work;
Above right Meanwhile, onshore, another phenomenon involving spring warming is underway. During winter, regular snowfalls build into a deep carpet of snow, covering the forest floor. Winds, blizzards and drifts pile snow higher around tree bases, so that the trunks resemble the chocolate flakes that are often pushed into the top of a soft ice-cream. Once the peak of snow accumulation has passed, the snow pack begins to settle, the layer becoming denser, firmer and more icy. Now the raised snow against tree trunks melts first, warmed by any sun hitting the tree until the trunk's radiated warmth melts the snow below the level of the main snow pack. The result is like that same chocolate flake standing in an empty ice-cream cone. At the top, at the level of the snow layer, the widest point of the cone may be a metre or more in diameter. The cone tapers down towards ground level and may sometimes be deep enough to swallow, temporarily, spring skiers (I speak from experience, having slid right down into a particularly deep one). Eventually, after a succession of warm, spring-like days, the ground will be revealed, and this is where the first spring flowers begin to appear, or where the dwarf bamboo first emerges;
Below Prolonged periods of low temperatures (below −15°C) bring hoar frost to the trees, 'diamond-dust' to the air, and frost flowers to exposed lake and river ice [ALL MAB].

Fur over Feathers

Largha Seal [YaM]

Steller's Sea Lions reach certain coasts of Hokkaidō in winter, but are rare elsewhere [YM].

Japan's list of coastal marine mammal species is impressive: Sea Otter, Northern Fur Seal, Steller's Sea Lion, Walrus, Bearded Seal, Kuril Seal, Largha (or Spotted) Seal, Ringed Seal and Ribbon Seal. There was once also Japanese Sea Lion, now sadly extinct. Of the living species, five are rare visitors, leaving four regular visitors. These occur mostly in winter, when they travel south with the cold current flowing past the Kuril Islands from the Bering Sea towards Japan, or with sea-ice drifting south down the Sea of Okhotsk.

Sea Otter, a relative of the otters, weasels and badgers in the family Mustelidae, ranges along coastal areas of the North Pacific and Bering Sea on both the North American side and the Asian side. It is closely associated with kelp and is considered a keystone species in the kelp beds of the near-shore ecosystem. In Japan, it is rare and found only around Hokkaidō coasts, primarily in the east. Sea Otter is a surprisingly large marine creature of the littoral zone, weighing up to 45 kg and measuring up to 110 cm in length, with a 30-cm tail. It has a distinctly rounded face and a thickly furred body, and it ranges in colour from blackish-brown to pale sandy-brown, varying in coat coloration depending on the individual and on whether it is wet or dry.

Sea Otters feed on a range of marine organisms, including shellfish, fish and, especially, sea urchins, which they obtain by diving in coastal waters down to as much as 90 m. They are commonly seen floating face upwards, and particularly so when nursing young. They may carry a large stone to the surface, resting it on the chest and using it as an anvil on which to crack open food items such as urchins. Because they forage on urchins and other grazing invertebrates of the kelp beds, they help to maintain those beds, which serve not only as vital fish nurseries but also as crucial absorbers of carbon dioxide.

Northern Fur Seals follow the cold Oyashio current south to northern Japan [NB].

Unlike pinnipeds, Sea Otter lacks blubber, so it is completely dependent on its fur for insulation. Its exceedingly thick fur is reputedly the densest of any animal's, the guard hairs covering an underfur with as many as 120,000 to 140,000 hairs per square centimetre, whereas Eurasian Otter has 'only' about 60,000 to 80,000 hairs per square centimetre. This dense coat requires extensive grooming to maintain its insulation and, although most grooming is done in the water, Sea Otter will occasionally emerge onshore to rest and groom.

The Sea Otter's population was estimated to have numbered up to 300,000 individuals, but hunting from 1741 to 1911 reduced the population to 1,000–2,000, which survived in small pockets of its historical range. Hunting and trade in its luxuriously dense fur led to the near extinction of Sea Otter during the latter half of the 19th century. Earlier, local hunting by Ainu[1] and Russian fishermen prior to the 19th century probably did little to dent the Sea Otter's population, but wider recognition of its presence in Kuril Island waters in 1872 led to its dramatic demise there. Ten thousand skins were taken from that region alone from 1872 to 1881[2]. Such over-exploitation led to a decrease in the numbers taken by hunters to around 1,000 from 1882 to 1891, to about 800 from 1892 to 1901, and to a mere 350 between 1902 and 1909. Hunting of the population in the Kuril Islands, and perhaps also in east Hokkaidō waters, is documented in Henry Snow's[3] book *In Forbidden Seas*, which is a potent reminder of how quickly plenty can be reduced to poverty when it comes to unmanaged and unrestrained harvesting of wild-animal populations.

International protection from 1911 led to a slow, but steady recovery in much of the Sea Otter's range. During the late 20th century, the expanding population in the Kuril Islands led to an increasing number of records off east Hokkaidō. The species has been seen from Abashiri eastwards around east Hokkaidō to Cape Erimo, in south-central Hokkaidō, but it is most likely to be encountered along the Shiretoko

1 The Ainu are the indigenous inhabitants of northern Japan and nearby areas of Russia.
2 Blakiston proposed a licensed quota system, but this was rejected by the Japanese government.
3 Henry Snow was a former clerk in Blakiston's company.

Sea Otter [YaM]

Sea Otters are slowly recolonizing eastern Hokkaidō from the Kuril Islands [WaM].

Sea otters are most likely to be found around kelp beds where they search for a favoured food — sea urchins; only occasionally do they come ashore [BOTH KS].

Sea Otters rely on their extremely dense underfur for insulation in cold water [WaM].

and Nemuro Peninsulas. Although still rare, Sea Otters are now regularly sighted around (and even on) the rocks off the rocky headland of Cape Nosappu overlooking the Nemuro Strait, around the Moyururi and Yururi islets off the southeast coast near Cape Ochiishi, and farther west around Cape Tōfutsu at Kiritappu. Previously, this species was a winter visitor from farther north, but it may now be breeding again in Japanese waters. It is, however, unlikely to expand far, as both kelp beds and sea urchins in Hokkaidō are routinely and extensively harvested for human food. Farther north, in the Kuril Islands, Sea Otters are sociable, living in groups of several individuals, sometimes dozens gathering together, despite females producing only one youngster a year.

The smaller Eurasian Otter used to occur widely throughout Japan. This medium-sized, brown, riverine mustelid once occurred in secluded rivers and along coasts near river mouths of the four main islands and also on offshore islands, including Tsushima. An aquatic, crepuscular carnivore, it foraged along rivers to their estuaries in search of prey, which included freshwater fish, crustaceans, invertebrates, birds, and small mammals such as voles. Eurasian Otter in Japan is generally designated as an extinct endemic subspecies *Lutra. lutra whitleyi.* Some Japanese taxonomists have considered it to be a full species, *Lutra nippon*, although there is little evidence to support this. Riparian habitat loss, disturbance and, especially, hunting from 1868[1] until 1927 led to swift range contraction and rapid population decline. As many as 1,000 individuals a year were hunted during the 1910s and 1920s, until between 1928 and 1953 only isolated local populations survived, and from 1954 onwards the species declined towards extinction. In 1960, Eurasian Otter was made a National Natural Monument, and in 1965 it was elevated to a Special National Natural Monument, but that recognition came too late to save it. During the 1950s, 1960s and 1970s only dead individuals were reported. By 1972, only 17 individuals were estimated to survive in the species' final stronghold in Shikoku. The last sight records seem to have been in 1979 and 1983, in Kōchi Prefecture. It was officially declared extinct nationally in 2012, having not been seen for more than 30 years. In 2017, however, Eurasian Otter footprints and droppings (confirmed by DNA analysis) were found on Tsushima Island, indicating either that a tiny population survives there or that a small number have colonized recently from the Korean Peninsula. Occasional reported sightings of Eurasian Otter in Hokkaidō or northern Honshū invariably refer to the introduced and invasive American Mink, which has established a feral population in those areas since the 1960s.

1 1868 is the single most significant year in recent Japanese history. One of the many things that the Meiji Restoration ushered in, temporarily at least, was greater freedom for gun ownership. This seems to have lasted until about 1872 according to Blakiston's notes.

Fat is a Feathery Issue

That young animals and birds typically start off small, and remain smaller than their parents for a long time, seems to be a dominant rule of life. Think of fox or badger cubs; think of young sparrows or bulbuls. They may appear like fluff balls, but from birth or hatching, and for several weeks or even months thereafter, they remain smaller than their parents. Of the Whooper Swans that migrate to Japan each winter, the cygnets, although capable of making the arduous migration, are still noticeably smaller than their parents six months after hatching. By that age, most young small birds (passerines) will be indistinguishable from their parents not only in body mass but also in feather length and in plumage colour. There is, however, one group that not only bends this rule, it shatters it.

The long-winged gliders of the ocean, the group of seabirds known to ornithologists as the Procellariiformes and to naturalists as albatrosses, shearwaters and petrels, do not play by the accepted rules. Among them, offspring routinely attain body masses that may be 60 percent higher than those of their parents. More than that, they accumulate fat at an astonishing rate, and then lose most of it before leaving the nest. This strange weight change has for a long time been assumed to be an insurance strategy. The parents feed far off at sea, where foraging conditions and food availability are notoriously variable. They may be away on foraging trips for days or weeks at a time. Some southern-ocean albatrosses take that to extremes. Breeding albatrosses from South Georgia are known to forage, for example, off the Brazilian coast, while non-breeders, or biennial breeders in their year off, may circumnavigate Antarctica on foraging trips. During the prolonged absences of their parents, the fluff-ball seabird chicks must wait out both hunger and whatever the fickle climate throws at them. Many of the albatrosses, shearwaters and petrels nest on remote islands in the Southern Hemisphere, where summer storms, even snow, are likely. The chicks may not eat for days and may experience chilling winds, rain or snow. Their fat layer then is their insurance policy against the weather and the delayed return of their parents – a reserve for them to live off. At least, that was the perceived wisdom, and it seemed to make sense. Recent research, however, has shown that it is only part of the story.

Streaked (*left*) and Short-tailed (*right*) Shearwaters [YaM]

Streaked Shearwater is Japan's most abundant breeding shearwater [JoH].

Once young albatrosses, shearwaters and petrels leave their nests they are on their own. There are no more free handouts, and no more family support; they must go it alone from day one. Their departure for life at sea has all the pitfalls of pushing the average person off the street into a light aircraft and saying 'Go fly this thing until you find a fuel supply!' The young seabird, such as a Streaked Shearwater of Japan or a Short-tailed Shearwater from the Antipodes, will have spent weeks cooped up in a burrow, its wings an unknown quantity until its instincts eventually drive it to begin stretching them at the entrance of the burrow. Until this stage, all food has come by parental delivery. A richly concentrated, partly predigested seafood chowder has been delivered to its door, in fact directly to its throat, with a regularity that has allowed it quickly to lay down an extensive layer of body fat. Then, one day, neither of its parents returns. Suddenly it is on its own. It must metamorphose from being totally dependent on the efficiencies of its parents' foraging to becoming a flying machine capable of navigating its way across oceans and finding its own food on the way. The requirements from day one are enormous. Is this not the point in their lives when the chicks are at the greatest risk? Should not this be the period for which they carry the fat, to see them through their first inexperienced days of learning to fly and learning how to recognize and catch food? Yet generation after generation of these birds puts on weight massively while in the nest, reaches a stage at which it dwarfs its parents, then slims right down *before* it leaves the nest burrow. Seen in that light, there is clearly something missing from the story; surely they should hold on to the fat for another week or so to see them through their flying and foraging education.

The remote island of South Georgia, far down in the South Atlantic between the Falkland Islands and the Antarctic Peninsula, has for long been the research territory of the British Antarctic Survey (BAS), and fascinating discoveries keep coming to light. Recent research there on Grey-headed Albatross, a relative of the three albatross species breeding in Japan, has revealed that chick growth has two distinct phases. During the first phase the internal organs develop, including those that process and use the rich food supply that the parents deliver to their chick. During the second phase, when the chick is between 60 and 80 days old, it develops the muscles and feathers required for flight. If the insurance-policy theory was correct, the chick would be at risk right from the beginning and it should accumulate fat from the very beginning. What researchers have found, however, is that the young albatross begins to lay down fat only during the second phase of growth, in other words after it is 60 days old. Suddenly, the insurance-policy theory seems even weaker.

Streaked Shearwaters breed on remote islands, where they are extremely susceptible to being preyed upon by introduced mammals such as cats [SuT].

What the BAS researchers have shown is that this later phase of growth has stricter limitations than the first one has. After all, if there is a food shortage during the first phase of a chick's life it will result simply in slower growth of its internal organs. Growth rate can increase once food supplies increase. During the second growth phase, however, the crucial flight feathers develop, and any interruption in their growth may weaken them and ultimately render the bird a less efficient flier. Adult seabirds replace their flight feathers slowly. They drop symmetrical pairs, one from each wing, or one from each side of the tail, and the replacements grow out slowly. The energy and nutrient demands of feather growth are enormous, and adult birds spread those demands over many months. The chicks do not have this luxury. They have 70 days at most in which to grow a full set of 60 wing feathers and 14 tail feathers to full size. Furthermore, these feathers are not a temporary set just to see them through; some of these feathers will last them for several years, so the young albatrosses must produce quality as well as quantity, and they must do it in a very limited time. Food shortage during this growth phase could be fatal. Irreparable damage to the flight feathers would render the young bird unable to fly.

Growing a full set of feathers from scratch requires enormous amounts of nutrients and energy. The fat layer that the chicks have accumulated by this stage helps to power this growth and ensures against food shortages that might otherwise be harmful. These birds are masterful gliders, shearing low over the waves, swooping up on the updraft so skilfully as to make those who fly hang-gliders deeply envious. Long wings, strong feathers, a light frame, and a low ratio of weight to wing surface are crucial for them as gliders, and the final phase for the chicks therefore is to *burn off* any excess fat. Any fat left after the feathers are fully grown will be extra unwanted baggage for a lean flying machine.

Once they have achieved self-powered flight, and once they have learned to navigate and to feed, the young petrels, shearwaters and albatrosses set off on mammoth journeys. The abundant Short-tailed Shearwaters that breed in burrows atop the cliffs of islands around New Zealand and Tasmania during the southern summer avoid the cold winters of the southern oceans by making their way in a huge loop all the way around the North Pacific. On their way they pass through Japanese waters, and there they can be seen in enormous numbers off the Pacific and Sea of Okhotsk coasts. Conversely, Japan's Streaked Shearwaters breed during the northern spring and early summer and migrate 3,500–5,400 km southwards to the tropical waters of the South China Sea, northern New Guinea and the Arafura Sea, some even reaching west Australian waters.

Short-tailed Shearwaters visit Japanese waters in enormous numbers as they migrate towards the Sea of Okhotsk and Bering Sea from their Southern Hemisphere breeding sites [MAB].

Tundra-like habitat can be found at high elevations in the alpine zone of the Japanese Alps [*above* MAB; *below* CC]; *inset* Spotted Nutcracker [JW].

Natural Ecology

The Habitats of Japan

The complexity of Japan's scenery (both sea-bound and intensely mountainous), its geographical spread and location, general topography and climate, and the range of soils, all combine to produce and support an astonishing variety of habitats, providing homes for a diverse array of plants and animals.

Japan's main terrestrial habitats include the upper slopes of towering volcanoes, such as Mt Fuji, and high montane ridges and cliffs, as in the Japanese Alps, down to low coastal areas including lagoons and swamps at sea level or barely above. The mountainous nature of the country helps to drive its regional and seasonal climatic variation and the distribution of biomes.

For its size, Japan has great climatic variation. Hokkaidō's subarctic winters with frigid seas and sea-ice contrast with the subtropical waters of the Nansei Shotō (the southwestern islands) and the Ogasawara archipelago. A steep altitudinal gradient, from sea level to 3,776 m, combined with a rich variety of habitats and species, creates much interest for the eager naturalist.

Japan's many rugged, sharply ridged and steeply flanked alpine peaks, some soaring to over 3,000 m, support habitats that include alpine flower meadows near their summits, with juniper and dwarf conifer forest lower down. In the northern half of the country, the lower flanks of the mountains support many deciduous broadleaf forest species, while in the southern half the same slopes support a wide range of evergreen broadleaf species. In the north, particularly in Hokkaidō, the forests take on a more clearly boreal aspect and support Brown Bears, while the coast in winter appears more subarctic, with Northern Fur Seals, Steller's Sea-Lions and Sea Otters offshore. In the south, the forests are clearly more subtropical, and in the far south, notably the Nansei Shotō, tree ferns and cycads (Japanese Sago Palm) grow wild. At certain of the river mouths there are mangroves, while offshore there are coral reefs, rich in tropical fish, along with marine turtles, rays, sharks and dolphins. In the past, Japan had extensive inland, riverine and coastal wetlands with numerous lakes, ponds and swamps in riverine plains and coastal lowlands. Although many of these wetland habitats have long since vanished, mainly through agricultural conversion, because of their fertility, some significant remnants do survive.

Wetland habitats include brackish coastal lagoons [MAB].

In Hokkaidō, alpine tundra can be found in the Daisetsu Mountains [MAB].

Temperate montane forest in northern Honshū is dominated by deciduous species [MAB]; *inset* Japanese Badger [IiM].

In Japan, the alpine tundra-like habitat, which occurs above the tree line, recalls the ecologically similar high-Arctic tundra beyond the tree line in northeastern Russia. Both are dominated by prolonged periods of freezing, with brief, intense summers. At slightly lower elevations and in the north, habitats are of a more boreal nature, dominated by mixed forests combining spruce, fir and a range of deciduous species, including oaks, elms, birches and alders. These regions experience extremes of climate, with prolonged winters, locally heavy snowfalls and frozen streams, rivers and lakes, and with extensive sea-ice in the Sea of Okhotsk reaching Hokkaidō.

At middle elevations, and dominating much of northern Japan, the habitats are essentially temperate, experiencing well-defined seasons, and with forest dominated by deciduous species and even some pines. Central and western Japan, which lie within the warm temperate zone, have more extensive broadleaf evergreen forest species and more pines. In certain regions, there are restricted areas of both lowland and upland grasslands and, in a confined area of Tottori Prefecture on the Sea of Japan coast of western Honshū, desert-like sand dunes exist.

Freshwater habitats range from montane streams and large rivers to brackish coastal lagoons and inland freshwater lakes. Lake Biwa, in Shiga Prefecture, Honshū, is one of the oldest extant lakes in the world. Offshore, there are coral reefs and seagrass beds in the south, while in the north there are kelp beds and sea-ice.

Alpine–Subarctic Habitat

Looking at Japan's habitats in a little more detail, we find that they vary from north to south in a pattern that is mirrored in elevation, from higher to lower levels. Northern Japan, especially Hokkaidō, experiences long subarctic winters, and short mild to warm summers. It has habitats typically considered boreal.

Long, subarctic winters are typical of northern Japan, here in east Hokkaidō [YM].

Permanent snowfields, stony tundra and Japanese Stone Pine can be found in the Daisetsu Mountains of central Hokkaidō [*left* MAB; *right* KM].

Above the tree line (about 1,500 m in the higher mountains of Hokkaidō, and about 2,500 m in central Honshū), there is rocky or stony, tundra-like high-alpine habitat. Few birds frequent this habitat in Hokkaidō, but in certain areas of central Honshū it is home to such species as Rock Ptarmigan [VU] and Alpine Accentor. This habitat is dominated by very prolonged periods of winter freezing, with only brief, intense summers. Above this zone there is only bare rock. Just below the tundra/alpine zone there is commonly a belt of hardy Japanese Stone Pine. This habitat supports a range of species, among them Spotted or Eurasian Nutcracker, Siberian Rubythroat, Japanese Accentor and Pine Grosbeak (the last in Hokkaidō only).

Much of the mid-elevation mountain slopes of Hokkaidō (as low as sea level in the east) and northern and central Honshū support mixed taiga/boreal-like forests. These include Erman's Birch, along with spruce and fir species. These habitats hold a range of bird species, including Eastern Buzzard, Hazel Grouse (Hokkaidō only), Black Woodpecker [VU] (Hokkaidō and northernmost Honshū), White-backed Woodpecker, Great Spotted Woodpecker, Japanese Pygmy Woodpecker, Brown-headed Thrush, Japanese Thrush, White's Thrush and Siberian Thrush, Japanese Robin, Red-flanked Bluetail and Sakhalin Leaf Warbler.

Northern Japan has very prolonged, severe winters dominated by cold airflows from the continent to the northwest and north, leading to low average temperatures, frozen lakes and rivers, locally heavy snowfalls and, along the shore of the Sea of Okhotsk, extensive sea-ice during late winter. In contrast, regions in the far south, especially the Nansei Shotō and the Ogasawara Islands, are subtropical year-round. These areas are affected by warm airflows from the south, southeast and southwest, and are warm to hot, with high average annual temperatures and high humidity throughout the year, and high average rainfall. Much of Central Japan belongs to the temperate climate zone, but with northern regions having longer winters and shorter summers and southern and western parts having shorter winters and longer, hotter and humid summers.

Japan also spans different climatic zones from east to west. The Sea of Japan coastal climatic zone is characterized by heavy winter snowfalls and mild summers, whereas the Pacific coastal zone typically experiences colder, cloud-free winter weather with little or no snow, but hot and humid summers.

A relict population of Rock Ptarmigan still survives high in the Japan Alps of Honshū. Ptarmigan are cryptically plumaged in each season [BOTH YM].

Mixed taiga- or boreal-like forest flanks the volcanoes of Hokkaidō. The Shikotsu Caldera is shown here, with Lake Shikotsu and the volcanic peaks of Tarumae and Fuppushi [MAB].

Mid-elevation forest in the mountains of central Honshū [MAB]

Cool temperate forest in northern Honshū; Katsura (*inset*) is a fairly common species in such forests [BOTH MAB].

Warm temperate mixed forest in central Honshū is home to many broadleaf evergreen trees and bamboos [*left* MAB; *right* ChM].

Cool Temperate Habitat

The dominant forest habitat of southwestern lowland Hokkaidō, and throughout northern and central Honshū at mid-elevations to lower elevations, is mixed temperate forest. The diversity of tree species here is considerable[1], and includes many familiar and unfamiliar species. The most recognizable, for those from outside the region, will be the pines, larches, oaks, elms, maples, magnolias, azaleas and mountain cherries. These forests support a high diversity of resident birds and summer migrant birds, among them various owls, cuckoos, thrushes, flycatchers and warblers, as well as Copper Pheasant, Japanese Sparrowhawk, Grey-headed Woodpecker (Hokkaidō only) and Japanese Woodpecker (Central Japan only).

Warm Temperate Habitat

The warmer areas, from central Honshū westwards to Shikoku and Kyūshū, support a very mixed forest, some of which is typically temperate deciduous, as in cool temperate forests farther north, but with an increasing proportion of broadleaf evergreen trees and tall bamboos to the west and south. These forests hold many of the same bird species as are found farther north, but with the addition (or higher densities) of species such as Grey-faced Buzzard, Ruddy Kingfisher, Japanese Paradise Flycatcher and, very locally, Japanese Night Heron [NT], Fairy Pitta and Dollarbird.

1 As much as three times the tree diversity for the same latitude band in Europe.

Subtropical forest with Itaji Chinkapin on Okinawa [KuM]

Subtropical forest on the island of Amami Ōshima is home to various endemic species including Lidth's Jay (*inset*) [BOTH TsM].

Subtropical-forest species include the wide-ranging Orange Oakleaf Butterfly (*left*) and the Ryūkyū Island endemic Okinawan Oak (*right*) [BOTH KuM].

Subtropical Habitat

The Nansei Shotō (or Ryūkyū archipelago) and the Ogasawara Islands support subtropical habitat and a very distinctive flora and insect fauna, with many endemic species. The stature of the forest trees in these areas is low, the canopy height surprisingly uniform, but tree-species diversity is high and the majority of these are broad-leaved evergreens.

The year-round climate in the northern islands is humid and subtropical while in the southern islands it is tropical. Summers are hot and winters are warm. Precipitation is very high, but largely concentrated in the early summer rainy season and the late summer to autumn typhoon season, when the islands may be battered by frequent storms.

Japan's subtropical islands support many resident avian endemic species and unique subspecies, among them Pryer's Woodpecker [CR], Owston's Woodpecker [EN], Okinawa Rail [EN], Ryukyu Green Pigeon, Ryukyu Scops Owl, Lidth's Jay [VU], Amami Thrush [CR], Ryukyu Robin [VU] and Bonin Honeyeater [VU].

Pryer's Woodpecker is endemic in the forests of northern Okinawa [ImM].

Grasslands

In certain parts of Japan there are restricted areas of grassland habitat, which may be either dry or wet. They include high-altitude meadow-like grasslands with considerable amounts of dwarf bamboo, these being commonly dominant around upper volcanic slopes and other areas exposed to strong winds. They include also diverse vegetation around alpine ponds or peat swamps, and reedbeds around wetlands at lower elevations. Grassland in Japan is not usually climax vegetation, but a stage of plant succession controlled by mowing, burning or grazing. The only truly natural grassland communities are those of windswept capes, windswept mountain tops and the alpine zone. A representative tall-grass grassland species is Maiden Silvergrass. These grassland areas may support Green Pheasant, Latham's Snipe [NT], Common Cuckoo, Meadow Bunting, Stejneger's Stonechat, Long-tailed Rosefinch and Chestnut-eared Bunting. Few bird species take advantage of stands of bamboo grass, but three that do are the native Japanese Bush Warbler, the summer visitor Asian Stubtail, and the introduced but now well-established Red-billed Leiothrix.

Japan's surviving reedbeds (once clearly a major habitat to judge from historical drawings and paintings) are home to a range of bird species, among them Eurasian Bittern [CR], Yellow Bittern, Eastern Marsh Harrier [VU], Red-crowned Crane, Ruddy-breasted Crake, Marsh Grassbird [EN], Oriental Reed Warbler, Black-browed Reed Warbler and Japanese Reed Bunting [VU]. The reed and grass fringes of wetlands support various warblers locally in Hokkaidō, examples being Sakhalin Grasshopper Warbler, Middendorff's Grasshopper Warbler and Lanceolated Warbler, and at the interface between such grassland areas and woodland or agricultural areas in Central Japan there may also be Green Pheasant and Chinese Bamboo Partridge.

Grassland at 1,400 m at Senjōgahara in Nikkō National Park, Tochigi Prefecture [MAB]; *inset* Green Pheasant [MC].

Japan's endemic Green Pheasant favours grassland habitats [AK].

A ubiquitous and prominent songster, Japanese Bush Warbler, is often found among dwarf bamboo [SP].

Bamboo species in Japan range from giants to dwarfs [MAB].

Dwarf Bamboo is the dominant ground cover in much of Japan [MAB].

Dry grassland habitat at Kirigamine Kōgen, Nagano Prefecture [MAB].

Wetlands

The native wetland habitats of Japan range from alpine ponds, peat swamps, montane streams, short fast-flowing rivers, freshwater lakes and reservoirs to fresh and brackish coastal lagoons and estuaries. While the first of these are mostly sterile from an avian perspective, as one descends towards lower elevations such species as Great Cormorant, Mandarin Duck, Long-billed Plover, Crested Kingfisher, Common Kingfisher, Brown Dipper, Japanese Wagtail and Grey Wagtail are increasingly likely, depending on the substrate. At 670 km^2, Lake Biwa, in central Honshū, is Japan's largest and oldest lake, formed about 5 Mya. From Hokkaidō to Kyūshū, there are many much younger volcanic caldera lakes. Lakes at low elevations in Hokkaidō and northern and western Honshū support considerable numbers of wintering or passage-migrant waterfowl, among them swans, geese and ducks. Coastal estuaries and mudflats attract large numbers of herons, egrets, spoonbills, gulls and ducks, and at certain seasons a wide range of migratory shorebirds, notably in western Japan. In southern Japan especially, estuaries and mudflats attract wintering Black-faced Spoonbill [CR] and Saunders's Gull [VU], both of which are globally restricted and rare species.

Mandarin Duck breeds along forested mountain streams at middle elevations [FT].

Native wetland habitats are diverse and include mangroves and mudflats in the southwestern islands [TsM]; *inset* Western Osprey, a fish-hunting specialist, is frequently found at river mouths and at mangrove areas in the south [MC].

Wetland habitats are among the most threatened in Japan [BOTH MAB].

Numerous shorebirds rely on coastal wetlands and mudflats [*left* NS] while birds such as the Black-browed Reed Warbler [*right* SP] favour damp areas with reedbeds.

Shingle- and rocky-bedded rivers at low elevations, here in Kyūshū [MAB], are the habitat of numerous fish species as well as Long-billed Plover [*inset* KT] and Japanese Wagtail.

Cultivation

Around 70 percent of Japan is mountainous, and about half of the remaining 30 percent has been developed for industry or urbanized, leaving only a small proportion of the country suitable for agriculture. Japanese crops include apples and citrus fruits, pineapples, potatoes and sugarcane, along with tea, rushes, sugar beet and silkworms. The most common crop, in terms of area, remains rice. Japan was once rich in inland river-plain wetlands and coastal wetlands, but most of these have been lost as a result of drainage. Extensive areas of wet rice cultivation do, however, provide alternative homes for certain wetland species, including insects, fish, amphibians and birds.

Agricultural areas, particularly rice fields, support many commensal species, and where rice fields are retained wet, or where they are farmed organically or at low intensity, as in the traditional *satoyama* system (which combines small-scale use of arable land and mountain foothills), they support many uncommon and several endangered species of amphibian. Common bird species here include the endemic Green Pheasant, the ubiquitous Black-eared Kite, a range of egrets and herons, various shorebirds on migration, Oriental Crow and Japanese Crow, White-cheeked Starling, buntings and

Cultivation has been ongoing in Japan for millennia: here in Kagoshima Prefecture, Kyūshū [*above left* MAB] and on Sado Island, Niigata Prefecture [*above right* and *below* SAKP].

Traditional *satoyama* agriculture combines small-scale use of arable land and mountain foothills [*left*; MAB; *right* ChM].

finches in winter. Very locally one can find Crested Ibis (Sado Island), Oriental Stork (Hyōgo Prefecture), Greater Painted Snipe, Hooded Crane [VU] and White-naped Crane [VU] utilizing farmland year-round or in winter.

The Urban Environment

Around 78 percent of Japan's 126 million people live in urban areas (45 percent in Tōkyō, Ōsaka and Nagoya combined). The urban environment is very densely developed, leaving little habitat even for commensal species. City parks, however, can attract surprising numbers of birds, especially during winter and on migration. Japanese Crow, Feral Pigeon (or Rock Dove), White-cheeked Starling, Brown-eared Bulbul and Eurasian Tree Sparrow are the most widespread resident species in the largely asphalt, concrete and glass desert of the urban environment.

Rice fields, here in Hokkaidō [MAB], provide Japan's staple diet and now offer the most widespread wetland habitat in the country. Some species, such as Greater Painted Snipe [*inset* KT] in Honshū, are at home in cultivated wetlands.

Some of the most crowded real estate on Earth is in Central Tōkyō [MAB].

Even in the urban setting some commensal species manage to make themselves at home [TK].

The Coast and Offshore Islands

Japan's coastline extends over about 29,000 km and experiences a great range of environmental conditions. Northern coastlines are arctic-like, blasted by wintry northern winds, lashed by frigid waters and strongly influenced by winter storms, sea-ice and the southward-flowing cold *Oyashio* current. Southern coastlines are subject to almost subtropical conditions, lapped by warm tropical seas and strongly influenced by typhoons, tropical storms and the warm, north-flowing *Kuroshio* current.

Rocky shores, especially, are the domain of Temminck's Cormorant and Red-faced Cormorant (eastern Hokkaidō only). Where there are sheltered bays, these may be frequented by winter gatherings of Whooper Swans or Bewick's Swans, and numerous other waterfowl; and where there are eelgrass beds Brent Geese [VU] may be encountered. Offshore, migrant seabirds such as loons (known also as 'divers'), grebes, sea-ducks and Streaked Shearwaters may be found, depending on the season. In warmer regions, there are Pacific Reef Egrets and Eastern Blue Rock Thrushes (north to southwest Hokkaidō). In the southernmost regions, particularly in the Nansei Shotō and Ogasawara Islands, there are

Japan's natural coastline is largely rocky, as here on Minami-shima, in the Ogasawara Islands [MAB].

Coastal erosion, winter storms, typhoons and tsunami all drive the construction of coastal defences [MAB].

A beautiful beach scene in northern Amami Ōshima [TsM].

Left The world's largest Rhinoceros Auklet colony is located on Teuri Island [MR]; *right* Tufted Puffin ranges around the Bering Sea and North Pacific and just reaches Japan in southeast Hokkaidō [WaM].

offshore coral reefs and seagrass beds, coral-sand beaches and, at some river mouths, small areas of mangrove forest.

The Japanese archipelago consists of several large, densely populated islands, a number of smaller populated islands, and thousands of very small islands and rocky islets most of which are unpopulated[1]. The majority of the islands close inshore are eroded fragments of what once were much larger islands, but some, and mostly the more distant ones, are oceanic – the tips of submarine volcanoes that have never been connected to the mainland.

The larger islands in the Sea of Japan, from Tsushima north to Rishiri, are well-known migrant traps, where numerous avian vagrants to Japan have been reported. Smaller islands provide breeding grounds for significant numbers of seabirds. Teuri-tō, off west Hokkaidō, for example, supports a very large Rhinoceros Auklet colony[2], while on islands off southeast Hokkaidō Leach's Storm Petrel and very small numbers of Tufted Puffin breed. Other islands, especially off western Honshū and off Kyūshū, provide breeding sites for the rare Japanese Murrelet. Farther south, subtropical islets are home to colonies of Brown Booby, Sooty Tern, Black-naped Tern [NT], Roseate Tern [NT] and Brown Noddy. Islands in the chain stretching from the Izu Islands to the Bonin Islands support important populations of Short-tailed Albatross, Black-footed Albatross, Bryan's Shearwater, Tristram's Storm Petrel, Matsudaira's Storm Petrel[3] and Bonin Petrel.

The Japanese archipelago spans climatic and biogeographical extremes and serves as a bridge between Northeast Asia and Southeast Asia. A major bird migration route, the East Asian–Australasian Flyway, extends along the continental coast of Asia, linking regions as far distant as southeast Australia and New Zealand with Kamchatka, Yakutia, Chukotka and Alaska. Many regular summer and winter visitors to Japan and numerous passage migrants move along this flyway or its various branches: following the line of the Japanese archipelago, branching in Kyūshū to the Korean Peninsula or through the main Japanese islands; branching again in Hokkaidō, some species migrating by way of the Kuril Islands and others via Sakhalin into northeast Russia. Kyūshū, in particular, is known as a 'crossroad of migration'. There, for example, in autumn, can be found Chinese Sparrowhawk migrating from Korea to Southeast Asia over Kyūshū to Amami Ōshima and Okinawa, while at the same time Oriental Honey Buzzards are migrating from eastern Japan to China by way of Kyūshū. Wherever the flyways cross water, wherever there are headlands, capes or offshore islands, there are regular places of landfall for tired migrants, and these make excellent places to search for common species and, at times, vagrants. These areas are especially fertile for the interested naturalist, connecting him/her to the larger processes of the natural world.

---

1 Throughout this book, the terms populated and unpopulated are used to refer to human populations, or to deities, and inhabited and uninhabited to refer to plants or animals.

2 Reputedly the largest colony in the world.

3 The only confirmed breeding site for this species in Japan is on Minami Iwō Island.

Teuri-tō, off the west Hokkaidō coast, has some of the highest cliffs in the country [TT].

Bonin Island Honeyeater is endemic in the remote Ogasawara Islands – here with Japanese White-eye (*centre bird*) [OVTB].

Subtropical broadleaf evergreen forest in northern Okinawa; *inset* Okinawa Robin [ALL KuM].

The Flora and Vegetation of Japan

Forest Types

Travel along the Japanese archipelago from south to north and you will encounter four main climax-forest types. These are: (1) subtropical broadleaf evergreen forest (in the Nansei Shotō and Ogasawara Islands); (2) warm temperate broadleaf evergreen forest (or laurel forests, with predominantly broadleaf evergreens, and some deciduous and needle-leaf evergreens, found in Honshū, Shikoku and Kyūshū); (3) cool temperate, broadleaf deciduous forest (in western Hokkaidō and Honshū, with predominantly deciduous trees); and (4) boreal, subarctic evergreen coniferous forest (in most of Hokkaidō and the highlands of Honshū, with a mixture of needle-leaf evergreens and deciduous tree species).

Transitional forest communities and special or local forest types also occur, depending very much on the soil type and the local climate, for example the small areas of mangrove forest to be found in the

Boreal mixed forest in Hokkaidō [BOTH US].

Japan's boreal forests support a wide range of fungi (*top left*) including Beard Fungus (*right, 2nd from top*), Fly Agaric (*right, 2nd from bottom*) and earthstars (Geostraceae (*bottom right*) as well as birds, such as the delightful white-headed form of Long-tailed Tit (*top right*) [ALL MAB except TOP RIGHT WaM].

southwest islands. There are also corresponding altitudinal forest zones: beginning in lowland broadleaf evergreen forest, one traverses through montane broadleaf deciduous forest, then subalpine evergreen coniferous forest, and finally, in alpine regions, scrub, grassland and rocky desert, which correspond to the arctic tundra. Many specific local non-forest vegetation types can also be found, such as those of coastal saltmarshes, and mobile and stabilized dune systems.

For its size, Japan's largely temperate Holarctic flora is rich and diverse, with more than 3,900 species of flowering plant (Compared with 2,900 for northeastern North America and 1,900 for New Zealand). In lowland coastal areas broadleaf evergreen oaks and chinkapins are found alongside Japanese Red Pine or Japanese Black Pine, while in the subcanopy layer there is a wide range of species, including yews, Japanese Bay Tree, camellias, viburnum, Japanese Spindle and Glossy-leaf Paperplant. In urban areas especially, the Ginkgo or Maidenhair Tree is widely planted, and elsewhere there are Japanese Zelkova, and of course innumerable cherry trees, mostly of the beautiful variety known as Yoshino Cherry (or *Someiyoshino*), which flower before they put out their leaves. In rural areas, the common cherry is the Mountain Cherry, which produces its flowers and leaves simultaneously. Flowering a little earlier, in April, and often bringing the first spring 'colour' to the forests are the perfumed, white-flowering magnolias, Kobushi Magnolia and Japanese Whitebark Magnolia, along with various yellow-flowering forsythias. The magnolias are an ancient lineage, believed to have evolved before pollinating bees did. Many magnolia species are pollinated instead by beetles, for which their perfume is a powerful attractant. Some even generate heat in their flowers to improve dissemination of the perfume. After the cherries, during late spring, come the various colourful azaleas and rhododendrons, ranging from white, orange and red to deep purple. In the warm temperate zone of the west there are Camphor Trees, Japanese Chestnuts and even podocarps (Fern Pine).

Montane forest covers much of Honshū [MAB].

Autumnal forests are rich in colours and fungi [ChM].

Ancient Trees, Ancient Scents – Camphor

Among the deep shades of the evergreen forests of central and southern Japan is a tree with a host of names and a host of uses. These forests are dominated by members of the laurel family Lauraceae, and this tree in particular goes by the name *Kusu-no-ki* in Japanese, and the name Laurel Camphor, Camphor Laurel, Gum Camphor or simply Camphor Tree in English. Camphor Tree is a species that ranges across the warm, humid southern parts of Japan and through adjacent areas of East Asia, particularly of the Korean Peninsula, Taiwan and continental China. Outside its natural range it has been successfully cultivated as an ornamental species, and as a wind-break and shade tree, in warm subtropical regions of the world, including India, southern parts of Europe, California, Florida and various other southern US states, and even in Argentina. As its scientific name indicates, Camphor Tree is a close relative of the Sri Lanka Cinnamon Tree, the dried bark of which we know as cinnamon, the spice used in food and drinks since ancient times.

The tree itself is an evergreen, shaped not unlike an American Linden or Basswood, and growing to a great size, reaching as high as 20–25 m, with a dark grey-brown, rugged main trunk that may attain 9 m in circumference and may also divide into several secondary trunks. Together, the main and subsidiary trunks support a dense glossy green canopy that casts a strong shade, which is particularly welcome in the city streets and parks where it has been planted. These impressive, scented trees may attain great age. For example, the dignified Camphor Tree that stands at the south side of the north gate of Shoren'in (shrine) in Kyōto is several hundred years old; it has been registered both as a giant and as a historical tree, and designated a Natural Monument of the city. Not to be outdone by Kyōto, the Camphor Tree east of the precinct at Yamada Shrine, in Tottori Prefecture, is said to have been planted more than 1,000 years ago, and is considered sacred by the people of the local town. Even older, however, is the giant Camphor Tree at Atsuta Jingu, in Nagoya, which is said to have been planted by the famous Buddhist priest Kōbō Daishi 1,300 years ago. And if you thought 1,300 years was old, well that, too, is beaten by the giant Camphor Tree of Kawago, in Takeo City, Saga Prefecture. Camphor Trees can be seen throughout Saga Prefecture, where the species is the prefectural tree, but the giant Kawago tree is its greatest living treasure. That particular tree, also designated as a National Natural Monument, is ranked as the third largest tree in Japan. It stands 25 m high, apparently has a deep root circumference of 33 m and has widely spreading surface roots that span 26 m from east to west and 33 m from north to south. This ancient tree is said to be more than 3,000 years old, meaning that it was here long before Japan was a nation. Takeo City seems well favoured by giants, as another 'Giant Camphor Tree of Takeo' at Takeo Shrine is ranked sixth nationally in size, while the 'Giant Camphor Tree of Tsukasaki' is ranked third in size in the prefecture.

Not only are Camphor Trees long-lived, but they are also astonishingly vigorous, surviving even the worst that humanity has thrown at them. The Camphor Tree at Sanno Shrine, Nagasaki, was designated a 'Natural Monument' by that city on 15 February 1969, because it had survived the atomic bombing of the city on 9 August 1945. On 3 November 1973, the Camphor Tree was made the official tree of Hiroshima City, in memory of all the trees that not only survived the atomic bombing but also recovered quickly and gave inspiration to the human survivors trying to rebuild their lives following complete devastation.

The Camphor's leaves are simple and alternate; the blades, often with wavy margins, are usually 4–10 cm long and 2–5 cm across. The central vein and two side-branching veins are particularly prominent. Cut a stem or bruise a leaf and you will immediately notice the strong aroma of camphor that emanates. Camphor Tree's flowers are individually small, yet noticeable because they occur in loose panicles of conspicuous greenish-white to cream flowers. The fruits consist of reddish or blackish berries much like those of the cinnamon tree, and they are often abundant on mature trees, attracting birds to feast on them during the winter.

The insecticidal properties of chemical camphor are particularly significant. Both the aromatic wood itself and the camphor (derived from the wood) provide protection against moths and other insects. It is not surprising that cabinets for storing natural-history samples are often made from Camphor wood to prevent damaging insects and arachnids from reaching the specimens. Research in North America has shown that camphor can be used very effectively to repel lady beetles (known also as ladybirds or ladybugs, depending on your background) without killing them. Some species of lady beetle are popular colourful creatures because they are beneficial in the garden and greenhouse, where they help to control aphids. They are sometimes, however, less favoured when they come indoors into homes and businesses in considerable numbers to overwinter. Adult beetles, using visual and chemical cues, usually choose the sunnier sides of prominent, exposed or light-coloured buildings as overwintering sites, making their way in through tiny crevices and cracks.

Camphor has aromatic qualities and is used medicinally in massage compounds to ease bruises, inflammation and joint pains, and in lip salve and inhalants. Locally, camphor numbs the peripheral sensory nerves and has mild antiseptic properties.

In fact, the herbal medicinal uses of camphor are many and varied. It is apparently helpful for colds, chills and various nervous, stomach and bowel complaints, as well as being utilized as a stimulant and even as a sedative liniment. Use it in moderation, though, because in large doses it is acrid and very poisonous.

Camphor Trees, such as this one in Saga Prefecture, Kyūshū, may live for several thousand years [SPTF].

Fresh green Camphor Tree leaves in early summer exude a distinctive scent [MAB]; *inset* the flowers are tiny, and only their abundance makes them conspicuous [NZUE].

Other Vegetation Types

The herb layer of Japan's forests is too diverse to describe here, but it includes numerous ferns and orchids, and, most distinctively, many Japanese forests have an undergrowth of dwarf bamboo species. The density of these dwarf bamboos makes the accessing of forests exceedingly difficult.

Many plant genera are endemic in Japan and many more have affinities with the Himalayan–Sino floral region. The flora of subtropical southern Japan has affinities with Taiwan and continental Asia beyond, whereas the flora of cool-temperate or boreal northern Japan has affinities with northeastern Asia. The alpine zone is particularly instructive: for example, the creeping pine known in Japan as Japanese Stone Pine, and elsewhere as Dwarf Siberian Pine, has a narrow distribution above 2,500 m above sea level in the mountains of central Honshū, whereas in northern Honshū it is found at around 1,900–2,000 m, and in Hokkaidō commonly above 1,400 m but as low as 200–300 m in some areas of Akan–Mashū National Park. This shift in altitudinal range from higher to lower from south to north is repeated among plants, insects and even some vertebrates in Japan.

Japan's island nature, the contrasting warm current from the south and cold current from the north, and the influence of saltwater and sea winds affect the natural distribution of maritime vegetation around the margins of the country. The conversion of coastal lowlands to rice-farming or industry means that only small areas of natural coastal vegetation survive outside Hokkaidō. The natural halophytic vegetation of salt-tolerant species is divisible into two regional groups: (1) that of northern Japan (Hokkaidō and Tōhoku), with species such as Common Glasswort, Sea Milkwort and Hoppner's Sedge; and (2) that of southern Japan (from central Honshū west to Kyūshū), with species such as Sea Blite, Asian Coastal Wormwood and Japanese Lawngrass. The natural vegetation of subtropical lagoons, river mouths and estuaries in the Nansei Shotō (and as far north as southern Kyūshū) consists of mangroves with species that include Narrow-leaved Kandelia, Mangrove Apple, Black Mangrove, Spider Mangrove and Grey Mangrove.

Throughout most of Japan the dominant plant species of freshwater marshes is the reed, sometimes known as Common Reed or Japanese Reed. It is also salt-tolerant and occurs around brackish marshes.

Another special foreshore habitat consists of dunes, which may be mobile or fixed. In warmer regions in the west, such as in the San'in region, and in the south, plants such as Saltwort, Asiatic Sand Sedge,

Coastal vegetation living under the influence of saltwater and sea winds includes Common Glasswort, here at its brightest in autumn in east Hokkaidō [ABTA].

Murain Grass and Beach Vitex colonize these areas and there may also be stable dune forests consisting of Shore Juniper and Japanese Black Pine. In the Nansei Shotō, the landward side of fixed dunes may be occupied by the widespread Screw Pine forest, while in the Ogasawara Islands the endemic Bonin Screw Pine is found. In the north, at least, we find another species named after the great naturalist of Bering's second expedition, Steller's Wormwood; and on the fixed dunes behind the beach and before the dune forests of Hokkaidō are thickets of the hardy Japanese Beach Rose, sometimes with Siberian Crab Apple. A colourful array of flowers brightens this habitat from late spring through summer. In addition to the reddish-pink of the wild roses, there are the nodding yellow flowers of Hokkaidō Daylily and Amur Daylily, the tall orange flowers of Thunberg's Lily, bright yellow masses of False Lupin interspersed with swathes of Asian Lily of the Valley and, in damp swales behind the dunes, the deep blue of Beachhead Iris. The dune forests themselves typically consist of Daimyo Oak and Mongolian Oak and are much battered and shaped by the prevailing wind and salt spray. Those fronting the forests are low in stature, shrubby, gnarled and crooked; they steadily increase in height inland but appear as if trimmed or clipped because of the severe conditions. Where they are mature, they shelter among them occasional Painted Maple, Prickly Castor-oil Tree, Bakko's Willow and Eurasian Aspen. Farther inland the climax forest may include Sakhalin Spruce, Sakhalin Fir and Erman's Birch, and in swampy areas and alongside streams Japanese Alder. Juniper, pine and oak forests are to be found among dune forests in northern Honshū, occasionally with stands of Japanese Black Pine and Japanese Red Pine alongside Japanese Lime, while Thujopsis (known also as Hiba) and Siebold's Beech are indicative of the oldest inland portions of stable dunes.

Differing communities of plants have adapted to the special characteristics, such as soil type, precipitation, exposure to wind or snow cover, of many other habitats. These include coastal cliffs, scree slopes, volcanoes, fumaroles and solfataras (sulphurous fumaroles), shingle-bedded rivers, riversides, swampy areas, ponds, lakes, marshes, fens and bogs. They include also the various elevations of mountainous regions (hilly zone, montane zone, subalpine zone, alpine zone and alpine desert), and so on, although there is no space to do them all justice here.

The large-scale and widespread conversion of lowland and coastal areas to industrial and domestic uses, and of wetlands to agriculture, combined with the use of agricultural chemicals (fertilizers and

Coastal and riverine mangrove forests occur only in the far south of Japan [TsM].

insecticides) and the policy since the late 1940s of felling natural forests and replacing them with plantations, has involved the destruction of natural plant communities and, in some cases, the complete loss of species. Long-established temple and shrine precincts have accidentally served as historical nature reserves, while the legal designation of some plant species, plant communities, forests, aquatic and alpine zones and localities as national monuments during the 20th century has helped in plant conservation. The national park and geopark movements have furthered those aims, and these parks are among the best places in which to explore Japan's natural flora and vegetation.

Waves of Colour

Imagine if it were possible to enjoy a live satellite feed of images of earth, so that on any given day it would be possible to zoom in and watch in real time how the seasons are progressing. Japan would be a marvellous place to have that currently impossible opportunity[1].

Because the country extends over many degrees of latitude and has an altitudinal range of over 3,000 m, seasonal changes wash back and forth, up and down the archipelago like an inexorable slow-moving tide of colour and expanding or contracting vegetation. If that imaginary view were possible, and recordable, we would be able to watch as the patterns of seasonal colour shift before our very eyes. The dark green hues of summer fade steadily and drain away from the north towards the south. The forests of the north would fade to oranges, reds and yellows, and as those colours flowed southwards a dusting of white would brush across the central highlands of Hokkaidō, then flow southwards in pursuit of the draining colour. In the south, patches of deep green would remain, giving contrast first to other patches of colour and then to the greyness of leafless trees. In spring, it would be possible to watch the waves of white and then hint-of-pink blossoms flicker and spread northwards like a running bushfire as magnolias, plums and then

1 We are getting closer to the possibility, however, with the combination of space-borne and ground observations.

Coastal dune forests are typically sculpted by onshore winds. Immediately inland, in the slacks between long-stabilized dunes, are coastal meadows rich in wildflowers [ALL MAB].

The tide of spring colour begins with subtle shades of cream, pink and green [US].

Golden Ginkgo leaves are emblematic of autumn [MAB].

The Cherry Blossom front is tracked on Japan's national weather news each spring [ChM].

cherries burst from bud into bloom. Following swiftly behind would be a wave of the freshest of delicate green, pursued by a wave of deepening, darkening green of foliage, and finally a late wave of colour as shrubs and trees put forth orange, pink and purple blooms. It would be like watching a living tapestry shift and change before our eyes.

In late autumn, in the early days of December, in the mountains of central Honshū, it is possible to intersect one particular wave of colour – the delightful, dreamy gold of Ginkgo leaves. That particularly beautiful wave reaches its southern limits here, petering out towards the southwestern region of Honshū. The bright yellow or pale gold of the Ginkgo's leaves, whether of a single tree standing proudly against dark evergreens or of an avenue of golden-leaved trees, is inspiring. It is as if a summer's worth of sunshine has been collected, distilled, concentrated, and at last returned, bringing not only colour but also mood-lifting brightness to gloomy early-winter days.

Living Fossils in the Heart of the Capital

A mature Ginkgo Tree, reaching 20–35 m in height, is a distinctive, dignified tree with a broad crown, widespread branches and curiously shaped leaves. While the tree's form may inspire, its history and its ancestry are worthy of considerable admiration. A member of the gymnosperms, or 'naked seed' plants, Ginkgo is sometimes called a 'living fossil'. This relict of ancient times can be seen as symbolic of changelessness, surviving from the distant past.

As the small fan-shaped leaves fall they twist and spin, ultimately carpeting the ground in drifts of fine-ribbed shapes that are more at home on earth than we are. Such golden cascades have been taking place each autumn and winter since before dinosaurs roamed the earth, and we are fortunate to be able to bear witness to this scene.

In Tōkyō, one can hardly avoid the official symbol of the capital. There the Ginkgo leaf shape is to be found almost everywhere, from roadside railings to the pages of official pamphlets, and the trees line many of the city's streets. I wonder how, as a symbol of changelessness, it came to be chosen as the emblem for a city that has reshaped and reinvented itself over and over again. You may enjoy delicious *ginnan* or Ginkgo nuts, which many Japanese-style eateries serve grilled on skewers and which many inns commonly conceal in bowls of *chawan-mushi* (savoury egg-custard).

Otherwise known as Maidenhair Tree, the Ginkgo's ancestors were first recognized in fossil form. Their leaves have been found in deposits dating back some 270 million years to the Permian period. They were once part of a diverse family of some significance, particularly during Jurassic times, when it contained at least six genera with many species. Now, this wholly unique family, the Ginkgoaceae, which arose during the Mesozoic Era, is considerably reduced to only a single genus with just one surviving species. The western world was ignorant of *Ginkgo biloba*'s survival until the German physician and botanist Engelbert Kämpfer (see *p. 13*), discovered it in Japan, where it had been grown from seed imported from China around 1190 CE. In China, this last remnant of a once great lineage of early seed plants had survived in cultivation at the hands of Buddhist monks, who had protected it in monastic, palace and temple gardens. Kämpfer carried seeds back with him from Japan to Europe, and trees planted in Europe in the early 1700s can still be found growing there today. It has subsequently been introduced as an urban street plant in many parts of the world, making it a surprisingly familiar tree for urbanites worldwide.

Herbalists know Ginkgo extract as an important natural ingredient in medicines to boost energy and stamina and to sharpen the mind. It is used professionally in the early stages of treating dementia – particularly Alzheimer's disease – to such an extent that it is the most widely sold plant medicine in Europe, and among the most popular herbal medications in the USA. Because Ginkgo is a potent antioxidant and free-radical scavenger, it is believed to possess important therapeutic powers that help to increase stamina, improve the circulation, add to our longevity, and counteract aspects of ageing such as degeneration of the retina, mental acuity and libido. Given the side-effects of the drug Viagra, people could have simply taken advantage of extract of natural Ginkgo.

We may admire Ginkgo for its medicinal value, for its food value, for being remarkably pest- and disease-free and for being tolerant of harsh conditions, including pollution, making it ideally suited for urban planting. Yet there is more than value here; there is an inspiringly beautiful tree. And one day, if my dreamed-of software arrives, perhaps we shall be able to watch in real time and admire as that golden wave sweeps down the country, followed by the northward sweep of pale green.

Ginkgo is an ancient relict tree that has become the official symbol of the nation's capital. It is a common urban planting and drifts of golden leaves brighten urban areas and temple complexes in autumn [*left* MAB; *right* & *below* ChM].

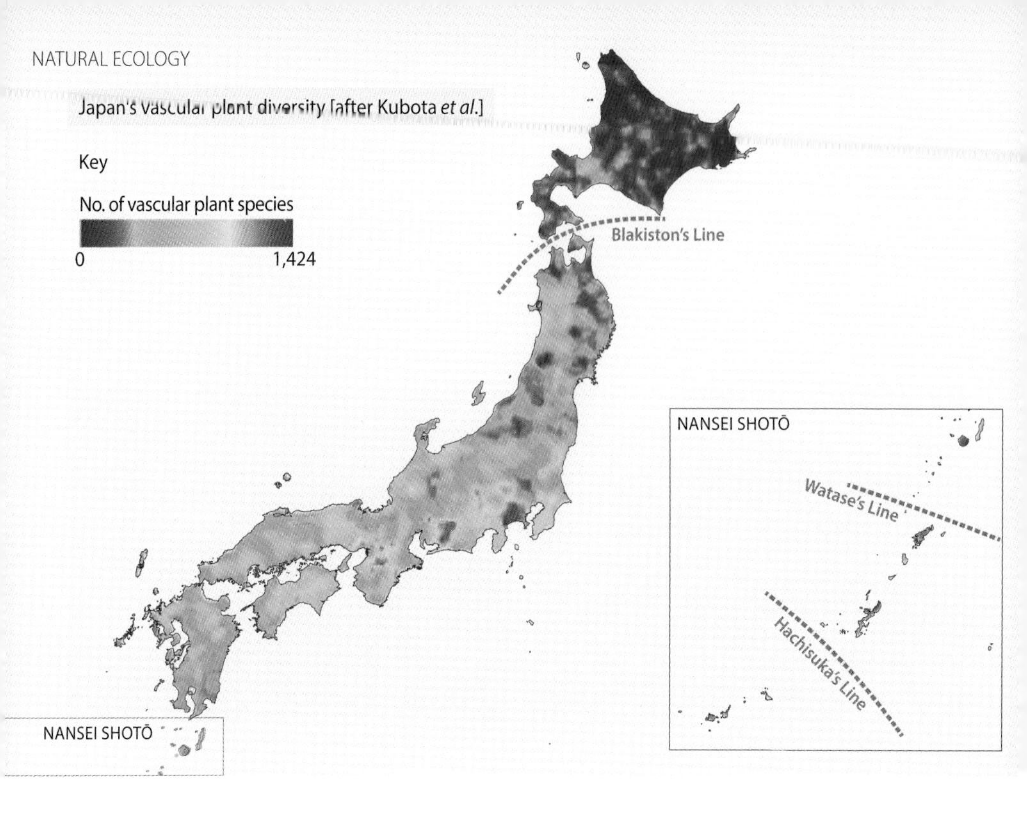

Japan's delightful landscapes support tremendous biodiversity [AOTA]; *inset* Japanese Giant Flying Squirrel [JH].

Japan's Biodiversity

Japan's largest known terrestrial mammal was the now extinct Naumann's Elephant [NNEM].

Japan's smallest mammal is the Least Shrew [IiM].

Given the size of the Japanese archipelago (378,000 km^2) and its location off a larger continental landmass, appropriate comparisons with Japan are those of the somewhat similarly sized archipelagos of the British Isles (315,200 km^2) and New Zealand (268,000 km^2). Size is definitely *not* everything, and the degree and length of isolation combined with geographical location and climate mean that these island groups are very different from one another. For example, more than 170 species of mammal, 50 of which are endemic, have been recorded in Japan (British Isles about 90; New Zealand 47, all but three marine), as have over 50 amphibians (British Isles 7; New Zealand 4, plus three extinct), more than 100 species of reptile (British Isles 11; New Zealand about 115), nearly 740 species of bird (British Isles 620; New Zealand about 340), just over 200 species of dragonflies and damselflies (British Isles 57; New Zealand 18), and 263 butterfly species (British Isles 71; New Zealand 62). Japan is among the world's most diverse zoogeographical regions.

Species diversity in Japan is considerably greater than that found in the British Isles, for example. One dragonfly specialist whom I met in east Hokkaidō explained how, during a 40-year study of a single marshland pool, he had recorded more than 40 species of dragonfly. This was at the time almost exactly the same number of species as were known from the entire British Isles! Examine other groups of organisms, such as trees or insects, and you find that Japan has more species than most of continental Europe[1]. This can be explained by the fact that Japan was spared many of the ravages of the last major ice age, whereas the British Isles were almost entirely scoured by ice. Even so, Japan's flora back then was markedly different from that of today, with open boreal woodland spreading across most of the main islands except Hokkaidō, which even supported steppe-tundra[2]. Thus, Japan is far richer in every way than either Britain or New Zealand. Yet Britain has the reputation of having an extraordinarily high number of birdwatchers and naturalists, and New Zealand is renowned for its natural splendour. Unfortunately, Japan's reputation as a powerhouse of natural biological diversity has been overlooked and understated.

1 Although perhaps we should consider Europe depauperate in species, with lower regional diversity than the same area of eastern Asia, because of its peninsular shape and the fact that Europe is isolated from the south by seas, mountains and a desert.

2 This term, derived from Russian ecologists, describes an ecological assemblage that has no modern analogue (and of which large herbivores such as mammoths were an integral part). During the Last Glacial Maximum it was the dominant ecosystem across northern Eurasia, and seemingly as far south as Hokkaidō.

Isolation and Endemism

Japanese Flying Squirrel [YaM]

Populations of animals and plants generally have more or less unimpeded flow of genetic material within them, as individuals choose mates and produce offspring. As soon as gene flow is restricted in any way by any kind of barrier, gene frequencies begin to vary in different parts of the population's range, leading to the evolution of distinct subpopulations. Subpopulations of animals and plants can form in many ways, including through geographical separation and ecological separation. Effective barriers to gene flow include oceans, air and water currents, habitat types and climate. For example, mountain-top plants are geographically restricted there because conditions for them downslope are unsuitable. In such cases individuals from the populations now separated on discrete mountain tops may rarely or never meet, and the degree of difference in the relative frequencies of different genes in the populations increases.

Changes in gene frequencies in isolated populations come about through a number of different mechanisms, which include random fluctuations and natural selection. In the latter, certain genes may confer reproductive advantage over alternatives, and so increase in frequency as those genes that are not advantageous decline owing to the decreased reproductive output of the individuals carrying them.

Sometimes individuals of two diverging populations can be told apart by their outward appearances. Biologists recognize races and forms of a species by such features as coloration pattern and relative sizes of different parts of the body. If the two populations are separated and do not interbreed for long enough, they can become genetically quite distinct. If individuals of each of these populations come together again, they may be unable to breed with each other as a result of behavioural, physiological or morphological differences that have evolved. If they are still able to breed, their hybrid offspring may experience decreased fitness, being not so well adapted to local conditions as are pure-bred individuals from either of the two populations. Genetic changes over time can result in the two populations being reproductively isolated, and they can then be thought of as separate species.

Geneticists today can map details of the genetic material in populations and quantify how distinct each population is. In some cases (such as Ryukyu Long-haired Rat, the Ryukyu spiny rats and Amami Rabbit: see *page 312*), the isolated species can become so different from their original population that they are classed as distinct genera, not just species. All the various taxa in biology (order, family, genus, species, subspecies, race, form) are today described usually in terms of genetic similarity. The longer an area has been isolated, the higher the taxonomic rank of its organisms is likely to be.

These genetic processes are very relevant to the biodiversity of Japan. The Japanese islands became separated from mainland parts of Asia as the Sea of Japan opened some time around 25–15 Mya, with their complement of species shared with the mainland. Over time, the species on what is now Japan experienced genetic drift away from the populations from which they were derived as a result of geographical isolation caused by the sea.

The situation is, however, more complex than this owing to climate cycles. Over the same time frame earth's climate has changed, with periods of extreme cold and much warmer periods. We are currently experiencing the Late Cenozoic Ice Age, which began 34 Mya. Its current phase is the Quaternary glaciation, which began 2·58 Mya. Within ice ages, the climate fluctuates between more severe glacial conditions and more temperate ones. These are referred to as glacial and interglacial periods. The earth is currently in such an interglacial period of the Quaternary glaciation, the last glacial period of the Quaternary having ended around 11,700 years ago. The interglacial in which we now live is the Holocene or Anthropocene epoch, which began approximately 11,600 years ago.

During the cold glacial periods huge amounts of seawater were locked up as ice, as global sea levels fell, and land bridges emerged between islands and from islands to nearby continental areas. During these periods, new colonizations of Japan could occur from different parts of the Eurasian continent. With each warming period ice melts and sea levels rise, cutting off the connections, and isolating the populations of strictly terrestrial species until the next ice age. As a result, the island archipelago of Japan has received repeated colonizations of new plants and animals over time. The original inhabitants of the early-formed islands have had longer to become more genetically distinct than have those animals and plants that colonized later. In the case of Japan, the landmass during glacial events was scoured by ice to only a limited extent, leaving many habitats to survive, and so there has been more time for the evolution of distinct or endemic forms, subspecies and species.

Amami Ishikawa's Frog [TsM]

The outcome is that Japan today exhibits a complex pattern of endemism, together with patterns of relatedness that are determined by the sources of those colonization events, and by the length of time during which stable climatic conditions have prevailed where the isolated Japanese populations now live. When plants and animals are confined in their distribution to the areas in which they evolved they are said to be endemic in or to that region. Their confinement to these areas may be the result of physical barriers to dispersal. They may also be recently evolved species that have not yet had time to spread from their centres of origin. As time passes, more species will evolve in the area. The proportion of the organisms in an area that is endemic is a clear guide to how long the area has been isolated.

Amami Rabbit [TsM]

Understanding Japan's Biodiversity and Biogeography

Japan harbours great natural diversity. By way of example, consider the carnivorous insects in the ancient[1] order Odonata (the dragonflies, or Anisoptera, and the damselflies, or Zygoptera). Japan hosts 203 species of Odonata and, taken as a whole, they provide a wonderful set of examples of biogeography[2] writ large in the Japanese archipelago.

Because the Odonata arose around 300 Mya, and before the continents as we know them were formed, some Odonata found in modern Japan have ultra-wide ranges that span North America, Europe and East Asia, including Japan. Others have wide ranges that extend from Europe and North Africa eastwards across Eurasia, including Russia, to Japan, while still more range from South Asia and Southeast Asia to Oceania, including Japan. Many have ranges that encompass China, Korea, the Russian Far East and Japan; some range from the south, such as Taiwan, north into the southernmost islands of Yaeyama, while yet others extend from the north, from the Russian Far East, south into Hokkaidō.

Among the 52 species (26 percent) that are considered endemic to Japan are some with fascinating nested[3] and overlapping distributions, revealing how the biogeography of this group is intimately related to the island nature of the country. Some species for example, while endemic, are widespread, occurring throughout all four main islands from Hokkaidō to Kyūshū (six species), whereas others are endemic to more limited areas of Japan and some to small island clusters, or even to single islands. There are endemic Odonata that are restricted to (from north to south): Hokkaidō and northern Honshū (1), Hokkaidō and Honshū (2), Central and northern Honshū (2), Honshū (1), Honshū and Kyūshū (1), Honshū, Shikoku and Kyūshū (9), Western Japan – central Honshū, Shikoku and Kyūshū (3), Western Honshū (1), Shikoku (1), Kyūshū (2), Shikoku, Kyūshū and the northern Nansei Shotō (1), Southernmost Kyūshū and northern Nansei Shotō (1), Northern Nansei Shotō from Amami Ōshima to Okinawa (4), The Amami Islands (1), The Okinawa Islands (1), Northern Okinawa (Yambaru) (4), The Yaeyama Islands (6), and The Ogasawara Islands (5).

If we consider a different order, the Hymenoptera for example, and look at just the ants of Japan, there are 273 species. They follow a very similar nested distribution pattern to that of the dragonflies, although with somewhat more complexity among the 135 (49 percent) that are endemic. There are for example endemic ants that are restricted to:

Hokkaidō only (1), Hokkaidō and Honshū (4), Hokkaidō, Honshū and Shikoku (1), Hokkaidō to Kyūshū (4), Honshū only (9), Honshū and Shikoku (1), Honshū and Kyūshū (4), Honshū, Shikoku and Kyūshū (15), Western Japan – central Honshū, Shikoku, Kyūshū and the Nansei Shotō (14), Shikoku only (1), Shikoku and Kyūshū (1), Shikoku and northern Nansei Shotō (2), Kyūshū only (3), Shikoku, Kyūshū and the northern Nansei Shotō (1), Shikoku, Kyūshū and the Nansei Shotō (1), Kyūshū and Nansei Shotō (4), Northern Nansei Shotō from Amami Ōshima to Okinawa (12), The Amami Islands only (8), The Okinawan Islands only (20), Nansei Shotō and Ogasawara (2), Okinawa and Ogasawara (1), The Yaeyama Islands only (18), Yaeyama and Ogasawara (1), The Ogasawara Islands only (7).

Similar analyses of other groups of organisms, whether butterflies, birds, arachnids or orchids, would reveal somewhat similar patterns of distribution, confirming how extraordinary Japan's biodiversity is, but calling for an explanation for how such distributions arose.

Islands of Endemism

Japan's main islands of Hokkaidō, Honshū, Shikoku and Kyūshū are home to a surprising array of creatures only hinted at here, from massive bears to diminutive chipmunks. While many of these species are shared with the Asian continent, with islands to the north or islands to the south, some of them, such as Japanese Macaque, Japanese Giant Salamander, Japanese Relict Dragonfly and Pryer's Petaltail (Dragonfly), are

1 The Odonata are a successful group that has survived since the Permian period (between 299 and 252 Mya).
2 Biogeography is the study of the geographical distribution of living organisms.
3 Nested distributions can be compared with Russian Matryoshka dolls, wherein a larger range overlaps a smaller range and that overlaps a smaller range still.

Wide-ranging dragonfly species occurring also in Japan include White-tailed Skimmer [*left* TaM], which ranges from Hokkaidō to Okinawa, and Northern Pygmyfly (or Scarlet Dwarf) [*right* OT] found from northern Honshū to southern Kyūshū.

Some dragonflies are endemic to tiny portions of the archipelago; examples include Japanese Matrona [*left* KuM] and Ryukyu Damselfly [*right* TsM], found only on the Okinawa and Amami Islands.

Japanese Relict Dragonfly [*left* KoM] and Pryer's Petaltail [*right* TO] are both unique to Japan.

Ryukyu Flying Fox is endemic to Japan's Nansei Shotō [KuM].

Iriomote Cat is one of a range of species found only in Japan's southwest islands [MEIW].

unique to Japan: they are found here and nowhere else in the world.

Ryukyu Flying Fox (*top*) and Amami Rabbit (*bottom*) [BOTH YaM]

Species that occur in a restricted geographical, or geopolitical, area are known as 'endemic'. Endemic in the biological sense refers to species known only from a particular, defined area, because that is where those species evolved and from which they have not dispersed. For biologists and naturalists it is a special term, since they know that they have to visit a very particular area to see the species. Endemic differs from 'indigenous' or 'native', which mean that the species occurs naturally in a given area, and could have a much broader distribution; it differs also from 'introduced' (or 'alien'), which means that the species originated from some other area. This biological meaning of endemic is very different from the meaning in medicine, where the word is used to refer to a disease or condition habitually present in a given area (and often even to indicate that a disease is widespread in an area).

For example, on Japan's subtropical southwestern islands, in the Nansei Shotō archipelago stretching between Kyūshū and Taiwan, and on the southern islands of Izu and Ogasawara (see *pages 299–351*), there are numerous endemic plants, insects, amphibians, reptiles and even birds and mammals – unique biological gems that can be seen nowhere else on earth, unless they have been introduced to these other areas.

A spectacularly colourful member of the crow family, Lidth's Jay, occurs only on the forested island of Amami Ōshima, which lies between Kyūshū and Okinawa, and nowhere else in the world. It shares its tiny world range with another island endemic, namely Amami Rabbit, and a host of endemic salamanders and frogs. Areas of great endemism are very attractive to visiting naturalists.

Throughout the forested parts of this archipelago one might bump into the small endemic Ryūkyū subspecies of wild pig (known as Wild Boar), or even the endemic Ryukyu Flying Fox. Among the more famous of the avian endemics of Okinawa are Pryer's Woodpecker and Okinawa Rail, but they are not the only ones. Confined to the particularly rich biodiversity hotspot of northern Okinawa known as Yambaru (or Yanbaru), the rail was first described in 1981, and soon came under threat from the introduced and highly predatory Small Asian Mongoose[1]. Although the woodpecker has been known for very much longer, it is a great rarity because of its dependence on mature forest trees. It has been under threat because of the loss of forest in this area through logging and development, and depredation of its nestlings and fledglings by snakes and crows.

On the volcanic islands of Izu and Ogasawara there are numerous endemic plant species, endemic insects, endemic snails, and a number of bird species known for having tiny and very restricted ranges indeed. Owston's Tit, a relative of the more widespread Varied Tit, occurs only on the southern Izu Islands, while on Haha-jima, a tiny island in the Ogasawara Islands group, Bonin Honeyeater, a relative of the white-eyes[2], can be found. Once occurring on nearby Chichi-jima, too, the honeyeater population there has become extinct, leaving the individuals on Haha-jima as the only ones in the world.

The long isolation of Japan's many islands has fostered the evolution of endemic plants and animals in almost all areas of the country.

---

1 Known also as the Javan Mongoose or Small Indian Mongoose.
2 It is also frequently called Bonin White-eye.

Japanese Deer is one of the most widespread native mammals in the archipelago [FT].

An array of salmon species, including Chum, Cherry and Pink, returns each year to spawn in the rivers of northern Japan [KS].

Patterns of Wildlife Distribution

The traveller on an imaginary journey down the archipelago from north to south is given an overview of the wildlife diversity to be found in Japan. Starting aboard the *Lilac*, a small ship that plies the route between the harbour town of Utoro and the tip of Hokkaidō's remote Shiretoko Peninsula, we ride seas that are home to the pretty, pied Dall's Porpoise and to Northern Minke Whale. The landward view to the east is to the wildest national park in Japan (Shiretoko National Park), still home to Japan's largest land mammal, the Brown Bear, and an excellent destination for naturalists. Hokkaidō's Brown Bears are among the largest of their kind in the world. It is difficult to travel here, even by boat, without sighting small groups of Japanese Deer. This wide-ranging Japanese species grows to its largest in Hokkaidō, and it and Red Fox are the mammals most regularly sighted as they move between the mountains and the shoreline.

In autumn, returning migrants vastly outnumber resident species of wildlife. Then, on the Shiretoko Peninsula, the river mouths boil, and the rivers shimmer with migratory salmon of up to six species (see *p. 174*) returning to spawn. Not surprisingly, this natural buffet attracts bears, for which Shiretoko is famous, down from the mountains. Only here, on this peninsula, is their natural range – from mountain slope to river mouth and seashore – still intact.

The Daisetsu Mountains, the great raised massif of central Hokkaidō, known to the indigenous Ainu (see *p. 149*) or Utari as *Kamuy Mintara* (the Playground of the Gods), are the next best home for bears. Although bears are in danger of being hunted to extinction in Japan, the lucky observer may find mothers with their cubs foraging for berry-bearing alpine plants amid brilliant paint-box autumn colours, with fiery mountain rowan trees flaming among deep green Japanese Stone Pines. The sighting of the diminutive, tail-flicking Siberian Chipmunks as they scurry among these high forests is, however, far more likely.

In Japan, Siberian Chipmunk is native only to Hokkaidō [MAB].

Ocean and Mountain Realms

Travelling the archipelago by coastal ferry allows observations of a wide range of marine species, including flying fish [NB].

Travel by ferry, wherever possible, offers a slow, reflective pace. So, to continue this imaginary single journey by ship for a while longer, the prolonged but comfortable ride south from Hokkaidō, down the Pacific seaboard of Honshū, not only offers time for reflection on the changes encountered during the journey, but also provides possible sightings of Northern Fur Seal, Pacific White-sided Dolphin and other cetaceans. Species possible along this route include Northern Minke Whale, Baird's Beaked Whale and Short-finned Pilot Whale, Orca, Risso's Dolphin, Harbour Porpoise and Great Sperm Whale. There is also a wealth of seabirds and other marine life, such as hammerhead sharks (family Sphyrnidae) and Green Turtles. My first sightings of silvery flying fish[1] with their shimmering elongated pectoral 'wings', taking off from the bow-wave of the ship as it entered warmer waters, was an inspiration. It surprised me so much that it led me off on a train of thought about the varied forms of expression that wild creatures have here in Japan and, ultimately, led me to write the book *Nature of Japan*.

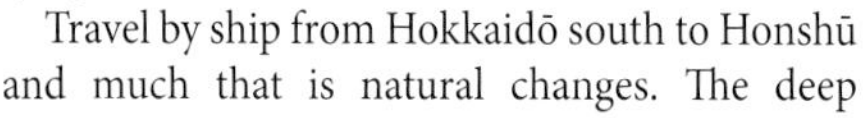

Travel by ship from Hokkaidō south to Honshū and much that is natural changes. The deep Tsugaru Strait separating Hokkaidō from Honshū (with the biogeographical line, known as Blakiston's Line [see *p. 16*], running through it) forms the southern limit of the Brown Bear's range. Instead, in the mountains of Honshū, there is a different bear, the smaller, elusive and now rather scarce Asiatic Black Bear, with a distinctive white crescent on its chest. The significance of this strait can be seen when examining Japan's mammalian fauna: although 21 species occur on both sides of this strait, 19 species are found in Hokkaidō but not on Japan's three other main islands, whereas 36 species occur on those main islands but not in Hokkaidō.

A more likely encounter than with a bear in the mountains of Honshū is with another strange inhabitant, the Japanese Serow. Seemingly part goat and part deer, this stocky creature is well adapted to forging its way through deep snow and wiry dwarf bamboo, and is frequently encountered by hiking and skiing visitors to Honshū's mountainous spine. The lower slopes of the mountains are home to Tanuki[2] and Japanese Macaques, reputedly the world's northernmost monkeys. On the Shimokita Peninsula, in Aomori Prefecture at the northern tip of Honshū, there truly are 'Snow Monkeys', as these macaques are often called. Although the species is now famed worldwide for its almost human habit[3] of resorting to thermal springs for winter warmth, this occurs only at a single location farther to the south in the mountains of Nagano Prefecture, Honshū. The forests that harbour Asiatic Black Bear, Japanese

1 Perhaps most likely to have been Barbel Flying Fish *Exocoetus monocirrhus* – just one of Japan's 3,000 or so species of fish.

2 Let's stop calling it Raccoon Dog. That is a name that causes unnecessary confusion; suggesting to some that it is related to the North American Raccoon *Procyon lotor*, which it is not. It is not really a dog, and it certainly is not a hunting dog for pursuing raccoons as some imagine. The Japanese name Tanuki is unambiguous.

3 A habit copied from humans.

The endemic Japanese Serow is typically secretive and shy, yet occasionally remains in clear view [MAB].

Japanese Macaques range as far north as the Shimokita Peninsula of Aomori Prefecture, at the northern tip of Honshū [NZFS].

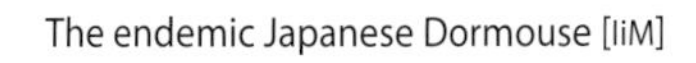

The endemic Japanese Dormouse [IiM]

The endemic Japanese Squirrel [FT]

Serow and Japanese Macaque are home also to the endemic Japanese Squirrel, the diminutive Japanese Dormouse and the cat-sized Japanese Giant Flying Squirrel. Hokkaidō, in contrast, lacks these species, but instead is home to Eurasian Red Squirrels and Siberian Flying Squirrels.

The rivers that drain western Honshū's peaks are home to an extraordinary creature: the record-breaking Japanese Giant Salamander. Some individuals have exceeded 1·5 m in length, making them the world's second largest amphibian. Both the giant salamander and the serow were hunted for food by the guides who led the English clergyman and Anglican missionary Reverend Walter Weston (1860–1940) into the Japanese Alps in the late 19th century. It was he who made pioneering hikes into these alps and helped to make Japan's peaks so famous, and now a mecca for mountain-hiking aficionados. But finding one of these rare salamanders today (now fully protected from the depredations of hunters and hikers alike) involves a diligent search in cold mountain streams at night. Not surprisingly, only dedicated naturalists can claim to have seen one in the wild.

Wetlands – Swamps and Wet Meadows

Rapid habitat conversion from the middle of the 20th century onwards has led to dramatic losses of Japan's coastal wetlands. At the same time, intensification of agricultural activities has depleted inland wetlands. Coincidentally, engineering practices of the period dictated that rivers be straightened, canalized and dammed, leading to the loss of riparian wetlands, too.

Some wetland species have managed to survive. Most famous is the deity of the marshes, Red-crowned Crane, which today thrives in east Hokkaidō. Although the related White-naped Crane has been lost as a breeding species in Japan, both Crested Ibis on Sado Island and Oriental Stork in Hyōgo Prefecture are making comebacks, thanks to intensive conservation activity.

Protecting wetlands, swamps and wet meadows for one species may assist the survival of many others, including wetland plants and associated insects, amphibians and birds. Now protected from hunting, Japanese Giant Salamanders prevail in certain areas of western Honshū, where we now have a greater understanding of the importance of naturally flowing rivers for this species. Across Japan several newts and at least 29 species of small salamander can be found, almost all of them endemic to

Wetland habitats, whether coastal or inland, have been drastically reduced in area over the last century [YS].

Wetland habitats are home to many unique species, such as Japanese Black Salamander [*left* MENR] and Japanese Clawed Salamander [*right* MOEN], both endemic to Honshū.

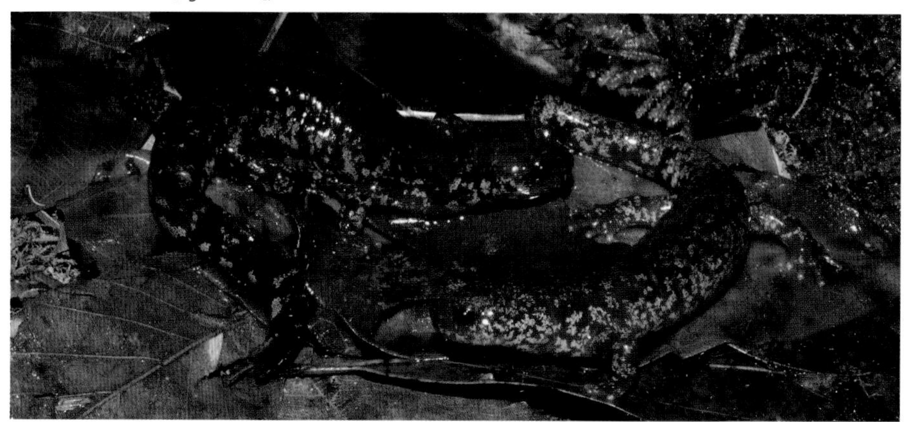

Two new species of salamander were discovered in Kyūshū as recently as 2019 [TA].

Damp woodlands and meadows in northern Japan are home to Sakhalin [*left* MR] and Middendorff's Grasshopper Warblers [*right* SP].

Japan, including Ezo (or Hokkaidō) Salamander and Japanese Clawed Salamander of Honshū. Much remains to be discovered in Japan, as revealed by the finding and naming of Mikawa Salamander, which was described in 2017 from the mountains of eastern Aichi Prefecture, in central Honshū, and the two new species (Chikushi Spotted Salamander and Stejneger's Salamander) described in 2019 from the mountains of northern Kyūshū.

Salamanders prefer damp pools, ponds, streams, damp woodlands and wet meadows, whereas the last-mentioned of those habitats, together with swamps, favours cranes, bitterns and snipe. Where there are extensive reedbeds in certain parts of Honshū, various localized bird species occur, including Marsh Grassbird and Japanese Reed Bunting. Rank grasslands in the north may hold Lanceolated Warbler and Middendorff's Grasshopper Warbler in summer. Swamp and woodland margins that have dense vegetation cover resound to the explosive, loud song of Sakhalin Grasshopper Warbler.

From Ubiquitous to Rare

Patterns of species distribution vary greatly in Japan. There are species that can be found almost everywhere, from the far north to the far south and at a considerable range of altitudes. At the opposite extreme are species that are not merely confined to a smaller area, or to a very particular and specific habitat, but may have evolved *in situ*, isolated on tiny fragments of the archipelago atop an isolated mountain, or restricted to a tiny island.

Whether in Kyūshū, in Shikoku or in Honshū, in an urban area or a rural area, there are five species of bird that are encountered almost everywhere and on almost every day. Perhaps the most conspicuous are Japan's two crow species. Between them, Oriental Crow, a relative of the western Carrion Crow, and Japanese Crow, a relative of the Asian Large-billed Crow, scavenge throughout the country. The former has a gently sloping profile and slender bill that continues to a fine point, whereas the latter exhibits an almost vertical forehead, a truly massive, more blunt-ended bill, and thuggish demeanour.

The huge Japanese Crow, often mistakenly identified as a Northern Raven by visitors to Japan, can be found building its nests in a wide range of environments, from rooftops and street trees to power pylons and forest trees. Some pairs even build their nests from abandoned (or stolen!) clothes-hangers from dry-cleaning shops. During winter, both crow species can be found roosting at night in concentrations of thousands in urban parks and street trees, enjoying the urban heat-island effect and the plethora of food available in cities. They are also at home, however, in rural farmland, foraging over newly harvested fields or around cattle barns and manure heaps, fossicking along the tideline for edible flotsam, or awaiting the return of the local fishing fleet at almost every harbour in the land. The intelligence of crows is remarkable. In Japan, Oriental Crows are renowned for cleverly using vehicular traffic to crack open or even smash walnuts, which the crows then eat.

Stroll in a residential suburb or drop by in any suburban park, and you cannot help but notice a cinnamon and brown dove wandering the paths and woodland edges: Oriental Turtle Dove. Its rhythmic, soporific "*de-de-po-po*" courtship song can be heard everywhere, from the forests of Okinawa to those of Hokkaidō, and it can be found foraging on the ground in almost every urban and suburban area of Japan.

While noting the duo of crows and the dove, it is hard to miss White Wagtail. This small, long-tailed, pied bird catches insects as it walks, bobbing its tail almost continuously as it does so. Largely unafraid of people, White Wagtail can be found from suburban supermarket parking lots to rural streamsides, roadsides and farmland. Equally widespread, and vocal, although a little more secretive, is Oriental Greenfinch. This large-billed finch consumes seeds of trees and common herbs, including dandelions. Its buzzing calls are distinctive; and in flight so, too, are its flashing yellow wingbars.

In contrast with these five ubiquitous species, other bird species have ranges that are restricted to single islands or tiny archipelagos; in other words, they are endemic (see *pages 90, 92*) and often rare. So, while some species are so common and widespread that they may be encountered during almost any birdwatching excursion, others are so localized and rare that it is necessary to plan a specific pilgrimage to a special location to find them.

Oriental Crow is ubiquitous throughout Japan [MaH].

The thuggish Japanese Crow is an intimidating scavenger and predator [MAB].

Both Eastern Spot-billed Duck and Mallard can be found throughout most of Japan [BOTH MAB].

White Wagtail is a ubiquitous ground-foraging insectivore [MR].

The most widespread seed-eating finch of Japan is Oriental Greenfinch [MR].

Oriental Turtle Doves may be found from montane forests to urban parks and from Okinawa to Hokkaidō [MR].

While some species range throughout the country, others, such as Lidth's Jay, are highly restricted and found only on one or two small islands [TsM].

Japan's Rodent Fauna

The commensal, omnivorous Brown Rat is found worldwide, including throughout Japan [IiM].

Among Japan's ubiquitous species are its rodents. The past connection of the proto-Japanese islands to the Asian continent, combined with the subsequent prolonged isolation of its many islands, has left Japan with a richly diverse rodent fauna. This fauna includes commensal species such as the ubiquitous Brown Rat (or Norway Rat), and specialist endemic arboreal species such as Japanese Dormouse. Japan's various squirrels are rodents, too, but they are treated elsewhere *(pp. 188–196, p. 218–220)*.

The omnivorous Brown Rat has a global distribution, so its presence in Japan comes as no surprise. It is readily distinguished by its long, mostly naked tail of 17·5–22·0 cm that is almost as long as the head and body together, 11–28 cm. It is one of the largest rodent species, weighing up to 500 g. Being happy in human company, it occurs throughout Japan in all residential and agricultural areas, particularly in coastal and lowland areas where there are concentrations of food waste at markets, fish markets, harbours, garbage dumps, sewers, storm drains, streams and riversides. Brown Rats are found also around crop fields and woodland margins. The species' range includes many offshore islands, which it may have reached naturally or been carried to by ships, and it occurs even on the isolated volcanic islands of the Izu and Ogasawara island chain, but no doubt it was introduced there accidentally. Rats burrow to create their nests or, alternatively, they use natural cavities or those in human-built structures. Able to breed at any time of the year, the females produce, annually, multiple litters of up to 18 young at a time, although they average 8–9 per litter.

At the other end of the size spectrum, Eurasian Harvest Mouse has a broad distribution across the Old World (Palaearctic), ranging from Europe across Siberia to eastern China, the Korean Peninsula, the Russian Far East and Japan. In Japan, it is absent from Hokkaidō and northern Honshū, but occurs from Niigata and Miyagi Prefectures south and westwards through central and western Honshū, Shikoku and Kyūshū. This small nocturnal and vegetarian mouse weighs just 7–14 g and measures only 5–8 cm in head-and-body length, with a very long (6–8 cm) prehensile tail. Its fur is a warm brown with an orange tinge, and it is pale buff or white below. It ranges from the lowlands to low mountains up to about 1,200 m above sea level, where it can be found in natural grassland and wetland habitats with tall vegetation, and in agricultural crops that provide similar habitats, such as abandoned fields. It builds a spherical nest of grasses (resembling a bird's nest) approximately one metre above the ground, in which it produces up to three litters (each of 2–8 young) in a year.

In contrast to the enormous global or Palaearctic distributions of Brown Rat and Harvest Mouse, several rodent species are endemic to Japan. Large Japanese Field Mouse and its sister species Small Japanese Field Mouse both occur throughout Japan.

Large Japanese Field Mouse is a medium-sized, terrestrial brown mouse weighing 20–60 g and measuring 8–14 cm in head-and-body length, with a tail 7–13 cm long. It occurs from the lowlands to the tree line in mountains, being found in mature forest and in areas with dense grasses, such as along riverbanks, overgrown fields and rice fields, from Hokkaidō to Kyūshū and on large offshore islands as far south as Yaku-shima and Tanega-shima. Its breeding seasons differ from region to region, encompassing the whole year, with multiple litters of 3–6 young per litter possible within a year. Herbivorous, it eats seeds, seedlings, stems and roots.

Small Japanese Field Mouse is a small, very long-tailed brown mouse found nation wide. It weighs just 10–20 g, measures 6·5–10 cm in head-and-body length and has a tail of 7–11 cm. Like its larger relative, it occurs throughout Japan, including on large offshore islands such as Sado, Yaku-shima and Tanega-shima, occurring from the lowlands to the alpine zone in mature forest with an accumulated

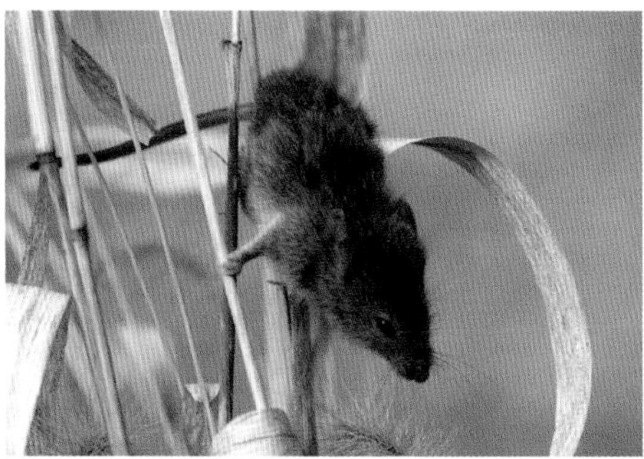

The diminutive Eurasian Harvest Mouse ranges from Europe to the Far East [IiM].

The endemic Japanese Dormouse is widespread but rarely seen [MM].

Large Japanese Field Mouse is endemic to Japan [MM].

The abundant Grey Red-backed Vole is a key prey species of owls, buzzards and Red Foxes [IiM].

litter of leaves and twigs. It is both terrestrial and semi-arboreal, nocturnal and crepuscular. It is also mainly herbivorous, eating seeds, fruits and green vegetation, but will also occasionally consume arthropods. The breeding season differs from region to region, with multiple litters (averaging four young per litter) possible within a year.

The most distinctive of all of Japan's smaller rodents is the nocturnal and mostly arboreal Japanese Dormouse. So special is this endemic dormouse that it has, since 1975, been recognized as a Natural Monument (Could any other country conceive of recognizing a rodent, albeit endemic, as a Natural Monument?). It has a unique broad, dark brown stripe down its back and a long, flattened furry tail. It is found in native mature forests from the lower montane zone to the subalpine zone, occurring from northern Honshū south to Shikoku and Kyūshū. It weighs 14–20 g (rising to 30–40 g prior to hibernation) and measures 6·8–8·4 cm in head-and-body length, with a tail 4·4–5·4 cm long. An omnivore, its wide range of foods includes seeds, fruits, buds, invertebrates and even birds' eggs.

During the breeding season, male Japanese Dormice occupy home ranges of up to 2 ha and females up to 1 ha. They build spherical nests of moss and bark, placing them in tree hollows and cavities or in the fork of a branch. They breed twice each year, in spring and autumn. The females give birth to 3–5 young in each litter. Like Siberian Chipmunk, they hibernate for up to six months of the year, although the length of hibernation depends on the temperature of the region and the altitude at which they live.

Japan's Ubiquitous Scavenger – Black-eared Kite

Overhead, silent as a spiderling's gossamer thread adrift on the breeze, a spread-winged kite tips its head and glances down from a cloudless sky. As the kite peers down, it twists its tail in a subtle unhurried motion. Time is on its side as it peruses the ground below. The gentle movement that begins at the base of its tail is barely perceptible, but it is transmitted along the length of the long, broad tail feathers and translates into an elegant twist at the fanning tail tip that sends the bird into a banking curve. The kite's outstretched wings enfold the warm gentle breeze and on it the kite quickly rises higher still, its twisting tail first appearing square-ended, now lightly notched, steering the way, and the bird drifts nonchalantly away above the farmland to a nearby woodland edge.

This scene could be anywhere in Japan, from Okinawa to Hokkaidō, for Black-eared Kite is Japan's commonest raptor, its ubiquitous avian scavenger. There are many other raptors in Japan, but none may be seen so easily and frequently and in so many different habitats.

Frequently seen with crows in pursuit, the kite's shape alone is sufficient to elicit a mobbing response from them. The unforgiving crows swoop and harry the kites both in the air and while the latter are perching. Although kites are most definitely classified as birds of prey, paradoxically they are not in fact predators. Unlike their mammal-, reptile- or bird-hunting relatives the eagles, hawks and falcons, kites are scavengers. They share piercing eyesight with their more predatory raptor cousins but, unlike them, they have weak feet and tiny, puny bills. Theirs is an aerial pursuit of the dead. You will find them soaring carelessly over almost any open area of farmland, drifting along casually above any road, river, estuary or stretch of shoreline.

Watch awhile and you may observe the kite's moment of recognition. Its seemingly endless silent gliding flight finishes abruptly in a tight banking turn; it makes a sudden spiralling descent to earth, and lands abruptly. There is a brief moment when it appears ungainly, but once it furls its great wings it regains its composure and soon it is scavenging at the corpse of a spent frog, a road-killed rodent or a tide-washed fish. Amphibians, reptiles, mammals and birds, along with earthworms, beetles and bugs, all are grist for the kite's mill – so long as they are dead.

Waxwings – Birds with Wanderlust

The waxwings epitomize the intimate relationship between birds and berries. Large numbers of two species, Bohemian Waxwing and Japanese Waxwing, visit Japan each winter. Waxwings are wanderers, remaining in a place only while favoured foods, berries on certain trees, remain plentiful, and then moving on quickly not just to a different tree but, more likely, a different district entirely.

Two species in particular attract them, Japanese Rowan and Japanese Mistletoe. The rowan has delightful fiery red autumn foliage and produces dense clusters, corymbs, of bright red berries. The mistletoe produces creamy white, pale yellowish-green or red berries and is a semi-parasitic plant that clings to other trees, where it photosynthesizes while drawing water and minerals from its host, on which it depends. The lives of these colourful winter wanderers are intimately entwined with that of the golden bough[1] itself, for the waxwings love not only the glistening pearly-white berries but also the faintly green ones and the waxen red ones[2]. Mistletoe and waxwings have together, it seems, co-evolved a mysterious, enchanting interdependent lifestyle, which to birders spells 'Winter'.

Waxwings are 'gulpers-and-dumpers': they gulp down berries whole and, after a lengthy period of digestion, they dump out the seeds. The mistletoe's berries are jelly-like and viscous, making the waxwings droppings similarly glutinous, ideally primed to stick to branches, where the seeds germinate into the next generation of mistletoes. Waxwings, like thrushes, are also strippers. While they do not peel off layers of feathers, they do gather in flocks and descend on berry-bearing trees, shrubs and mistletoes, stripping them entirely of their fruit before moving on.

Whereas Bohemian Waxwing has a Holarctic distribution, occurring both in North America and across northern Eurasia, Japanese Waxwing has a very restricted range in East Asia, making it a much sought-after species. Japanese Waxwing differs most noticeably in having a red, rather than a yellow, tip to its tail.

---

1 The mistletoe was considered sacred by the Druids of Europe, where it became known and revered as the Golden Bough.

2 Japanese Mistletoe produces berries of either colour, in contrast to European Mistletoe, which has only white berries.

Black-eared Kite is the most widely encountered raptor in Japan. It scavenges a great diversity of dead prey [*left* MAB; *right* ImM].

Bohemian Waxwings migrate into Japan for the winter and may be found feasting on the berries of street trees [SP].

Despite its name, Japanese Waxwing does not breed here; it is an erratic and highly mobile winter visitor much sought after by visiting birders [SP].

Japanese and Bohemian Waxwings form mixed flocks [SP].

An Abundance of Starlings

Chestnut-cheeked Starling [YaM]

A widespread and abundant bird of Japan is White-cheeked Starling. Like Common Starling in Europe (or where it has been introduced in North America), White-cheeked Starling is territorial during the breeding season but, after breeding, gathers in increasingly large flocks. These frequently forage on the ground, probing into short grass and leaf litter for invertebrate prey, but during autumn and winter they are often to be seen in flocks in fruiting trees, consuming apples or persimmons as they ripen.

A second species, Chestnut-cheeked Starling, is a summer visitor to central and northern Honshū and Hokkaidō and a migrant through western and southern Japan. Like its larger cousin the White-cheeked Starling, Chestnut-cheeked Starling is also a hole nester. Both starlings favour natural cavities, such as those made by the many woodpeckers of Japan, and they will also use nestboxes.

During spring and autumn migration, both species may be seen together in mixed flocks. During the breeding season, however, White-cheeked Starling is more likely to be found in urban and suburban areas and farming regions, while Chestnut-cheeked Starling is more a bird of mature woodland and forest.

Outside the breeding season, White-cheeked Starlings are commonly seen in flocks clustering along roadside telephone wires and power lines or performing spectacular flock manoeuvres as the birds head to their communal roosts. At such times, their flocks, known as murmurations, move spectacularly, resembling swarms of bees or clouds of smoke. In the related Common Starling these murmurations have been shown to be a successful anti-predator strategy, reducing predation through dilution and predator-confusion effects, meaning that an individual is less likely to be eaten when with others than when it is alone.

White-cheeked Starling is a common species throughout the country [TsM].

Chestnut-cheeked Starling is a migrant and summer visitor [SP].

Common Starling is, despite its name, only a scarce winter visitor [ImM].

A Flight of Butterflies

Butterfly diversity in Japan is a delight. Whereas the islands of the UK and Ireland host 71 species and those of New Zealand 62, an astonishing 263 species have been recorded from the Japanese archipelago thanks to its diverse range of local climates and habitats. As with Japan's other groups of organisms, some are very wide-ranging throughout the country in their appropriate habitat, while others are restricted to very much smaller areas or groups of islands. Palaearctic species familiar in Europe, such as Dark Green Fritillary, Silver-washed Fritillary and Small Tortoiseshell, are familiar also in Japan. They each occur only in Hokkaidō and parts of Honshū. Then there are those that occur in the eastern Palaearctic such as Oriental Hairstreak, which here ranges from Hokkaidō to Kyūshū. Some tropical species range from India to Japan, one example being Orange Oakleaf (see *p. 65*), but here it occurs only in the subtropical southwest islands from the Amami Islands to the Yaeyama Islands.

More restricted globally are those butterflies known only from East Asia such as Zephyrus Hairstreak, which occurs from Hokkaidō to Kyūshū. Other East Asian species, such as Long-tailed Spangle, East Asian Fritillary and Green Hairstreak, likewise range here from Hokkaidō to Kyūshū, while Many-striped Hairstreak occurs in Hokkaidō, Honshū and Shikoku. There are also those species that are endemic to Japan, such as Ezo Hairstreak, which is found from Hokkaidō to Kyūshū.

The Japanese archipelago is rich in butterfly species such as the Many-striped Hairstreak (*above top*), Long-tailed Spangle (*above middle*), East Asian Fritillary (*above bottom*) and Ezo Hairstreak (*below*) [ALL TK].

A Diverse Array of Thrushes

Brown-headed Thrush [YaM]

No fewer than 21 species of thrush have been recorded in Japan. While several are vagrants or accidentals, others are regular summer visitors, winter visitors, or spring and autumn migrants. Some, such as Dusky Thrush, are abundant migrants.

The summer migrants include several exciting Asian species. Among these are Siberian Thrush, White's Thrush, Brown-headed Thrush and Japanese Thrush, mostly elusive birds of montane forest, more likely to be heard singing their simple songs than to be seen. The migrant Eyebrowed Thrush is common and appears for just a few weeks of the year, and it, too, is very elusive. Two of Japan's winter-visitor species, Pale Thrush and Dusky Thrush, are very different.

Pale Thrush is frequently encountered in central and western Japan, whereas Dusky Thrush is more a bird of northern Japan. Both will forage for invertebrates in the open on the ground, and may be found in urban and suburban parks, extensive gardens and, in the case of Dusky Thrush, agricultural fields. They forage also on berry-bearing and fruit-carrying trees and, like the waxwings, they may appear suddenly, strip berries for a while and then move on. Look for them wherever there are fruiting orchards and windfalls on the ground.

Dusky Thrush is an abundant bird breeding across northeast Asia and noisy flocks arrive in Japan during October, when they are often first noticed through their frequent rapid and strident "*chek-chek-chek*" vocalizations in flight.

White's Thrush is the largest thrush species in Japan [TsM].

Dusky Thrush – an abundant winter visitor [JW]

Eyebrowed Thrush – an elusive migrant [MJ]

Siberian Thrush – an elusive summer visitor to montane forest [JoW]

Pale Thrush – a common winter visitor to southern Japan [MAB]

Brown-headed Thrush – a migrant within Japan [MC]

Japanese Thrush – a summer visitor [WaM]

Seed-shatterers, Megaflocks, and Auspicious Birds

Japanese Grosbeak [YaM]

A look at Japan's resident and migrant passerine birds reveals great diversity. Finches are renowned seed-eaters, among them Japanese Grosbeak and Hawfinch, which have the strongest bite of almost any small bird. The Hawfinch's scientific name literally means seed-shatterer. It has much larger jaw muscles than most other birds of its size, giving it sufficient bite pressure to enable it to crack cherry and plum stones – quite a dramatic feat and one that makes it a seed predator, not a disperser of seeds.

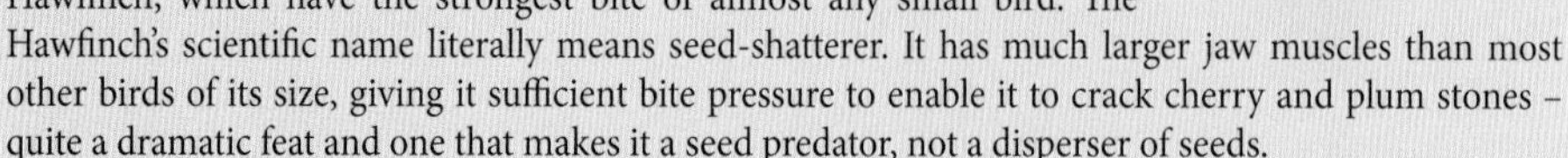

Hawfinch is a common resident of northern Japan and a common winter visitor throughout the main islands of the archipelago. Its impressive larger relative, the almost thrush-sized Japanese Grosbeak, is a resident of the south and west and a summer visitor in the north. The most widespread of Japan's finches, however, is the olive-and-yellow Oriental Greenfinch, which in winter can be seen in large flocks. Even so, greenfinch flocks are sometimes dwarfed by those of the visiting Brambling, a distinctive finch with an orange 'shawl' which sometimes appears in flocks of many tens of thousands of individuals.

In Japanese culture, the colour red, which is associated with the sun and the national flag, is considered an auspicious colour denoting various attributes, including strength, passion and self-sacrifice. Among Japanese birdwatchers, red birds excite particular passion. Of the finches, therefore, it is not Oriental Greenfinch or Hawfinch that is considered exciting, but the red-coloured Pine Grosbeak, Grey-bellied Bullfinch, Asian Rosy Finch, Long-tailed Rosefinch, Pallas's Rosefinch, Common Redpoll and Red Crossbill.

Bramblings, readily identified by their distinctive orange chestband occasionally occur in megaflocks [*left* ImM; *right* FT].

Japanese Grosbeak is a thrush-sized finch [WaM].

A seed-shattering Hawfinch [SP]

Common Redpoll is a common and widespread winter visitor [SP].

Eurasian Siskins are common and widespread in winter and may occur with redpolls [SP].

Pine Grosbeak is typically found around Japanese Stone Pines at high elevations in Hokkaidō [WaM].

The scissor-like bill of Red Crossbill (and other crossbills) is an adaptation for eating coniferous-tree seeds [SP].

Pallas's Rosefinch is a scarce winter visitor to Japan from the Asian continent [ImM].

Long-tailed Rosefinch winters in the south and breeds in the north of Japan [WaM].

A Bevy of Buntings

Meadow Buntings are typical of rural and agricultural areas [YaM].

More than 20 of Japan's resident and migrant passerines are buntings. These are akin to the finches (see *p. 112*), but typically have a smaller head and a longer tail. Several of those recorded in Japan are vagrants, but many are resident or regular visitors.

Meadow Bunting is common throughout Japan, although only a summer visitor in Hokkaidō. Chestnut-eared Bunting is resident in the southern half of the country and a summer visitor in the northern half, preferring rank grasslands and rough meadows, but it is now uncommon and declining. Rustic Bunting and Elegant Bunting are both winter visitors, generally more common farther south in the country. Yellow Bunting [NT] is a scarce local summer breeding endemic, Grey Bunting a more common summer visitor in montane areas of the north and a winter visitor in the south, but it is Masked Bunting that is Japan's most familiar bunting. This common migrant and resident (it breeds in the north, winters in the south, and is resident in the middle) is a constant presence in lightly wooded areas, gardens, hill forests and areas with rough vegetation. While Masked Bunting remains abundant, another of Japan's buntings has almost entirely disappeared. Yellow-breasted Bunting [NT] was once a common summer visitor to reedbeds and swampy grasslands of the northern wetlands. Its attractive, although repetitive, song was once a constant sound in the background. Yellow-breasted Bunting migrates to wintering grounds in southern China, where it has been almost entirely eliminated. Today it is a very rare visitor to Japan.

Elegant Bunting is a winter visitor to southern Japan [YM].

Rustic Bunting is a common winter visitor to the southern half of Japan [TK].

Grey Bunting favours montane forest habitats for breeding, but winters in the lowlands [SP].

Chestnut-eared Bunting is restricted to rank grasslands and rough meadows [SP].

Yellow Bunting is a scarce endemic breeding species that winters outside Japan [KT].

Japan's most abundant bunting species is Masked Bunting [WaM].

A once common breeding visitor in northern Japan, Yellow-breasted Bunting is now almost extinct in the region (and in much of its world range) [KT].

Life on the Edge

With one of the most extensive coastlines of any country for its size, it comes as no surprise that Japan is a land of fishermen and that the Japanese diet contains a great variety of seafood. Much of the coastline is rocky, forming tidal pools that are packed with invertebrate life, so much so that Red Foxes often scavenge along the rocks at low tide. In certain areas in the north seals haul out, although mammalian life in general is scarce around the coast. Shorebirds gather at muddy foreshores and along sandy beaches where they are undisturbed, while egrets and herons may be found hunting in pools and channels among the rocks. On craggier sections of the shore, the colourful Eastern Blue Rock Thrush may be encountered.

Among the strangest of all avian phenomena is the gathering of White-bellied Green Pigeons at the coast. This forest-dwelling species is cryptic and secretive, foraging quietly on tree fruits in the forest canopy from Hokkaidō to Kyūshū, but at certain sites many individuals visit the coast, flocking down to rock pools where they drink saltwater! The purpose of this remains unclear. Perhaps it serves to neutralize some aspect of their main diet.

Where grassy-topped islets are isolated by craggy rock shores, Japan's most widespread resident seabird, Black-tailed Gull, is likely to be found breeding. In Japanese it is known as the 'sea cat' (*Umineko*) from its distinctive mewing call, which is inevitably heard by anyone visiting islands where these gulls breed. *Umineko* breed in colonies from northern Hokkaidō to western Kyūshū, favouring islands, islets and even isolated rocks. They use grasses growing nearby to build their nests, in which they lay their clutches of two or three densely mottled olive-green and brown eggs. Approach their nesting site too closely and they will soon begin to swoop and call. They do not press home their attacks as some terns do, but, because they nest in large colonies, the panic soon spreads to other pairs and, when they all

Most of Japan's extensive coastline is rocky [UNSP].

White-bellied Green Pigeons are typically at home in forested habitats yet they regularly visit rocky coasts, where they drink saltwater [TK].

The shoreline offers rich pickings to Black-eared Kites and Red Foxes [YM].

Eastern Blue Rock Thrush is more frequently seen along rocky coasts than anywhere else in Japan [ImM].

Coastal scenery in central Honshū [JW]

Harlequins and other sea-ducks

Shallow bays, fishing harbours, and the churning waters off Japan's northern rocky coastlines in Tōhoku and Hokkaidō all share a particular feature: they attract some of the most appealing and dramatic of Japan's waterfowl, the sea-ducks. None comes prettier than Harlequin Duck, which occurs abundantly around northern coasts, where it dives for shellfish and crabs in shallow water.

Frequently, in the same areas as the Harlequins, there are flocks of crooning and moaning Black Scoters, Siberian (or Stejneger's) Scoters and Long-tailed Ducks. Their calls carry even over the roughest of seas. As much at home in the surf zone as in calm water, these birds seem unperturbed by the pounding waves of autumn and winter storms and brighten any visit to the coast. Joining them are tousle-crested Red-breasted Mergansers and, occasionally, the latter's more freshwater-loving larger cousin the Common Merganser (or Goosander). Where the scoters, Harlequins and Long-tailed Ducks dive deeply to forage for prey on the bottom, the mergansers dive to chase their fish prey through the water column, in much the same way as the cormorants and grebes of Japan.

In winter, the smaller, more slender grebes can also be found diving offshore and often around sea-ducks and cormorants. The commonest is Red-necked Grebe, although at times in autumn and during winter Slavonian Grebe and Black-necked Grebe may also be common, the latter often resorting to fresh or brackish lagoons and moving in densely packed mobs as they forage.

In addition to the sea-ducks, a wide range of waterfowl visits Japan's freshwater wetlands each winter (see box on page *224*).

Harlequin Duck – male (*front*), female (*back*) [YaM]

The pretty Harlequin Duck is a scarce breeding species but a common winter visitor in northern Japan [*above* MC; *below* MR].

Black Scoters – common in winter around northern coasts [SP]

Male Long-tailed Duck – a winter visitor typically found offshore around northern Japan [MAB].

Male Common Goldeneye – a common winter visitor found offshore and at rivers and lakes [MAB].

Common Merganser (Goosander) – most frequently found at rivers and inland lakes [JW].

Red-breasted Mergansers – common winter visitors typically found offshore [PP].

take to the wing, their massed aerial presence is often sufficient to panic a potential egg or chick predator.

There is another behaviour to be aware of when visiting gull colonies. When the birds are disturbed they fly up, panic-induced peristalsis takes over – and splat! So, when you are standing beneath a swirling flock of outraged gulls, it is not a good idea to look up. Better instead to retreat and observe the gulls from a greater distance. They will soon settle, and treat you to views of their nesting behaviour – changing over at the nest, delivering food to their young, and so on. The adult has a very distinctive long beak with a prominent black band and deep red tip. If there are young in the nest and the parents are busy visiting, watch the chicks closely: they will peck at the red part of a parent's bill. All the larger gulls have either a red spot or a band, and the chicks use it as a pecking target. The pecking stimulates the parent to regurgitate the chick's next meal. It is as effective as pressing the button on a Japanese vending machine!

Almost all gulls are daytime hunters, and most are rather generalist feeders. Black-tailed Gulls are no exception, being as willing to snatch a morsel thrown from a tourist boat as they are to snaffle the offal discarded by fishing boats at sea. The tideline, too, is their scavenging ground, and here just about any recently animate object is considered potential food. Despite the amount of food available from garbage, gulls in Japan have not yet adopted the urban habit to anything like the degree that they have in the UK.

The red, black and yellow bill pattern of Black-tailed Gull is unique. Nestlings peck at the pattern when begging for food [*above* winter MC; *below* summer JD].

Black-tailed Gulls breed on craggy grassy-topped islets [BOTH TaM]].

Japan's Large Gulls

Black-tailed Gull [YaM]

Two East Asian gull species, Black-tailed Gull and Slaty-backed Gull, breed commonly in Japan. They are joined during spring and autumn migration seasons and in winter by great numbers of several other large species, such as Glaucous Gull, Glaucous-winged Gull, Vega Gull and Common Gull.

Visit any river mouth or fishing harbour in winter, and gulls are likely to be evident. They are far less common away from the coast than is the case in many other countries. If you visit any offshore island around northern Japan in summer, you will find colonies, sometimes very large ones, of Black-tailed Gull. The same can be said for Slaty-backed Gull on islands around Hokkaidō, but in the last four decades some Slaty-backed Gulls have adopted a very different nesting habit.

Whereas Slaty-backed Gulls were once confined to nesting on grass-covered headlands and clifftops, they have learned that rooftops offer relatively safe havens. Not surprisingly, they prefer flat or gently pitched roofs, especially those with snow guards, as these provide extra stability for their nests. Research has shown that they prefer sites near the edges of roofs where they can see their surroundings more easily and take off more readily. From the centre of Sapporo, Hokkaidō's capital city, to the fishing ports of the Shiretoko Peninsula, Slaty-backed Gulls can now be found nesting commonly on flat-topped high-rise buildings and harbour-warehouse roofs. They have even begun to nest on exposed breakwaters and harbour walls, although in such locations the nests and their contents are prone to being peeled away by strong winds as there is nothing to anchor them in place if the nesting adults take off.

Slaty-backed Gull is a widespread and common large gull in northern Japan, where it has taken to nesting on rooftops in some areas [BOTH MAB].

Glaucous Gull (*left*) and Glaucous-winged Gull (*right*) winter around northern coasts, harbours and river mouths [BOTH SP].

Commensalism in Farmland and Town and the Mystery of *Entotsu Suzume*

The plain brown cap and black cheek spot are the distinguishing marks of both male and female Eurasian Tree Sparrows. Despite its name it is at home nesting in buildings, during winter it gathers into large flocks [BOTH MAB].

The common, resident sparrow throughout East Asia, including Japan, is not, as is supposed by many visitors, House Sparrow[1], but rather it is Eurasian Tree Sparrow. This occurs throughout the country from rural farmland to towns and cities, although it has declined considerably in recent decades. Along with the crows and common rodents (rats and mice), in East Asia Eurasian Tree Sparrow is a commensal species, living alongside humans[2].

Its close relative, Russet Sparrow, a winter visitor to southern Japan and a summer visitor to the north, is more of a woodland species, although both have adapted to using human structures for nesting. Old-style Japanese domestic, agricultural and even commercial architecture was constructed mostly from wood, providing many holes, niches and interstices that could be used by sparrows and other birds for nesting. My memories of early travels in northern Japan include numerous weathering, grey wooden-clad buildings, a simple style of domestic architecture dating back to the 19th century and the settlement of Hokkaidō by the Japanese. Rickety, creaking and draughty, these older buildings provided ample access points for cavity-nesting birds, including sparrows. More modern, better-sealed construction now prevents this access. Sparrows have discovered, however, that snow poles and snow arrows have no caps, and many electricity poles offer narrow spaces into which both sparrow species can now enter and nest.

One kind of sparrow, which is far less common than it once was, is *Entotsu Suzume* (literally, 'chimney sparrow'). Neither a species nor a subspecies, this sparrow may nevertheless be unique to Japan. When almost all houses in northern and colder regions of Honshū and Hokkaidō used smoky wood-, coal- or coke-burning stoves, the narrow metal chimneys of those stoves offered warm rooftop cavities into which sparrows could retreat during winter. I have seen several of them myself. They take on a dirty, dusky appearance – as well they might – from direct contact with the chimney soot.

In contrast to Eurasian Tree Sparrow the sexes of the migratory Russet Sparrow have differing plumage – male (*front*), female (*back*) [YaM].

1 Sometimes called the English Sparrow by residents of North America.

2 Whereas the House Sparrow is the common commensal sparrow from Europe east to South Asia (and in countries where it has been introduced), from Southeast Asia to Japan it is replaced in that niche by the Eurasian Tree Sparrow.

People and Wildlife

Islands and Extinction

Islands experience greater extinction rates than continental areas do. Random events that can lead to the extinction of small, isolated populations are more common on islands. In addition, those same taxa are susceptible to humanity's destructive habits. In recent centuries mankind had caused the extinction of several endemic species in Japan, initially through hunting, then more particularly through habitat destruction and, most recently, through the introduction of alien predators and competitors. The Ministry of Environment's recent publication, the *Red Data Book of Japan*, is a saddening catalogue of pending biological disaster, for it reveals that 36 percent of all mammal taxa, 18 percent of reptiles, 19 percent of birds, 32 percent of amphibians and 24 percent of freshwater fish are Extinct, Endangered, Vulnerable, Rare or Localized. That around one-quarter of all Japan's vertebrate populations are at risk is an all too common outcome of economic development, as seen both here and around the world, especially during the latter half of the 20th century.

Among others, the Japanese populations of Eurasian Otter and Grey Wolf have already been hunted to extinction (see *p. 262*), but these are by no means isolated cases.

Brown Bear, with an estimated population of fewer than 2,000 individuals in Hokkaidō, has been hunted at an unsustainable rate, scientifically sampled unsustainably, and its habitat has been greatly reduced. Tinned Brown Bear meat can even be found on sale at tourist spots in east Hokkaidō. Extinction is perhaps only a few decades away for the species. Asiatic Black Bears, Japanese Serow, even Japanese Macaques, are similarly under enormous hunting pressure, primarily because any conflict between people and wildlife is seen as best remedied by eliminating the species said to be causing the problem.

Compounding the stress caused by habitat loss, the introduction of alien species has also had a serious impact. The craze for pets, particularly exotic species, combined with a lack of experience in their care and welfare, has led to many being abandoned. This exacerbates the problems faced by wild populations of native species. Abandoned exotics can have disturbing ecological impacts, particularly if these are predatory snakes or, as in one river near Tōkyō, predatory fish in the form of piranhas. The dangers of such releases are enormous. Even more dangerous is the uninformed deliberate introduction of alien species, such as Small Asian Mongoose, American Mink, North American Raccoon, Largemouth

Escaped or released alien species such as American Mink (*left*) and North American Raccoon (*right*) have spread widely causing considerable damage and long-term ecological consequences [BOTH IiM].

'black' Bass and Bluegill, without thought for their impacts on native species and long-term ecological consequences.

Witness to the fact that there is a mismatch between the image of the Japanese as having nature (albeit simplified and idealized) as part of their cultural identity and the stark reality that, during forays into the countryside, one is far more likely to bump into a dozen fishermen or hunters, butterfly- or beetle-collectors than to meet a single naturalist armed with binoculars or hand lens. Furthermore, the most visibly nature-oriented visitors to the countryside these days are photographers and, with some exceptions, they seem driven more by a desire for images than for knowledge and understanding of their wildlife subjects. The current emphasis lies in nature to be exploited. Do Japanese people really love nature, as we are led to believe? Indeed, they *do* share a deep, powerfully cultural empathy with the symbols of nature in Japan, and for some the spirit of Japan is in nature. In the modern world, however, such empathy with nature seems of no more practical value than an intrinsic religious or philosophical belief in human goodness. Is there in fact a fundamental environmental commitment? Indeed, there is, but one must delve back into ancient Japan to find that commitment. The Japanese people believed that all natural landscapes, including the sun, the mountains and the fertile lowlands, were created and populated by deities known as *kami*. *Kami* are emblematic of 'nodes' or 'concentrations' of spirit associated with what we see as physical objects, whereas spirit itself is continuous, indestructible, and cannot be differentiated. This is at the root of the problem; because spirit has these qualities, it does not matter if you destroy any of its physical representations, the spirit just flows into something else. You want to chop down a tree – fine, propitiate the local 'knot' of spirit associated with the tree and then cut it down. Its spirit will just flow into something else. The belief in a continuous spirit world denies the uniqueness of individual natural objects, such as trees or whales, and of individual natural categories (species, landscapes). Furthermore, they believed that the will of these *kami* controlled the cultural domain. People therefore provided shrines and made offerings to the *kami* in order to placate their reckless domination. From this perspective, culture lay in the hands of nature, not vice versa. Understanding this concept of nature as superior to culture helps to explain the true Japanese geographical concept of landscape, rather than the concept absorbed as western sciences came to permeate Japan. While a traditional perspective thus promotes Japan as being a nation of 'nature-loving' people, this is in strong contrast to the changing reality of modern life with its anthropocentric hierarchical ideology. This contrast deserves much closer attention and analysis.

Although blessed with an enviable diversity of land forms, habitats and wildlife, Japan is in the unenviable position of being poised on the brink of losing a large proportion of its natural riches, particularly that proportion which distinguishes it so markedly from other regions of the world. The single-minded post-war pursuit of a larger gross national product has sacrificed the natural environment and gross national happiness. The human influence on the landscape is a powerful one. Some habitats, particularly wetlands, are now in very short supply or seriously endangered, many of them having been destroyed for 'development' and many of those remaining not being protected.

Extinctions in Japan are set to escalate over the next few decades. A century hence, naturalists reflecting on the Japan of their recent past may contemplate a natural fauna that was once very much richer, unless attitudes and actions change quickly. Evidence for such change is now apparent – in the requirement for environmental-impact assessments, in the spread of nature schools, in the protection of wetlands of international significance (Ramsar Sites), and in the attention given to such issues in the media.

Japanese Wolf [*left* WUFE] was hunted to extinction. The spread of westernized agriculture and the drive to exterminate predators led to the extinction of the Hokkaidō Wolf [*right* NHHJ].

Bluegill (*above*) and Black Bass (*below*) have both been introduced to Japan for sport fishing, at considerable cost to native species [BOTH PUDQ].

Red-crowned Cranes have recovered remarkably well from near-extinction a century ago [JW].

Conservation Successes

During the 18th century and the first half of the 19th century, Japan's human population remained steady at about 30 million. Following the Meiji Restoration of 1868, the population increased as the country strove to build a modern nation state. It grew to 60 million by 1926 and passed 100 million in 1967, after which the annual growth rate slowed to about one percent during the 1960s, 1970s and 1980s. Between the 2010 and 2015 population censuses, the population declined by one million from a peak of 127·09 million, the first decrease since such censuses began in 1920. By 2017 the population had fallen further, to 126·71 million, and it is expected to shrink year by year from now on as the post-war baby boomers age and die. Although many areas of Japan remain densely populated, especially the island of Honshū, the population is unevenly spread. In 2015, the average population density measured 340·8 persons per square kilometre. As stated earlier, however, most of Japan's human population and its entire agricultural and industrial economy are concentrated in the coastal plains and inter-mountain basins, which make up less than 30 percent of the country.

Changes in human population size and distribution have had a significant negative impact on natural habitats, and the species they can support, throughout Japan.

Protected Areas

The Japanese government makes considerable efforts to protect and preserve elements of the country's diversity. Not only are landscape features, such as the sand dunes in Tottori Prefecture or the volcanoes of Kirishima in Kyūshū, protected as geoparks, but ecosystems, environments, habitats and particular collections of species, as well as individual species, are protected through conservation efforts in a nationwide network of nature reserves, 52 Ramsar Sites (wetlands of international significance amounting to 154,696 ha), 34 national parks and 58 quasi national parks spanning the country from northernmost Hokkaidō to the southern Ryūkyū Islands and the Ogasawara Islands. To qualify as a national park each area must offer outstanding scenic beauty, distinctive physiography and geology, and flora or fauna of high scientific value. So far, more than 5·6 percent of Japan's land area qualifies under these criteria and is currently designated as national park.

The designation of 34 national parks indicates the value placed on Japan's delightful landscapes, such as here at Bandai–Asahi NP [SHUT].

Japan's National Parks and Ramsar sites – *see regional maps for National Park names*

Ramsar sites

1 Sarobetsu-genya
2 Kutcharo-ko
3 Tōfutso-ko
4 Notsuke
5 Fūren-ko & Shunkuni-tai
6 Kiritappu Shitsugen
7 Akkeshi-ko & Bekambeushi Shitsugen
8 Kushiro Shitsugen
9 Akan-ko
10 Uryūnuma Shitsugen
11 Miyajima-numa
12 Utonai-ko
13 Ōnuma
14 Hotoke-numa
15 Izu-numa & Uchi-numa
16 Shizugawa-wan
17 Kaburi-numa
18 Kejo-numa
19 Oyama Kami-ike & Shimo-ike
20 Hyō-ko
21 Sakata
22 Yoshigadaira Wetlands
23 Oze
24 Oku-Nikkō Shitsugen
25 Watarase-yusuichi
26 Kasai Marine Park
27 Yatsu-higata
28 Sakata
29 Tateyama Midagahara
30 Katano-kamoike
31 Nakaikemi Wetlands
32 Tokai Hilly Land
33 Fujimae-higata
34 Kushimoto Coral Communities
35 Biwa-ko
36 Mikata-goko
37 Lower Maruyama River
38 Nakaumi
39 Shinji-ko
40 Miya-jima
41 Akiyoshidai Groundwater System
42 Kuju Botatsuru & Tadewara Shitsugen
43 Higashiyoka-higata
44 Hizen Kashima-higata
45 Arao-higata
46 Imuta-ike
47 Yakushima Nagata-hama
48 Man-ko
49 Kerama Shotō Coral Reef
50 Kume-jima streams
51 Yonaha-wan
52 Nagaru Amparu

NANSEI SHOTŌ

Key
National Park
Ramsar site

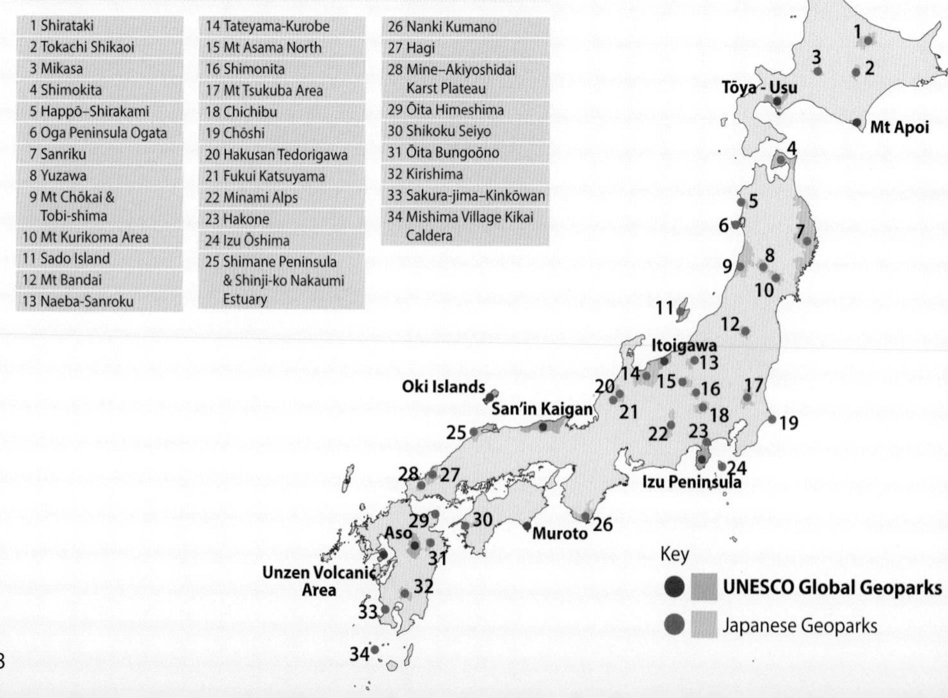

Japan's national parks[1] have been designated relatively recently and long after the islands became densely populated and developed a history of land ownership. Consequently, national parks have been created not necessarily where the government owns the land, but where the need for nature or landscape preservation is recognized. Out of necessity, any preservation of natural beauty is achieved by means of mutual cooperation between public and private stakeholders.

Geoparks

During the 1990s a geopark movement began in Europe, and by 2004 a Global Geoparks Network had been established. The International Year of Planet Earth in 2008 was the catalyst for introducing the concept and activities of geoparks to Japan, and the movement took off rapidly. In 2015, UNESCO recognition led to all existing global geoparks being incorporated into the International Geoscience and Geoparks Programme. The development of the geopark network in Japan began only in 2007, yet so rapidly has it expanded, in recognition of its geodiversity and geoheritage, that by 2019 it had nine UNESCO Global Geoparks and no fewer than 35 National Geoparks, with 16 further sites eager to join.

Japan's geodiversity is inescapable. The influences of geology on daily life and society in Japan are so great that the Japanese people have developed a special relationship with their dynamic landscape and the earth sciences. Japan actively promotes its geological natural heritage and the earth sciences to the public as a tourism resource. Its geoparks are spread from Hokkaidō to southern Kyūshū and they include a wide range of natural heritage sites, from the caldera and volcano of Tōya and Usu in southwest Hokkaidō to offshore islands such as the Oki Islands, off western Honshū, the San'in Kaigan Geopark (see *p. 214*) and the Unzen Volcano Area Geopark, in western Kyūshū . The purpose of a geopark includes display and education, promoting geotourism, hazard awareness, and exploration of the deep-rooted links between Japanese society, history and culture and the geological underpinning of the nation.

World Heritage Sites, Natural Monuments and Conservation Efforts

Five natural locations and 20 cultural sites are protected by UNESCO World Heritage Site (WHS) status, the five natural ones being: the forested mountains of Shirakami-Sanchi (1993), the isolated island of Yaku-shima (1993), the Shiretoko Peninsula (2005), the Ogasawara Islands (2011) and Amami Ōshima, Tokuno-shima, Northern Okinawa and Iriomote (2021).

Significant and iconic species themselves, such as Red-crowned Crane and Amami Rabbit, may be protected today as either Natural National Monuments or Special Natural Monuments, although often such actions are necessary because of past negative human impacts.

Through Japan's rapid modernization and industrialization after the Meiji Restoration of 1868, combined with its post-1945 recovery and reconstruction after World War II, there has been a significant loss and degradation of its natural habitats. Especially hard hit were coastal and lowland habitats, most particularly coastal wetlands, estuaries, bays, shallow coastal lagoons and lakes, most of which were drained for use as farmland, converted for domestic or industrial use, or simply swallowed up by the cancer-like growth of urban sprawl. The need for lumber for reconstruction, along with the impact of spreading transport infrastructure consisting of railways, highways and airports, resulted in rapid and visible inroads into Japan's beautiful landscape. Although largely mountainous and forested, much of this forest now consists of forestry plantations.

Against this backdrop of habitat loss, Japan has also experienced industrially induced pollution of air and water, sometimes triggering terrible illnesses (such as Minamata disease, caused by mercury pollution in southwest Kyūshū in the 1950s). Furthermore, given its anachronistic stance on lethal cetacean research

1 Japan's legal system provides for three kinds of nature parks: national parks, quasi national parks and prefectural nature parks. The former fulfil the strictest criteria and currently number 34. Areas that approach (but do not fulfil all of) the required criteria but which may be upgraded in the future are quasi national parks (58), and have an area equalling about half that of the national parks. Together, natural parks protect about 15 percent of Japan.

Crested Ibis is unusual for such a species in that it undergoes a dramatic transformation between its breeding [*left* MAB] and non-breeding plumages [*right* SAKP].

and the commercial harvesting of whales, Japan seems an unlikely country from which to learn about conservation. Nevertheless, there have been several success stories, each involving a large avian species. In each case, the species was brought within a hair's breadth of extinction through overexploitation for its feathers, through hunting for its meat, or through habitat pollution and habitat loss.

Japan's Red-crowned Crane population reached a critical low in the early 20th century when fewer than 30 birds survived in the Kushiro Marsh area. A century later, legal protection, habitat protection and a winter feeding programme have brought this population back to more than 1,750[1] individuals, which are now spreading northwards and westwards through east Hokkaidō, and even beyond, from their stronghold in eastern Hokkaidō. Oriental Stork followed a similar pattern. Having lost suitable nesting and foraging habitat in western Honshū, it, too, has made a comeback, numbers in the wild exceeding 100 again as a result of successful captive breeding and release in Hyōgo Prefecture.

Crested Ibis is the ultimate 'Japanese species', given its scientific name of *Nipponia nippon*, yet it suffered habitat degradation and loss until the last few birds survived only on the island of Sado, where they finally died out. This ibis has been 'resurrected', like Oriental Stork, thanks to captive breeding and release (of birds donated by China); and now more than 350 live on Sado. In the cases of the crane, the stork and the ibis, considerable efforts were made to involve the local agricultural communities and in each area they have responded with gentler, more bird-supportive farming methods. These farmers have reaped benefits through being able to sell the value-added rice that they grow, and the regions have benefited through bird-related tourism.

The recovery of Short-tailed Albatross from presumed extinction in the mid-20th century to a current population of over 6,000 on its remote island stronghold, on Tori-shima, is especially heartening. Survival has depended on the natural homing ability of the few survivors of massive depredation for the feather trade and then volcanic eruption, combined with human efforts to stabilize their nesting slopes on the island and to translocate young birds to a different island to seed a new population.

The enormous efforts (both in time and in financial support) made by government departments, public and private organizations and individuals for the continued survival of each of these four species indicate a strong and successful desire to prevent species extinction in the future[2].

1 This is the currently estimated population. Winter counts are made at a fixed number of established sites, but since winter feeding was reduced during the 2010s many cranes no longer visit the count sites. Numbers counted at the established sites therefore have begun to fall, with approximately 1,350 found during the most recent count.

2 Very significant efforts are also being made to reduce or remove alien predators from islands, such as Okinawa, Amami Ōshima and the Ogasawara Islands.

Oriental Storks are thriving once more thanks to a captive-breeding and release programme [BOTH TsM].

Short-tailed Albatross has made the most dramatic recovery of all Japanese species, from near-extinction to over 6,000 individuals in less than a century [TR].

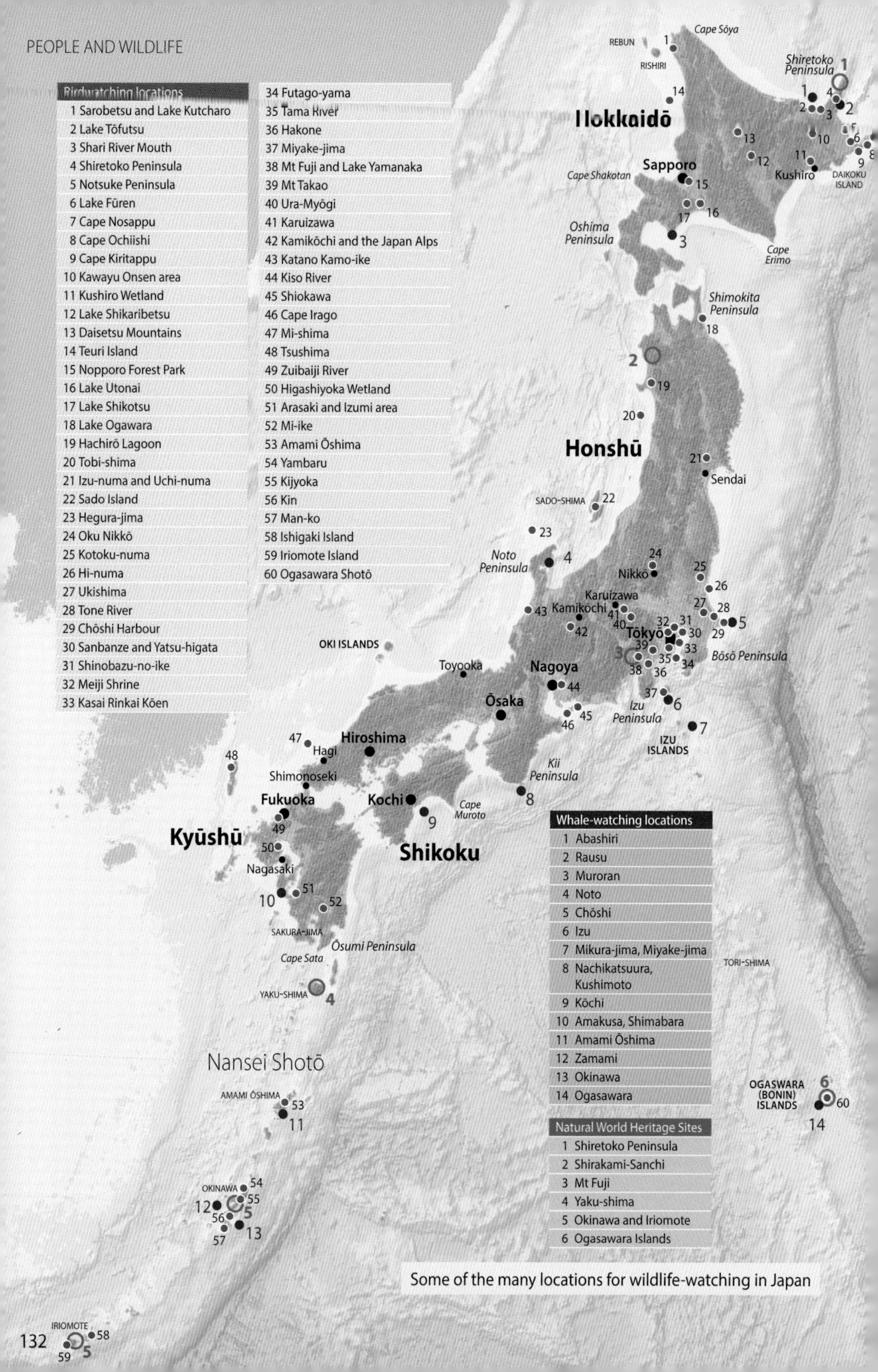

Birdwatching locations
1 Sarobetsu and Lake Kutcharo
2 Lake Tōfutsu
3 Shari River Mouth
4 Shiretoko Peninsula
5 Notsuke Peninsula
6 Lake Fūren
7 Cape Nosappu
8 Cape Ochiishi
9 Cape Kiritappu
10 Kawayu Onsen area
11 Kushiro Wetland
12 Lake Shikaribetsu
13 Daisetsu Mountains
14 Teuri Island
15 Nopporo Forest Park
16 Lake Utonai
17 Lake Shikotsu
18 Lake Ogawara
19 Hachirō Lagoon
20 Tobi-shima
21 Izu-numa and Uchi-numa
22 Sado Island
23 Hegura-jima
24 Oku Nikkō
25 Kotoku-numa
26 Hi-numa
27 Ukishima
28 Tone River
29 Chōshi Harbour
30 Sanbanze and Yatsu-higata
31 Shinobazu-no-ike
32 Meiji Shrine
33 Kasai Rinkai Kōen
34 Futago-yama
35 Tama River
36 Hakone
37 Miyake-jima
38 Mt Fuji and Lake Yamanaka
39 Mt Takao
40 Ura-Myōgi
41 Karuizawa
42 Kamikōchi and the Japan Alps
43 Katano Kamo-ike
44 Kiso River
45 Shiokawa
46 Cape Irago
47 Mi-shima
48 Tsushima
49 Zuibaiji River
50 Higashiyoka Wetland
51 Arasaki and Izumi area
52 Mi-ike
53 Amami Ōshima
54 Yambaru
55 Kijyoka
56 Kin
57 Man-ko
58 Ishigaki Island
59 Iriomote Island
60 Ogasawara Shotō
Whale-watching locations
1 Abashiri
2 Rausu
3 Muroran
4 Noto
5 Chōshi
6 Izu
7 Mikura-jima, Miyake-jima
8 Nachikatsuura, Kushimoto
9 Kōchi
10 Amakusa, Shimabara
11 Amami Ōshima
12 Zamami
13 Okinawa
14 Ogasawara
Natural World Heritage Sites
1 Shiretoko Peninsula
2 Shirakami-Sanchi
3 Mt Fuji
4 Yaku-shima
5 Okinawa and Iriomote
6 Ogasawara Islands
REBUN
RISHIRI
Cape Sōya
Shiretoko Peninsula
Hokkaidō
Sapporo
Cape Shakotan
Kushiro
DAIKOKU ISLAND
Oshima Peninsula
Cape Erimo
Shimokita Peninsula
Honshū
Sendai
SADO-SHIMA
Noto Peninsula
Nikkō
Karuizawa
Kamikōchi
Tōkyō
Bōsō Peninsula
OKI ISLANDS
Toyooka
Nagoya
Ōsaka
Izu Peninsula
IZU ISLANDS
Hiroshima
Hagi
Shimonoseki
Kii Peninsula
Fukuoka
Kochi
Cape Muroto
Kyūshū
Shikoku
Nagasaki
SAKURA-JIMA
Ōsumi Peninsula
Cape Sata
YAKU-SHIMA
TORI-SHIMA
Nansei Shotō
AMAMI ŌSHIMA
OGASWARA (BONIN) ISLANDS
OKINAWA
IRIOMOTE

Some of the many locations for wildlife-watching in Japan

Wildlife-watching

Key Wildlife-watching Areas

Japan's diverse habitats provide tremendous opportunities for wildlife-watching. This diversity can be explored on foot or by some other mode of transportation, through the window, or even from an outdoor hot spring.

The southwestern islands of Iriomote, Ishigaki, Okinawa and Amami Ōshima offer particularly good opportunities to seek out remarkable insect life, including numerous endemics, along with endemic amphibians and reptiles, and a number of endemic birds. The mountainous island of Yaku-shima, 75 km south of southernmost Kyūshū, offers the highest peak, Mt Miyanoura (1,936 m), between the Japan Alps (up to 3,193 m [Kita-dake]) and the high peaks of Taiwan (up to 3,952 m [Yushan]), and thus it presents a tremendous range of habitats supporting an unexpected array of species. This botanical paradise holds an astonishing range of insect life, including southern tropical species at sea level and northern species in forest at higher elevations on the same island. It also hosts some of the oldest trees in the world, such as the *Yaku Sugi*, or Yakushima Cedar, one of which – the *Jōmon Sugi* – is reputedly more than 7,200 years old. The same island boasts Japan's southernmost Japanese Macaques, living alongside one of the southernmost populations of Japanese Deer, and by hiking high into the mountains it is possible to see the famous Yakushima Azalea, which is endemic to the island.

In Kyūshū, the isolated volcanic range of the Kirishima-Kinkōwan National Park (Japan's first national park and part of the Japanese Geoparks Network) offers majestic natural features. It is home to more than 20 peaks, including Kirishima's highest, Mt Karakuni (1,700 m), and sacred Takachiho-no-Mine

Isolated Yaku-shima has a rugged alpine landscape in a subtropical zone [MAB].

(1,574 m) (which features in Japan's founding myth[1]), along with crater lakes, waterfalls, rivers and hot springs. Kirishima's scenery includes Mt Shinmoe, which erupted spectacularly, for the first time in approximately 300 years, in January 2011 (I happened to be admiring the same mountain when it erupted again in sub-Plinian[2] fashion on 4 February 2011). To the south, standing in the middle of Kinkō Bay, is spectacular Sakura-jima (1,117 m), one of the world's most active volcanoes. Kinkō Bay itself is the remnant of the enormous Aira Caldera formed by a massive eruption around 30,000 years ago.

Ancient cedars on Yaku-shima [MAB]

Kyūshū's location, its proximity to the warm, north-flowing Kuroshio ocean current and its abundant rainfall (said to reach an astonishing 400 cm a year) provide a rich environment for plant life, including for the island endemic Kyushu Azalea, and the rare Kirishima endemic Kirishima Crab Apple, which flowers during early May. Many plants there have Japanese names that include the word Kirishima, such as Fragrant Witch Hazel, known as *Kirishimamizuki*, Kyushu Azalea, which is known as *Miyamakirishima*, and Kirishima Thistle, known also as *Kirishimahikodai*.

Northwest from Kirishima are the world-famous wintering grounds of cranes at Arasaki, on the coast near Izumi City. Almost 15,000 cranes of four migratory species, but predominantly Hooded Cranes and White-naped Cranes, gather spectacularly each winter, from November to February, attracting birdwatchers from around the world. Farther north still are the wetlands of the Ariake Inland Sea. These provide important wintering grounds for the rare Saunders's Gull, and resting and foraging sites for a host of East Asian shorebird species.

Small islands in the Sea of Japan, from Tsushima (off Kyūshū) in the southwest to Rebun and Rishiri (off Hokkaidō) in the north, provide migratory traps for birdwatchers and birds alike, as here almost any migrant bird from East Asia may appear in spring or autumn.

Japan's main island of Honshū offers a host of wildlife-watching locations, too many to mention. The forests around Mt Fuji and the woodlands in the Karuizawa and Tateshina areas are all within easy reach of Tōkyō, and support a range of Japan's common and endemic mammal species, including Japanese Badger and Japanese Giant Flying Squirrel, as well as endemic and migratory birds, including the much sought-after Copper Pheasant and Yellow Bunting. In addition, the three ranges of the Japanese Alps support important populations of Asiatic Black Bear, Japanese Serow and Japan's last remnant Rock Ptarmigan population.

The islands stretching south from Tōkyō, first the Izu Islands and then the Ogasawara Islands (described later in this book), support numerous endemic plants and a range of endemic or range-restricted birds, including Izu Thrush [VU], Owston's Tit and Izu Robin.

---

1 In the Age of the Gods, when deities ruled the heavens, they saw 'an island floating in a foggy sea', known today as Kirishima ('fog island'). On orders from Amaterasu Omikami, the sun goddess, nine deities descended to earth and took their first steps on what is now called Takachiho-no-Mine. They went on to found Japan and the imperial dynasty.

2 Plinian eruptions (sometimes called Vesuvian eruptions) are volcanic eruptions resembling that of the city-destroying Mount Vesuvius eruption in 79 CE, described by Pliny the Younger; sub-Plinian eruptions produce lower volumes of magma at a lower rate with unsteady eruptive columns reaching up to 20 km high.

The endemic Yakushima Azalea [AJ]

The isolated volcanic range of the Kirishima Kinkōwan National Park [TOOI]

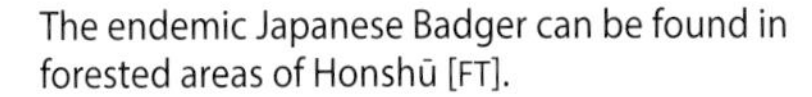

The endemic Japanese Badger can be found in forested areas of Honshū [FT].

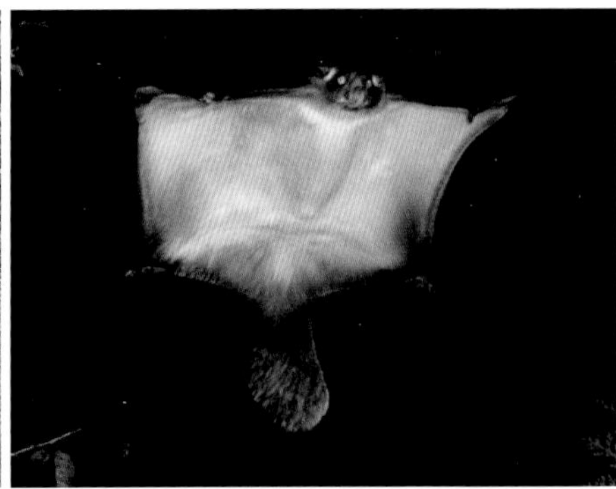

Like the badger, the Japanese Giant Flying Squirrel is largely nocturnal [FT].

Various wetlands and river mouths, ranging from those in the southwest to those in the north of Honshū, support enormous numbers of wintering waterfowl, and few are more famous than the gathering at Izu-numa and Uchi-numa, a pair of interconnected lakes (which rarely freeze in winter) in the alluvial plain of the Hasama River, in Miyagi Prefecture. They are famous for the spectacular carpet of Lotus flowers covering the lake in August, and for the enormous roosting and feeding flocks of Greater White-fronted Geese [NT] numbering tens of thousands that gather here during late autumn, winter and early spring, and which attract birdwatchers from far and wide to witness them. Few sights are more stirring than seeing clouds of birds departing at sunrise or returning at sunset.

The wildlife delights of winter in Hokkaidō are now so famous that photographers from around the world are drawn here each year. With a backdrop of gorgeously snowy landscapes they capture images of dancing Red-crowned Cranes against spectacular gatherings of Steller's Eagles on floes of sea-ice, long-furred Red Foxes foraging for voles across snow-covered meadows, herds of Japanese Deer in the woodlands. They also find flocks of elegant Whooper Swans swimming through drifts of mist and ice. During those winters when sea-ice reaches Hokkaidō's northern shores and the Nemuro Strait, visitors should take a trip out by boat to the ice edge to witness eagles hunting. Each winter, the period from late November to early March, offers wonderful birdwatching and wildlife-watching. In spring, summer and autumn, Brown Bears can be viewed along the shores of the Shiretoko Peninsula, Orca and Northern Minke Whales can be seen offshore, and many thousands of migratory Short-tailed Shearwaters, on

The largest gathering of wintering cranes is at Arasaki, near Izumi, in southwest Kyūshū [JW].

their long journey from antipodean islands to the Bering Sea, carpet the water.

Summer in Hokkaidō offers wonderful wildflower-viewing opportunities on mountains across the island, but especially in the Daisetsu range. During summer it is also possible to see Northern Pika, Siberian Chipmunks, and summer-visiting birds such as Siberian Rubythroat. The forests and calderas of Akan–Mashū National Park are spectacular at any season and serve as reminders that the closer any area is examined the more there is to find.

Owston's Tit is an endemic of the Izu Islands [ET].

Wildlife-watching Tips for Japan

Japan is without doubt overpopulated, but it feels more seriously so because the 126 million human inhabitants are crowded into the small proportion of the country that is flat, providing suitable conditions for coastal ribbon development and inland urban sprawl. Even within these coastal lowlands and urban areas, however, there are places well known for their local wildlife. Just remember that to seek wildlife there generally means to be surrounded by numerous other people, unless one visits early in the morning on weekdays. If your schedule permits, aim to avoid weekends and national holidays, both those of Japan and those of adjacent Asian countries as Japan has become a popular tourist destination.

Even in the more populated parks, and at much-frequented riversides and shores, the same golden rules for wildlife-watching apply as those in the very wildest of places. Move slowly, seek

The majority of Japan's Greater White-fronted Geese winter in the Izu-numa area of Miyagi Prefecture [MAB].

Whooper Swans are locally common winter visitors in northern Japan [MAB].

out quieter paths and trails, often scan ahead and around with your eyes and your binoculars, and listen carefully for the songs and sounds that may give away the presence of your quarry.

Take every opportunity to visit the interior of the country, away from the coasts, and seek out quiet valleys and mountain trails. As so much of the countryside is experiencing human depopulation, this will become increasingly easy as the years go by.

Siberian Rubythroat is one of several attractive summer visitors to Hokkaidō [SP].

Note that most species of wildlife avoid humans, their domesticated stock and companion animals. Many mammals are more nocturnal where there is considerable human activity, and even most birds are more active early and late in the day. You can greatly increase your chances of wildlife-watching success by being out from before dawn and again until after dusk, avoiding the warmest part of the middle of the day as it is often too hot for much wildlife activity, especially in summer (although this time may be utilized to look for butterflies and dragonflies). Even nocturnal species tend to be more active in the hours immediately following sunset and immediately preceding sunrise, but be aware that the daylight timing in Japan can be a little confusing for those unused to it. Since 1888, Japan has observed Japan Standard Time, a single standardized time throughout the entire country and throughout the year. There is just one time zone, and no daylight-saving time, despite Kyūshū and the Nansei Shotō being very far to the west of Tōkyō, and most of Hokkaidō being well to the east. Japan's time zone leaves it out of sync with areas of Russia immediately to the north and northwest. This means that in midsummer first light may be around 03:00 and last light a little earlier than 20:00, while in midwinter first light may be around 06:30 and last light before 16:30. Take this timing into account when planning your excursions, and perhaps use a sunrise–sunset app on your phone to plan your viewing. Japan's seasonality and great length mean that you will experience considerable local variation in conditions. Make sure that you are suitably prepared with protective clothing against the extreme cold, wind, rain, snow and heat.

In addition to being out early and late, and moving slowly, remember that most species respond to colour (birds especially), movement or recognizable patterns. Therefore, while camouflaged clothing may be a little too extreme for the average naturalist, do at least wear muted or neutral colours when in search of birds and mammals; insects seem not to react to colours.

Mammals, although limited in their colour vision, are extremely sensitive to smells, and wind direction is therefore important. Wherever possible, approach mammals from downwind, as they will likely be less concerned by your presence until they see you.

Watching wildlife is greatly enhanced by means of binoculars, and I strongly recommend the using of a pair with 8× or 10× magnification and objective lenses of between 32 mm and 50 mm. A telescope (30× to 60× magnification) mounted on a tripod is also of great help when watching wetlands and coastal areas, and even in forests. A telescope will provide more satisfying and closer views than binoculars will. Using such optical equipment means that it is not necessary to approach wildlife very closely. Photographers, however, even when armed with the longest of lenses, will no doubt want to approach their subjects more closely in the hope of better images. Even so, do be aware that such photography may actually harm and permanently disturb wildlife. Please consider the welfare of your subjects and use field craft and stealth; use a hide or blind to conceal yourself so as not to disturb your quarry. Watch your subject's behaviour carefully and ensure that your presence (at whatever range) is not causing it or them undue stress.

Nature-watching in Japan

Japan offers the naturalist traveller a surprising range of wildlife spectacles to enjoy throughout the year. Natural-history tours of Japan, pioneered in the early 1980s for international visitors, offer an opportunity to see more of the country than just the traditional cultural centres.

Nature-watching is practical at any time of year (although avoidance of the early-summer rainy season is a good idea). Begin in winter in the Nansei Shotō, exploring subtropical forests in search of endemic plants and birds, and then visit Arasaki, in Kyūshū, to witness the phenomenal sight of 10,000–15,000 Hooded Cranes and White-naped Cranes at their traditional gathering. No winter visit to Honshū is complete without wandering in the beautiful mixed forests, perhaps around Karuizawa, and, of course, the hot-spring monkeys of Hell Valley, in Nagano Prefecture, are an absolute must. Japanese Serow may be found here and, if not, travel north to the Shimokita Peninsula to watch the Snow Monkeys (Japanese Macaques) and Japanese Serow near the village of Wakinosawa. On the way drop in at Izu-numa, a large shallow lagoon in Miyagi Prefecture, where most of Japan's geese spend the winter. At dawn and dusk, this is as powerful a spectacle as the cranes in Kyūshū. Continuing to Hokkaidō, a visit timed to see the Sapporo snow-festival in early February is perfect for continuing to Akan–Mashu National Park to watch Whooper Swans swimming in ethereal morning mist. Sea-ice on the Sea of Okhotsk, sea eagles, Japanese Deer, Red Fox and iconic Red-crowned Cranes are all to be seen by the visitor while making a circuit including the Akan–Mashū, Shiretoko and Kushiro Shitsugen National Parks.

In summer, a similar journey timed to travel north with the flowering season of the azaleas is blissful. Summer flowers and summer birds bring the forests alive with colour, in contrast to the strong monochrome of winter. While the Arasaki crane flock, the swans and the sea eagles will all have migrated north, in their stead there are flycatchers and warblers, cuckoos and shrikes to look for, and the botanical diversity of the forest itself exceeds that found in the whole of Europe.

Birdwatching is a wonderful hobby that can fill idle moments or whole days at a time. It has two major benefits compared with other outdoor activities. First, the equipment required is minimal – a field guide, a notebook and a pair of binoculars – and, secondly, it can be done anywhere, not only in wild places but from a balcony or in a garden.

Wildlife photography is very popular, drawing hundreds to sites with cranes, monkeys or shorebirds [MAB].

During spring and autumn migrations birdwatchers flock to wetlands and islands in the Sea of Japan, such as Hegura-jima [MAB].

In the city, visit a park, pond or grove of trees and there will invariably be birds. Outside the city there may be fields, streams, woods or mountains to explore; and if near the coast, then sandy and rocky shores, headlands, estuaries and bays all provide opportunities for birdwatching.

From city high-rises to busy airports, from urban parks to remote mountains, birds can be seen, and watching them reveals much about the distinctions between species and about their choice of habitats. Most of all, it reveals interesting elements of their behaviour. In urban environments, the commoner, rather widespread generalist species tend to be the norm, although unusual species can appear on migration. In contrast, once the city has been left behind, different communities of species will be found, depending on the habitat.

The key to birdwatching is being sharp-eyed and questioning. Even if the only species to visit your balcony feeder are the ubiquitous Eurasian Tree Sparrow and Brown-eared Bulbul, there are hours of entertainment to be had in learning about their behaviour. At the opposite extreme, if you travel to the various regions of Japan you will find that the species differ from island to island, and journeys from Hokkaidō to Okinawa during winter and summer may reveal 300 or more species of bird.

For some species timing is crucial. Visiting east Hokkaidō during February and again in July will reveal completely different species in the two months. On the other hand, Okinawa and the Nansei Shotō can be visited at almost any time of year, as most species there are resident.

There are clubs and groups to join to add to your birding fun. Once caught, the birdwatching bug provides life-long enjoyment.

Whale-watching in Japan

When it comes to whales, Japan is best known for its resumption of commercial whaling, much of which takes place now in Japanese waters, where Northern Minke Whale is one of the main targets (The cultural, historical, scientific, sociological and political aspects of this subject are worthy of a whole chapter in itself!). As a result, Japan's position as an excellent destination for whale-watching has been overlooked, not only by most visitors but also by those in Japan tasked with promoting inbound tourism. In fact, Japan offers exciting and excellent whale-watching opportunities from the Shiretoko Peninsula in northeast Hokkaidō to the southwest islands (Nansei Shotō), and from Shikoku to the Ogasawara Islands. Small-boat captains, some of them former fishermen or even former whalers, now provide an array of opportunities for the public to watch a diverse set of cetaceans in Japanese waters, and an increasing number of visitors now recognize these opportunities, whether in the Kerama Islands, off Okinawa, or off Chichi-jima, in the Ogasawara Islands.

It may come as a surprise that the Japanese public, especially youngsters, love whales, dolphins and porpoises. The growth in this interest combined with local declines in fish stocks has encouraged the development of a new industry, which has spread quickly. Whale-watching is a huge and growing industry internationally. This fledgling industry began in Japan in the tropical Ogasawara Islands in 1988, and since then the number of watchers of whales, dolphins and porpoises has increased dramatically. Today, there are companies specializing in whale-watching in many parts of Japan from the Kerama Islands off Okinawa to the Ogasawara Islands in the Pacific off Tōkyō, from Kōchi Prefecture in Shikoku to Muroran in west Hokkaidō, and in the Nemuro Strait off east Hokkaidō. The whale-watching industry continues to grow, becoming increasingly valuable in terms of revenue and creating recreational opportunities.

The whales, dolphins and porpoises found in Japanese waters range in size from the tiny Indo-Pacific Finless Porpoise, which averages only 1·55 m in length and weighs just 30–45 kg, to the record-breaking and very rare Blue Whale, the largest living mammal on earth, which reaches up to 30 m in length and weighs 100–170 metric tons. In terms of numbers, Pacific White-sided Dolphin and Dall's Porpoise are

One of the great wildlife spectacles involves the dawn and dusk flights of more than 15,000 Hooded Cranes in Kyūshū [MAB].

The silent and elusive Japanese Serow can be found in many forested, mountainous areas of Honshū [AK].

Urban and suburban parks and gardens support a range of common species, such as Eurasian Tree Sparrow [ImM].

Japanese White-eyes feed on flower nectar and fruit, particularly mikan placed specially for them [TsM].

The widespread Brown-eared Bulbul frequently visits berry-bearing shrubs in gardens [JW].

There is considerable regional variation among Japanese bird species. This Brown-eared Bulbul is from Japan's Ryūkyū Islands [KuM].

Orca, also known as Killer Whale, are frequently seen from late winter into summer off the Shiretoko Peninsula [TC].

Whale-watching is a popular pastime from Hokkaidō to Okinawa and the Ogasawara Islands [GOKA].

the most abundant and most easily seen. Japan hosts many species in the Order Cete[1], especially in the Family Balaenopteridae, the rorquals, such as Northern Minke Whale. Some species are largely pelagic while others frequent coastal waters. Few sights are as awe-inspiring as a Humpback Whale launching itself out of the water in a full breach off the Kerama Islands or emerging to spy-hop, or as a pod of Orcas keeping pace with their watchers in the Nemuro Strait, off Hokkaidō.

One of Japan's prime areas for whale- and dolphin-watching is Cape Muroto, the southeastern tip of Shikoku, where Great Sperm Whales can be seen during the peak season of March to May, and dolphins year-round. At least 14 cetacean species have been seen there.

Although found all around the Japanese archipelago, Northern Minke Whale, the smallest of the *Balaenoptera* genus, is most easily seen off east and north Hokkaidō, especially in the Sea of Okhotsk off Abashiri. This slender-bodied small baleen whale with a pleated throat and a small dorsal fin is sometimes known as the Lesser Rorqual. It measures 7–10 m in length and weighs 5–10 metric tons. Its dorsal surfaces are black or dark grey and its ventral surfaces are white. Streamlined for speed, it has a tall and clearly falcate[2] dorsal fin located two-thirds of the way along the body towards the tail. It rarely shows its broad, smooth-edged and slightly concave tail flukes, even when diving. Often visible beneath the water are the pectoral flippers, which are short and pointed with a prominent, but individually variable white band across the mid-section. Typical of the baleen whales, its head is sharply triangular, with a single head ridge and 50–70 throat grooves, giving it tremendous elasticity when sieving huge amounts of water for prey. It feeds mainly on krill and small fish, such as Walleye Pollock and Pacific Herring, but takes whatever species are most common.

1 I have continued to use the terms Cete and Cetacean here, even though in the latest taxonomy the Order Cetacea/Cete has gone because it has been shown that the whales, dolphins, porpoises and beaked whales are not so closely related as once thought; in fact they are polyphyletic, having different origins. Instead, the various families have all been absorbed within the Order Artiodactyla, but that is, as yet, a rather unfamiliar grouping for most of us non-taxonomists.

2 Meaning bent or curved like a sickle.

Pacific White-sided Dolphins are common in northern coastal waters [GOKA].

Dall's Porpoise off the Shiretoko Peninsula [GOKA]

Bryde's Whale off Kochi Prefecture [OT]

From a variety of sites along Japan's Pacific Ocean coast, cetacean-watchers have the possibility of going in search of nearly 70 species. Humpback, Great Sperm, Northern Minke and Baird's Beaked Whales, along with pilot whales, Dall's Porpoise, Pacific White-sided Dolphin, Common Bottle-nosed Dolphin and Risso's Dolphin, as well as Orca, can all be easily seen. Although a number of these species can be seen from commercial ferries and tourist boats plying coastal waters, the chances of seeing them on commercial whale-watching tours are greater.

The second largest pelagic cetacean in the world, Fin Whale (18–22 m long and weighing 40–60 metric tons), is found also around the Japanese archipelago from Hokkaidō to the Nansei Shotō. Although it remains scarce, it, too, can sometimes be seen from whale-watching boats. A unique feature of this enormous whale is that it is the only vertebrate known to be asymmetrically coloured, the left side of the lower jaw being dark and the right side white, while the colour pattern is reversed on the tongue. Grey Whale, at 13–14 m and 15–35 metric tons somewhat larger than a Northern Minke Whale, is rare but is occasionally seen off Japan's Pacific coast. The massive Great Sperm Whale, while ranging all around the Japanese coasts, is most frequently seen off the Pacific coast, particularly over deep-water areas. This slow-moving deep-diving whale with a box-like head has just one blowhole, situated far forward on the left side of the head. Unlike the large baleen whales, which produce exhalations that rise vertically to varying heights, the blow produced by the sperm whale is low, bushy and projected forwards at a low angle. Great Sperm Whale differs further in having a distinctive lower jaw that is extremely narrow relative to its head and comes armed with two rows of simple teeth. Although measuring only 12–18 m in length, it can weigh up to 50 metric tons. Dives to depths of 2,800 m have been recorded for this whale, and individuals commonly remain submerged for an hour or more while hunting their main prey of squid, octopus and fish.

Fin Whale can sometimes be seen off Hokkaidō's Sea of Okhotsk coast [MAB].

The small Northern Minke Whale is the most abundant baleen whale in Japanese waters [CC].

Indo-Pacific Bottlenose Dolphins favour the warmer waters off Kyūshū [JoW].

Humpback Whales are the most visibly active of all the large whales [*left* OVTB] and can even be found on decorative drain covers on Zamami Island [*right* MAB].

The Nemuro Strait, off the Shiretoko Peninsula, is the best area to look for social Orcas [SF].

Spinner Dolphins are best seen off the Ogasawara Islands [OGVO].

A perennial draw for whale-watching visitors is the chance of sighting the social, even gregarious pod-forming Orcas that occur in the Nemuro Strait, off the Shiretōko Peninsula. They can be seen around all Japanese coasts, but the currents flowing through the shallow channel off east Hokkaidō seem to have attracted the highest density of these animals in the country. This high-speed marine predator is the largest and most distinctive of the marine dolphins. Males range in length from 6·7 m to 9·8 m, while females are smaller at 5·5–6·6 m; they weigh 2·6–9·0 metric tons, and the triangular dorsal fin of a large male may be up to 1·8 m tall. This top predator feeds on a very wide range of prey, including whales, seals, seabirds, turtles and fish.

Of the more than 30 small cetaceans found in Japanese waters, six marine dolphins are frequently encountered while whale-watching. The dashing Dall's Porpoise, with its distinctive white flanks, is typical in cold northern waters, while in warmer subtropical southern waters there are Indo-Pacific Bottle-nosed Dolphins and Spinner Dolphins; then, in the mild waters off the main islands, and especially around Kyūshū, Common Bottle-nosed Dolphins can be seen. Perhaps the most abundant and widespread small cetacean species around Japan is Pacific White-sided Dolphin. It ranges around all Japanese coasts and is most particularly encountered off the Pacific coast. One of the smallest of all cetaceans in Japan is Harbour Porpoise. This species frequents coastal areas and river estuaries and, in consequence, is often both the first and the last cetacean to be sighted during whale-watching excursions.

In the tropical waters of southern Japan, particularly the Ogasawara Islands and the Kerama Islands, it is possible to visit the calving grounds of one of the most dramatically active of all cetaceans – Humpback Whale. Both archipelagos offer the relaxed atmosphere of a subtropical lifestyle, but the Kerama Islands have the added flexibility of very ready access. Humpback Whale is a semi-social long-distance migrant, and the Japanese islands lie at the southern extreme of that migration route. Each year the whales migrate from food-rich summer feeding grounds in the cold waters of the Bering Sea, beyond the Aleutian Islands, to their calving grounds in the warm subtropical waters around Japan's southern island groups. They arrive there to give birth in late December, and remain until March. When considering record weights, mother Humpback Whales weigh around 30 metric tons and give birth to a one-ton baby!

Perhaps in years to come, the wide diversity of whales around the archipelago will make Japan a top destination for international whale-watchers, making it as well-known as Hawaii and Alaska are for Humpback, as Vancouver Island is for its Orcas, as New England is for its Humpback, Fin and Northern Minke Whales, and as Kaikoura (New Zealand) is for its Great Sperm Whales. Perhaps future visitors to Japan will talk as glowingly of their whale-watching trips here as they will about their visits to see cherry blossoms, autumn maples or ancient shrines.

Nature Photography Tips and Hints

Whether your interests include scenic beauty, seasonal colours, geological features, historical culture or wildlife, photographing your favourite subjects has to be one of the best ways of remembering pleasant times past.

Despite there having been three different forms of recording media available (traditional 35 mm film, advanced systems, and digital) in recent decades, the ultimate decision comes down to framing the picture, and that means choosing your lens, not your medium, and developing a 'good eye'.

Good light is the key to a good photograph, and choosing the moment therefore is crucial, but wildlife photography imposes different restrictions from those set by other subjects. After all, a wild animal will rarely wait for the right light or remain to pose once disturbed. For wildlife photography, a gentle approach is required – no loud noises and no bright or garish clothing. Approach your subject steadily, slowly, and take 'security' pictures as soon as possible, in case your subject is spooked.

Wherever possible, get down to your subject's level, and even below its own eye-level, if at all possible. Place your subject in a setting so that its environment contributes to the frame. The fast, unpredictable movements of mammals and birds require fast shutter speeds and a fast 'motor drive'. Long focal-length lenses (200–400 mm or more) are preferable. For landscape, scenery, geological and geographical features, trees and flowers, short focal-length lenses and a tripod are a great benefit. They do not merely reduce camera shake, but they make it easier to select and frame the perfect picture. For wildlife, too, a tripod is a boon. The movement of the subject may be troublesome enough without adding the movement of the camera.

All good wildlife photography depends on a combination of knowledge and luck. Knowledge to choose the right place at the right time, and to be able to predict the behaviour of your subject; and luck that the light will remain ideal when its behaviour is revealed.

Perhaps the final axiom is the simplest: the best photographs are taken when you are carrying your camera, so do not leave it behind. If all else fails, use the camera in your pocket – on your smartphone!

Humpback Whale off the Kerama Islands showing its diagnostic tail shape and unique fluke pattern [MAB].

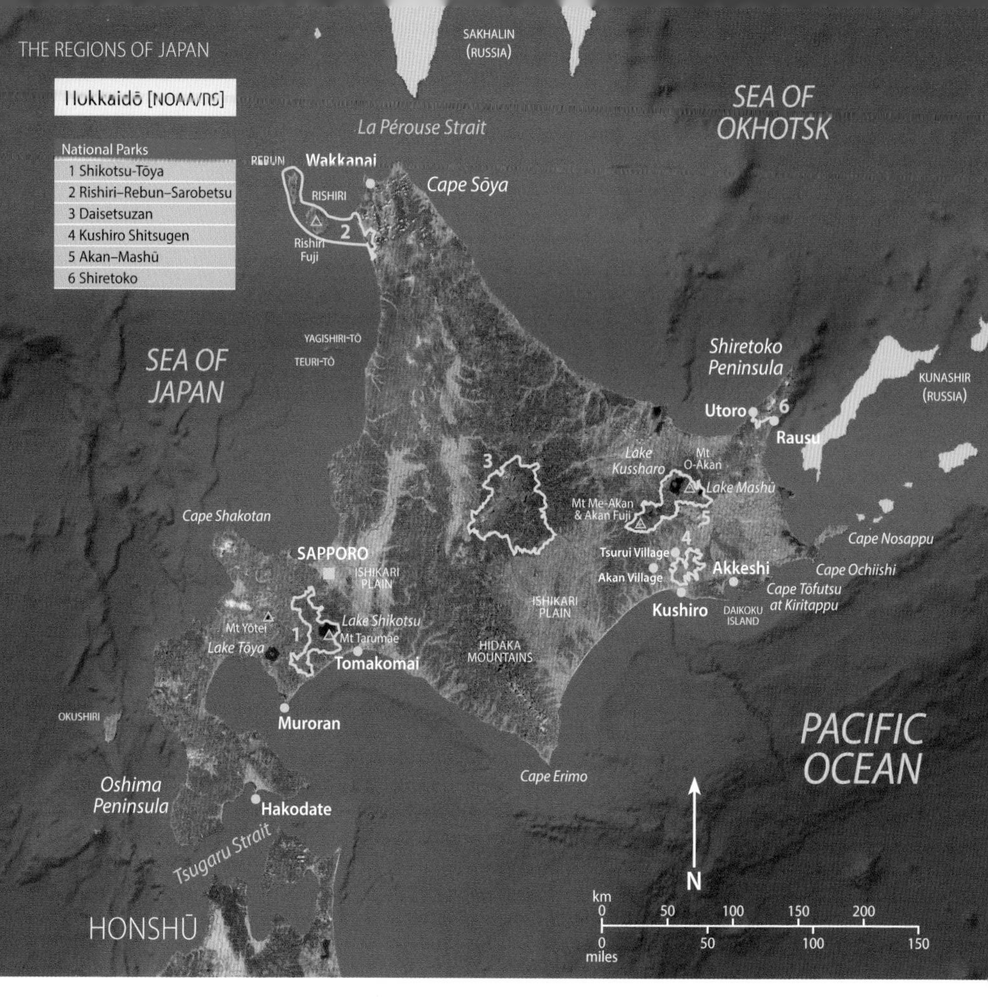

Much of Hokkaidō is covered with montane forest, here at mid-elevation [US]; *inset* Hazel Grouse [WaM].

The Regions of Japan

Hokkaidō – Land of Ice and Snow

Hokkaidō, Japan's rough-cut diamond of the north, is the country's subarctic zone.

Despite having a population of about 5·5 million people, it has the important reputation as being Japan's wild frontier. Hokkaidō represents just over 20 percent of the nation's total land area, but has less than five percent of its total population. Until 1800 its Ainu population outnumbered the Japanese, although now they constitute a very small minority.

Yezo and the Ainu People

Hokkaidō was formerly known as Yezo (or Ezo). It was to this remote northern island that the indigenous Ainu people of Japan were increasingly restricted by the northward movement of ethnic Japanese until, by the 12th century, Yezo was their only homeland. Although Japanese made early forays here from Honshū from the year 659 onwards, and occupied many localities in the southern peninsula from the late 16th century, most of Yezo was seen as suitable only for refugee warriors during times of civil war or for banished criminals.

Yezo was perceived as remote, inhospitable and cold, but Japanese influence continued to spread until, by the Meiji Restoration in 1868, Japanese people had settled in many areas, particularly at ports and the mouths of rivers, where they controlled fishing and other extractive industries. In 1869, Yezo was renamed Hokkaidō and colonization accelerated under official government schemes. Hokkaidō is now as modern and efficient as the rest of the country, but with a population of less than six million in an area constituting about a fifth of Japan, and with most of that population concentrated in the Sapporo area, much of Hokkaidō appears uncrowded and has a more rural feel than any other region of the country.

Fishing, farming, forestry and mining are the main industries, and tourism draws several million people northwards each year. For the Japanese, this is their 'wild west' frontier; to the overseas visitor on the other hand, Hokkaidō's broad expanses of dairy agricultural land, its root crops and its corn fields are reminiscent of familiar, tamer European or North American landscapes.

Although Hokkaidō has little of cultural or historical interest for the international traveller other than the lifestyle of the earlier Jōmon and later Ainu inhabitants, there is much to attract outdoor enthusiasts. Hokkaidō's extensive forests, mountain ranges, lakes and nature parks provide many opportunities for camping, hiking and hot-spring bathing, and help to make it Japan's premier region for wildlife-watching.

Hokkaidō is situated at the same latitude as southern France, yet, because of the cold ocean current from the north (instead of Europe's warming Gulf Stream) combined with the prevailing winter winds that blow in from Siberia, the winter climate is subarctic, with temperatures averaging below zero, and sometimes reaching minus 30°C. Hokkaidō's long cold winters give it a guaranteed snowy season, with snow especially heavy in the west, that attracts local and overseas skiers and snowboarders alike to the delights of perfect powder snow. Hokkaidō's summer temperatures approach 30°C, making conditions pleasant and rarely too hot during the months from May to September.

For the Ainu, the wild animals that they encountered, hunted and depended on for their livelihoods were deeply significant and recognized as a pantheon of beneficent gods. They believed in them as deities and communed with them through intermediaries. Some, such as the salmon, *Kamuycep*, and bear, *Kimun Kamuy*, were particularly significant because they provided important food at different times of the year. Wild areas were seen as the homes of the gods, and so the great Daisetsu mountain range, with its stunning array of volcanoes, craters, alpine meadows and boreal forests at the heart of

Hokkaidō, became known as *Kamuy Mintara*, meaning the garden or playground of the gods, because it was there that their benevolent god-spirits, especially Brown Bear, roamed.

Other gods, too, shared their lives. The cranes that bred each year in the remote corners of once vast wetlands were known as *Sarurun Kamuy*, the deity of the marshes. These mysterious birds could be heard trumpeting from their reedy homes, although for much of the year they remained invisible among the tall stems.

Kotan Kor Kamuy was a deity of a different type. This defender or protector of the village was a giant owl, Blakiston's Fish Owl. Like the crane, it can still be found in Hokkaidō, but, in contrast to the crane, its numbers remain perilously low. Arguably the largest owl in the world, it nests in cavities in enormous trees in riverside forest, especially near the conflux of two streams where fish abound. Its traditional habitat was exactly where Ainu settlements were situated, close to fresh water with plentiful fish. In the past, the deep sonorous calls of this nocturnal deity would have rung out during the winter months over many an Ainu village. For much of the year this owl, true to its name, eats fish, but as winter bites, and ice covers the streams and rivers where it hunts, food-finding becomes difficult and it broadens its diet. The owl then becomes more of a generalist, eking out an existence by feeding on small flying squirrels, small birds, and mice, voles and other rodents that risk a visit to the surface.

Glaciation, Sea-level Rise and Fall

Hokkaidō is a frontier land that, until about 60,000 years ago, was connected to the Asian continent to the north and northwest across what is now the shallow La Pérouse Strait[1], via Sakhalin, and across the even shallower Strait of Tartary[2] to the current continental coast by way of a northern land bridge. During interglacial periods sea levels rose and submerged the land bridge, in a process that may have been repeated several times. The presence and absence of various periglacial features[3] in Hokkaidō strongly suggest that northernmost Hokkaidō experienced continuous permafrost during the most recent glacial period, whereas most of the rest of the island experienced discontinuous permafrost.

Because of the past existence of the northern land bridge, Hokkaidō shares many of its natural floral and faunal characteristics with those of the Okhotsk coastal region. From a natural-history perspective Hokkaidō is very different from the rest of Japan. Although surrounded by channels and seas, it is the currently deep channel to the south, the Tsugaru Kaikyo or Tsugaru Strait, that is most significant. The Tsugaru Strait connects the Sea of Japan with the Pacific Ocean, and has two narrow necks approximately 20 km across at its eastern and western ends. While only 140 m deep in the east and 200 m deep in the west, at its deepest it reaches 449 m.

While this strait's exact age remains contentious, sea levels would, given its current depth, need to have fallen considerably for a land bridge between Hokkaidō and Honshū

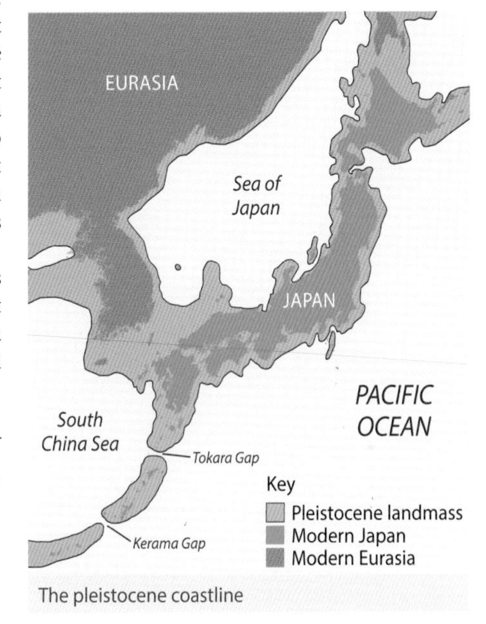

The pleistocene coastline

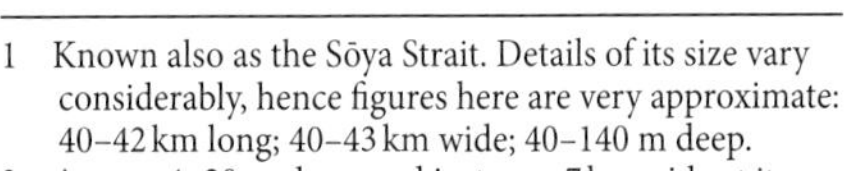

1 Known also as the Sōya Strait. Details of its size vary considerably, hence figures here are very approximate: 40–42 km long; 40–43 km wide; 40–140 m deep.

2 A mere 4–20 m deep, and just over 7 km wide at its narrowest point.

3 Such as the richness of soil wedges and palsa scar-like features and the lack of typical ice-wedge casts and pingo scars.

to have existed, although some do postulate one as recently as 18,000–20,000 years ago. For most of the time when Hokkaidō was connected northwards to northeast Asia, it is likely that the deep Tsugaru Strait separated Hokkaidō from Japan's main islands to the south. This explains why Hokkaidō's fauna, especially its mammals, shares more northern affinities than southern ones, and why those same species failed to colonize Honshū, Shikoku and Kyūshū. Many northern species reach their southern limits today in Hokkaidō. Conversely, many species range northwards through Japan to the northern tip of Honshū, but have been unable to cross to Hokkaidō. Therefore, from a natural-history perspective, Hokkaidō is a unique place.

Climate and topography dictate which species can live in any region, and Hokkaidō is no exception. Prevailing northerly winter winds create weather here that is more arctic than temperate. For months each year snow blankets the island, lakes and rivers freeze, and sea-ice drifts against the northern shore. Only hardy species can overwinter here. Summer is dramatically different, with southerly winds bringing migrant birds, a wealth of wildflowers in bloom, and temperatures often above 30°C. Winter reigns for six months of the year in the mountains and for four in the lowlands, but the brief summers are glorious.

Hokkaidō is mountainous at heart, and has extensive boreal forests and coastal wetlands, craggy cliffs and offshore islands. Hokkaidō's forests are dominated by a mixture of species, including subarctic conifers such as Ezo Spruce and Sakhalin Fir, and hardy deciduous trees, including Mongolian Oak, Father David Elm, Manchurian Ash, Späth's Ash, Katsura[4], Japanese Wingnut, Japanese Walnut, Erman's Birch and Siebold's Beech, while the undergrowth is typically dominated by dwarf bamboo (otherwise known as bamboo grass), the mix depending on the elevation. At lower elevations the forests support more temperate hardwoods, such as oaks, maples, elms, limes and magnolias, along with Prickly Castor-oil Tree, Koshiabura

4 Sometimes called the Caramel Tree because of its scent.

The strangely patterned leaves of Kiwi Vines may be partly white or partly pink [MAB].

Climbing Hydrangea bloom in Hokkaidō during June and July [US].

Mixed lowland forest at Nopporo Forest Park [MAB]

and Amur Cork Tree with a shrubby undergrowth of Forked Viburnum, Oval-leaved Blueberry and Japanese Rhododendron, and a ground cover of evergreen dwarf bamboos such as Kuril Bamboo and Broad-leaf Bamboo. Woody vines such as Climbing Hydrangea, the kiwi vines Hardy Kiwi and Variegated-leaf Hardy Kiwi, and Silver Vine bring an added dash of colour to the forests. Hokkaidō is famous for its small surviving population of Brown Bears, its plentiful Red Foxes, abundant Japanese Deer and winter hordes of clamouring Whooper Swans, and most of all for its breeding population of grace embodied – the Red-crowned Cranes of east Hokkaidō.

Hokkaidō, situated on the Pacific Ring of Fire, is Japan's northernmost island and forms the southern border of the Sea of Okhotsk. With both sea-ice and active volcanoes, Hokkaidō is truly a land of fire and ice. Seas and straits separate Hokkaidō from Russia to the north, west and east, while the deep Tsugaru Strait to the south separates it from the very different island of Honshū. Beautiful mountain ranges with dramatic peaks, spectacular volcanoes, gorges and lakes, all contribute to making Hokkaidō the part of Japan where nature is most conspicuous.

Kotan Kor Kamuy – Lord of the Owls; Protector of the Village

An enigmatic denizen of Hokkaidō's riparian forests is the world's largest owl – Blakiston's Fish Owl. This magnificent bird, an Ainu deity, was for many years the victim of forest clearance and the tidying-away of large dead trees from riverine habitat. These are huge birds and need huge cavities in which to nest, and there simply were not enough cavities or birds surviving. A conservation initiative, begun with a simple nestbox project in the early 1980s and subsequent supplementary feeding, has been successful, but has borne only very slow-growing fruit. These enormous long-lived birds, living in Japan's harshest northern environment, produce *at best* only one or two young a year. Their potential for population recovery is therefore low. Furthermore, suitable habitat for them is limited, leaving the young facing a difficult time once they are pushed out of their parents' territories. These slow, low-flying creatures fall easy victim to traffic accidents where their riverine habitat runs close to main roads – such that the help which they receive through the nestbox project and supplementary feeding is mostly cancelled out by accidental mortality.

Their extreme rarity, combined with the extreme environment, has made them the focus of wildlife-watchers and bird-photographers from around the world. At two sites it has been possible to manage human access, making photography possible without interfering with the birds.

Blakiston's Fish Owl is monogamous, and pairs remain together in their year-round territories for many years. As dusk falls they prepare to leave their day roosts, and within an hour or so of dusk

Although winters are becoming increasingly mild, they are still long and subarctic in nature [MAB].

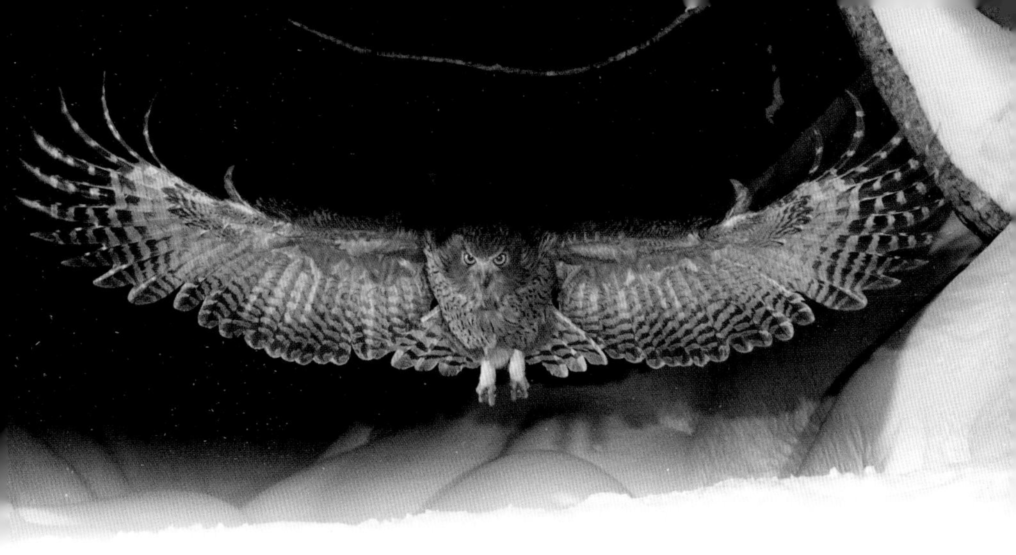

The size of a sea eagle, the enormous Blakiston's Fish Owl is dramatic [FT].

Blakiston's Fish Owls inhabit lowland and mid-elevation rivers with plentiful fish [YS].

Pair-members remain together in their territory throughout the year and duet in courtship [WaM].

During daytime Blakiston's Fish Owls roost in the open, typically close to a large tree's trunk [MAB].

The Blakiston's Fish Owl population is supported by an ongoing nestbox project. Each breeding season the pair produces just one or two chicks [YS].

they start calling, producing a deep, sonorous duet. The male gives a double note, followed so quickly by the female's single even deeper note that the sound may appear to be that of a single bird calling. Eventually, the pair will set off around its territory, visiting favoured fishing haunts, or swooping on any vole, brown frog or flying squirrel that risks exposure. These owls breed exceptionally early in the year, so that by February the female will already be incubating and, with luck, within a few months contribute another one or two young owls to the tiny population of Hokkaidō.

A fledgling Ural Owl soon after leaving its nest [WaM]

Blakiston's Fish Owl is not the only owl of Hokkaidō. Occurring throughout the island is another resident, the very pale subspecies of the widespread Ural Owl. This mostly nocturnal vole-hunting specialist occupies the night shift, a niche filled in the daytime by the Eastern Buzzard. The Ural Owl roosts and nests in natural tree cavities and can sometimes be found basking in the sun at the entrance of its roost site.

In the milder southwestern and western parts of Hokkaidō the Northern Boobook and Oriental Scops Owl occur as migrants. Their calls liven up summer evenings. Meanwhile, in the main islands of Japan, Ural Owl is the largest wide-ranging species, while Northern Boobook and Oriental Scops Owl are the commonest although mainly from central Honshū westwards. They are joined by several other scarcer species, including Japanese Scops Owl and both Long-eared and Short-eared Owls.

Single Ural Owls, and pairs, roost during the day in large tree cavities [MAB].

A Mountainous Heart

At the heart of Hokkaidō lie the mountains of Daisetsu. Established as a national park in 1934, Daisetsu-zan's 2,310 km² area makes it the country's largest national park. Hokkaidō's highest peak, Asahi-dake (2,291 m), is located here. The raised Daisetsu plateau, ringed with peaks, gorges, waterfalls and steam vents, is wild and tempestuous in winter, when it is cloaked in wind-sculpted ice and snow. It takes on a kinder face in summer, when it is carpeted with alpine flowers and when it rings to the sounds of birds, such as Siberian Rubythroat and Japanese Accentor.

On the forested lower flanks of the Daisetsu mountains, herds of Japanese Deer browse the dwarf bamboo. Red Fox may be encountered at any level, as may the summer-active Siberian Chipmunks. In areas with jumbled loose rocks, a small rabbit-like creature, Northern Pika, announces its presence with a piercing whistle. Lord of all, though, in the garden of the gods, is Brown Bear, for Daisetsu is one of that species' last remaining strongholds in a country with a low tolerance of nature that is untamable. At higher levels in the mountains of central Hokkaidō, mixed forests of deciduous broadleaf species with Erman's Birch, spruce and fir give way to increasing numbers of conifers and Japanese Rowan. Leaving the tree line behind, you enter the alpine zone, where there are scattered patches of brilliant colours from late June onwards. Up on the plateau itself, the very much more limited vegetation includes dwarf Japanese Stone Pines, along with alpine shrubs including Lingonberry (Cowberry or Mountain Cranberry), and alpine snow meadows carpeted with an array of alpine flowers, including Deer Cabbage, Hog's Fennel, Wedgeleaf Primrose, Aleutian Avens, Shrub-like Penstemon, the delicately bell-like Clubmoss Mountain Heather and Kuril Bamboo. Some of the species present here in the alpine flora of these mountains can be found also on higher mountains in northern and even central Honshū, yet they share their affinities with the tundra of far northern Asia, being Pleistocene relics of a Beringian flora, while the birds, such as Siberian Rubythroat (which migrates here to breed each summer), have come from Southeast Asia, making this a rarefied meeting ground of north and south. The autumn colours here are from a lavish palette indeed.

Alpine flower meadows on the tundra-like raised Daisetsu plateau are especially rich with wildflowers during June, July and August [US].

The bizarre spherical algae, known as Marimo, lives in Lake Akan [MAB].

Looking east from Daisetsu, the peaks of Akan, in east-central Hokkaidō, are visible. They are situated in an enormous protected area (914 km^2) known as Akan–Mashū National Park. Many believe this to be Japan's most attractive national park. It is certainly one of the oldest, having been established in 1934. The western portion of the park contains beautiful Lake Akan and is dominated by the paired peaks of Mt O-Akan (1,370 m; the male) to the southeast and Mt Me-Akan (1,499 m; the female) to the southwest, with its satellite Akan Fuji (1,476 m). The Me-Akan composite volcano, the highest point of the Akan complex, is comprised of nine overlapping cones with three summit craters. It is highly active, with the most recent eruption in 2008, while Me-Akan constantly spews steam and gas. Lake Akan is famed for scenic views and its bizarre filamentous green alga that grows into velvety green balls known as Marimo, which can reach a diameter of 34 cm. The hike up from Akan over the peaks and down the other side to Lake On-ne-tō affords spectacular views at any season, especially in autumn, when paint-box colours add to the scenery. O-Akan is a more serious hike, but also manageable in a day.

The forest in the Akan–Mashū National Park rings to the calls of six species of woodpecker, including the huge Black Woodpecker and the diminutive Japanese Pygmy Woodpecker. Many other forest birds occur here, too, along with Red Fox, Eurasian Red Squirrel, Siberian Flying Squirrel and Siberian Chipmunk. Haunting night-time wavering whistles here are the loud penetrating calls of rutting male Japanese Deer, which roam the forests in search of females.

The Mashū Caldera, with its super clear lake and extensive forests, is beautiful at any time of year [LMTA].

The expansive scenery of the Akan–Mashū National Park looking west towards distant Mt O-Akan [MiH].

The enormous Kussharo Caldera with Lake Kussharo is at the heart of the Akan–Mashū National Park [MiH].

Summer scenery in the Akan–Mashū National Park [MAB]

Erman's Birch forest is especially attractive during winter [MAB].

Steam vents and sulphur springs break out at the surface on the flanks of Mt Iwō [MAB].

Geothermal activity abounds in this region. In addition to active Me-Akan, there are outdoor hot-spring baths at Kussharo Kotan and Wakoto (marvellous by day and by night), steaming fumaroles on the flank of Iwō-zan (between Kussharo-ko and Mashū-ko), and onomatopoeic 'bokke', small areas of bubbling mud, beside Lake Akan, as well as hot-spring resorts at both Kawayu and Akan.

East of Akan, over the pass towards Teshikaga, an attractive drive in itself, there are splendid views back to the volcanoes. To the northeast lies Kussharo-ko, an enormous caldera lake with a 57-km perimeter. A year-round beauty spot, this enormous lake freezes over in winter, when the weird singing sounds of the ice and the snaking pressure ridges of ice across the lake are special features. Thermal vents keep tiny portions of Kussharo-ko ice-free, and thus flocks of Whooper Swans remain throughout the winter, looking ethereal in the early-morning mist rising off the tiny strips of open water. The swans keep up their soft contact calls throughout the change from day to night, altering their tempo only when a disturbance breaks out between families.

A few kilometres farther east lies beautiful Mashū-ko. With a shoreline of 19·8 km, a surface area of 19 km^2 and a volume of 2·7 km^3, Lake Mashū is one of the greatest scenic spots in all Hokkaidō. When the weather is kind, it is truly spectacular. The crater's steep internal cliffs encircle, and rise 200 m above, the lake, while the pumice-covered floor of the crater lies 211 m below. The water of the lake is of an astonishingly clear blue[1], and the lake has no visible outlets. Lake Mashū is the only lake registered as a GEMS/Water Baseline Monitoring Station in Japan. The panoramic view from the Mashū crater rim takes in Mt Shari to the northeast, with the Shiretoko Peninsula beyond, and Kussharo-ko to the west with the Akan volcanoes farther off to the southwest.

Mashū-ko is one of the essential elements of a visit to Hokkaidō, although many visitors are thwarted by the fogs that sometimes rise from the deep clear waters to fill the crater in summer. In early winter, however, with cold clear skies, there is much to watch for, and the rim provides the perfect viewing point.

1 Known locally as Mashū Blue.

Among the very first flowers to emerge after early snow melt are those of the huge Asian Giant Skunk Cabbage [US].

Black Woodpecker is Japan's largest woodpecker species [SP].

For the naturalist willing to spend time watching, it is an ideal spot to search for birds of prey. Raptors are usually late to rise. For them, riding currents of rising warmer air an hour or so after the sun has risen makes for easy flying. Across the far side of the lake to the east lies the rugged 857-m peak of *Kamui-nupuri*, the mountain of the gods (known also as Mashū-dake). Here in winter occasional pairs and family parties of Northern Ravens can be found dogfighting and displaying their aerobatic skills, running rings around Japanese Crows. The caldera rim, flanked with Erman's Birch and an understorey of dwarf bamboo, is often traversed in autumn and winter by Peregrine Falcon [EN], Eurasian Sparrowhawk [NT], Northern Goshawk [VU], Black-eared Kite, Eastern Buzzard, Rough-legged Buzzard, and Hokkaidō's two large eagles – White-tailed and Steller's – moving between the Sea of Okhotsk coast and the lowlands, capes and lakes of southeast Hokkaidō. The eagles especially linger on occasions, to roost in trees near the waterline and hunt across the lake. The forested outer flanks of the caldera are home also to Japan's largest woodpecker, Black Woodpecker.

The Enchanting Red Fox

Red Fox [YaM]

Secretive, sly and a shapeshifter (if Japanese folk tales are to be believed), Red Fox is an elusive, nocturnal creature throughout most of Japan. But be on your guard, for there are foxes about! Beware the shape-changers and take care to avoid possession! Foxes are usually far craftier than to allow themselves to be seen and are quite unwilling to reveal themselves in their true forms, for they exist on many planes, as godly messengers, as tricksters, deceivers, and as wild and wily animals. These are deceptive, far-ranging, and strange creatures.

There is much more to the 'humble' fox than meets the eye, and proverbs attempting to explain the lore of the wicked, deceiving fox abound. There is something mesmerising, almost bewitching about them. If you try to pass a fox without stopping and looking, you will discover what I mean.

In real life, we are dealing with an out-and-out opportunist, willing to try to eat anything ranging in size from small insects such as beetles and grasshoppers, earthworms, and even berries and fruits in the autumn, to larger animals including rodents and occasionally birds and hares.

It is the Red Fox's skills and abilities at shape-changing, into people and trees, that have been most often extolled. In 1889, the most extraordinary tale of all was circulating, about this craftiest of all the shape-changers that achieved new heights in assuming the shape of a steam locomotive on the Tōkyō–Yokohama railway! The phantom train appeared before the driver of a real train, but, when the driver put on a burst of speed in order to catch up with the phantom, the latter disappeared and all that the driver ever found was a fox, crushed beneath his wheels.

The moral of the fox's tale is to be ever watchful, cautious but polite, for you cannot be sure just which kind of fox you are meeting, and never curse the fox for, as the medieval proverb says, '*The fox fares best when he is cursed*'.

The endemic subspecies of Red Fox that occurs in Hokkaidō is quite different from that in Central Japan, being frequently diurnal, and in its skill in catching prey it is a match for the more famous Arctic Fox. During winter it is common to see a fox patrolling vast snowfields during daytime. After trotting back and forth across a field the fox suddenly pricks up its ears, its whole body becomes tense, it rushes forward, and then, suddenly, it performs a four-footed leap high into the air, pouncing down with all four paws bunched on one spot so as to punch down through the snow crust. With its neck arched and its face pointing straight down, its leap turns into a mad digging scrabble and its foreparts disappear deep into the snow layer; moments later it appears with a vole in its mouth, having burst through the roof of a vole's winter quarters.

The Red Fox is a scavenger, nocturnal in most of Japan, but frequently diurnal in Hokkaidō [MAB].

Secretive, sly and a shapeshifter (if Japanese folk tales are to be believed), Red Fox is a character from folklore, and appears frequently in art and at special shrines [MAB].

Caught in the act! Like Arctic Foxes, Red Foxes in Hokkaidō have perfected a leap-and-pounce strategy, punching down through the snow crust to catch their prey [RW].

Voles are an important prey item for the Red Fox in Hokkaidō, where the foxes hunt them beneath the snow in winter [MH].

Red Fox roam extensively in search of prey. Their conspicuous tracks are distinctively linear [MAB].

Red Fox cubs are endearingly photogenic [WaM].

Kimun Kamuy – Lord of the Land, Hokkaidō's Brown Bear

Even though Japan is a busy crowded country, it also, amazingly, has many wilderness areas. These are so large that not one but two species of bear still occur here. The two provide an interesting example of biogeography and show how species may be isolated by the various channels between the islands of the archipelago.

In the mountain forests of Honshū one might, if lucky, come across a roughly overturned log or a patch of disturbed leaf litter, both indications that an Asiatic Black Bear has been foraging there. Those even more fortunate might see a lumbering blackish shape foraging across a high alpine slope between the Japanese Stone Pines. In Hokkaidō one can see its much larger cousin, Brown Bear. Forays into the Daisetsu Mountains or along the shore of the Shiretoko Peninsula may provide sightings. In Japan, the ranges of these two species seem to have been forever separated by the depths of the ancient Tsugaru Strait between Hokkaidō and Honshū.

Each bear species holds territory and some Asiatic Black Bears live within sight of Japan's capital, Tōkyō, while Brown Bears have entered Hokkaidō's capital, Sapporo. Human–Brown Bear conflict is an issue much on the minds of Hokkaidō residents, not only in the city but also island-wide. Such conflict is overinflated and overdramatised in the public mind, and inevitably the bears come off worse. In 2017, according to the Japan Bear and Forest Society, 108 people were injured in bear attacks and two were killed (out of 126 million). Meanwhile, nationwide, 3,779 bears (out of fewer than 10,000) were killed by humans. To put those statistics into context, it is estimated that, in Japan, more than 20 people die annually from hornet stings, while traffic accidents claim several thousand more lives each year. The public perception of risk belies reality.

Brown Bear, a Holarctic species ranging across Eurasia from Scandinavia to Kamchatka and Chukotka and across North America, is Japan's largest terrestrial mammal. Sometimes referred to as an endemic subspecies, *Ursus arctos yesoensis*, Hokkaidō's bears range in weight mostly from 150 kg to 250 kg, the very largest males exceeding 300 kg and measuring 200–230 cm in length. This large, stocky bear is found across Hokkaidō. It has a broad rounded head, short and erect rounded ears, and a long narrow muzzle. The limbs are powerful, the feet large with broad pads and long claws, but the tail is very short. Males and females differ in size, and males can be considerably larger than females. This bear's fur

Brown Bears are at home along the coast and in the mountains in Hokkaidō [FT].

Brown Bear, Japan's largest terrestrial mammal, thrives on the Shiretoko Peninsula [BOTH IA].

is long and varies in colour from pale tawny to dark blackish-brown, and most individuals have a paler neck and upperside. Typically, the legs are darkest.

Although omnivorous, Brown Bear is largely herbivorous, eating a wide range of plant species in great quantities. It favours fresh grasses and herbs in spring, and berries, fruits and nuts in summer and autumn. Its diet also includes a wide range of animal foods, particularly invertebrates, such as sandhoppers on beaches, along with earthworms, ants and bees. It will scavenge at Japanese Deer carcasses and, where its range still reaches the sea coast, those of beached seals and cetaceans. In autumn, it takes spawning salmon where still-active spawning runs occur within its territory.

These bears are crepuscular and diurnal. The males are solitary, whereas females are typically accompanied by their cubs of the year, or one-year-olds. Although active for most of the year, Hokkaidō's bears spend the winter from mid-December until late March in a state of torpor. The start and end of this so-called hibernation depends somewhat on location, food supply and weather conditions. They make their den underground on a slope, sometimes beneath a tree. It is in their dens, during winter 'sleep'[1], that the females give birth to one or two cubs.

Brown Bears naturally range from the coast to the alpine zone, and favour mixed deciduous and coniferous forest and alpine grasslands. They have become scarce, mainly as a consequence of human activity and persecution, and nowadays they generally avoid, or are absent from, lowland and coastal areas except where these are protected, such as along portions of the Shiretoko Peninsula. Today they are more easily found in mountainous areas wherever they can avoid disturbance.

Culturally, Brown Bear is a fear-inducer for Japanese people[2], yet to the indigenous Ainu of Hokkaidō it is revered as their deity of the mountains – *Kimun Kamuy*. Traditionally, the gifts of the bear god were vitally important in the Ainu calendar and included medicines made from the gall bladder, warmth provided by the bear's furs, its fat, and finally food in the form of its meat. Hunting with traps, snares and arrows, and bringing a bear to bay with wits and strength must have been a daunting and dangerous task.

1 The topic of torpor/hibernation is one of great interest at present. Although bears were previously thought only to enter torpor and not go into full hibernation, the emerging consensus is that they really do hibernate; it is just that their large size means that they do not cool so much as do, for example, small squirrels. Hibernation, it appears, is a continuum, from deep hibernation in small species such as Siberian Chipmunk to shallow hibernation in large species such as Brown Bear.

2 Perhaps this seemingly irrational fear has been driven largely by news of the extraordinary Sankebetsu Incident. In late 1915, seven settlers were killed by a single Brown Bear in the worst bear attack in Japanese history. The incident was not researched until the 1960s, and then was sensationalized in the 1980s. It transpires that the bear had been shot and injured before it made the fatal attacks while searching for food.

Each season, before bear-hunting could begin, the Ainu sacrificed a young bear so as to return its spirit to the world of the gods in the *Iomante* festival. Amid feasting and revelry, the young bear was entertained with tales, shot with blunted arrows, and then crushed to death. Its spirit was then led, with the flight of an arrow, away from the world of men and back to its own spirit world. Every bear trapped or killed through the ensuing season would also be treated with respect, and eventually another cub would be captured, reared and kept to provide the essential spiritual link with the world of the gods for the following year.

By the final decades of the 20th century, this event, as with so many aspects of our once natural world, had been stripped of any solemnity, reduced and trivialized. The bears were no longer hunted by Ainu, no longer treated as a god, but shot out of fear and mistrust by Japanese hunters. With the sickening ease and detached sense of safety that a high-powered rifle brings, the bears fall unceremoniously. The meat and paws perhaps 'grace' a restaurant table, and the gall bladder, worth more gram for gram than the rest of the carcass put together, will enter the lucrative and often illicit folk-medicine market. Much less dignified than the slaughter of the adults is the incarceration of their orphaned cubs in notorious bear farms, several of which still exist in Hokkaidō. Brown Bear skins are also stuffed and mounted in mercenary poses outside souvenir shops. They stand there mutely as temptations for visitors to enter, to fish yen from their pockets, and to trade them for useless souvenirs. What an ignominious position for a deity – how low our gods have been brought by commercialism. Yet, rather astonishingly, despite the persecution which it suffers at human hands, somehow the magnificent Brown Bear still hangs on in Hokkaidō.

Brown Bear is Japan's largest terrestrial mammal and although generally wary and shy, they should always be treated with respect. Females produce just one or two cubs each year [BOTH WaM].

Beach-washed carcasses are important sources of food for bears on the Shiretoko Peninsula [*left* YM; *right* GOKA].

Shiretoko Peninsula

To the northeast of Akan–Mashū National Park lies the Shiretoko Peninsula. Named *Sir etok* by the Ainu, it means 'End of the Earth'. This remote, admonishing finger of land protrudes 65 km northeastwards from the irregular diamond shape of Hokkaidō, separating the Sea of Okhotsk from the Nemuro Strait. Shiretoko is now Japan's wildest national park (386 km^2). The forests here consist of a mix of spruce, fir, Mongolian Oak, birches, maples and rowans. Consisting of a well-forested mountainous ridge of volcanic peaks, dominated by Mt Rausu (1,661 m) and Mt Iwō (1,562 m), the Shiretoko Peninsula supports one of the healthiest populations of Brown Bear left in Hokkaidō. Sightings are rare unless you ride north by boat towards the cape from Utoro or Rausu, where the bears forage along the coastal strip during spring, summer and autumn. In summer it is possible to see also Northern Minke Whale, dolphins and porpoises, along with various seabirds such as Spectacled Guillemot [VU] and Temminck's Cormorant. Several pairs of White-tailed Eagles nest along the peninsula and, in winter, their numbers are swollen by hundreds more arriving from Russia. In winter, they are dwarfed by the world's largest and most magnificent eagle, Steller's Eagle, best seen north of Rausu.

North of Utoro lie the pretty Shiretoko Five Lakes, reflecting the mountains on clear days. A two-kilometre trail here makes an attractive walk, and lovers of *onsen* (hot springs specially used for bathing) will not want to miss the hot waterfall known as Kamuiwakka. From June to October, the high pass across the peninsula from Utoro to Rausu is open, and the view from the pass eastwards to the island of Kunashiri is dramatic. This road passes through the subalpine zone dominated by dwarfed Japanese Stone Pine trees, and from below the pass a hiking trail strikes off southwards for Lake Rausu. Another trail, beginning just above the town of Rausu, heads north for Mt Rausu and Mts Iwō and Shiretoko and the cape beyond. Whereas Mt Rausu is a manageable, though strenuous day-hike, the journey to the cape requires several days and careful planning. The long, cold winters and short summers make hiking possible only from June to September.

In winter, the weather is wild and tempestuous, the landscape obliterated by swirling clouds of storm-blown snow. In summer, east-coast fogs are forced through the mountain gap to batter and ruffle the

The rugged Shiretoko Peninsula offers some of the wildest cliff, forest and volcanic scenery in Japan [MAB].

The creamy-white blooms of Asian Skunk Cabbage are an early sign of spring [ABTA].

Kamchatkan Trilliums bloom before the forest trees are in leaf [US].

Sea of Okhotsk coast with dampness – this truly feels like the far end of the earth! When high pressure dominates, however, and frost crisps the snow, it becomes a magical, exciting winter land. In spring, the creamy perfection of the unfortunately named Asian Skunk Cabbage and the Kamchatkan Trillium replace the snow's whiteness. With the brief vibrant flourish of summer, emerald-green becomes the dominant colour, broken here and there by the orange of tall lily spikes and the pinks of wild roses. The full palette of brilliant colour takes over as the splendour of autumn pours down the wooded hillsides like an overflowing paint-box. In from the salt sea come Pink or Humpback Salmon and Chum or Dog Salmon, thronging the brackish river mouths, fighting their way through life's great transition, and surging up the sweet rivers to spawn. Then, as autumn's rush fades into winter's bleakness, the eagles return.

Spectacular scenery, deep snow, sea-ice and magnificent wildlife combine to make Shiretoko National Park one of the most exciting areas of Japan. Its very remoteness helps to retain its role as a haven for wildlife.

Hokkaidō's Wetland Heaven

Hokkaidō's landscapes are not all mountainous. Although the majority of Japan's national parks from Hokkaidō to Okinawa are indeed mountainous, in southeast Hokkaidō, north of the coastal city of Kushiro, lies Japan's wetland supreme. The low marshland and swamp forest of Kushiro Shitsugen National Park is a rare exception. This enormous mire of peat, pools and waving reed fronds is an overgrown delta formed as the result of siltation from the outpourings of volcanoes in eastern Hokkaidō. Six thousand

Japanese Huchen is the largest of the salmonids here and spends its entire life in freshwater [KS].

years ago this area was an inlet of the sea, and from 4,000 years or so ago sea retreat and steady silting of the delta of the Kushiro River have led to the complex freshwater ecosystem that we see today. Many springs, streams and rivers feed into Kushiro wetland, but its main source is the Kushiro River, which drains from Lake Kussharo in Akan–Mashū National Park, and draws to it many smaller streams since the whole area declines towards the southeast. It is well worth taking the long canoe ride from north to south, leaving behind the stunning scenery of Akan's volcanic peaks and entering a slow, quiet world where enormous Japanese Huchen, growing to a metre in length and weighing up to 30 kg, native Hokkaidō Brown Frogs and many dragonfly species live among the greatest expanse of marsh that remains in Japan.

On cool summer nights, the calling of the frogs from below, the bizarre sounds of Latham's Snipe displaying above and, on misty summer mornings, the calls of Common Cuckoo and Black-browed Reed Warbler all bring the marsh to life. In winter, these sounds are replaced by those of brittle reed stems rattling, frosted river ice cracking, and the soughing of the wind. In the brief interlude between winter and summer, impressive cries echo across the marshes, the powerful duetting of *Sarurun Kamuy*, the graceful Ainu deity of the marshes.

Hokkaidō (or Ezo) Brown Frog, restricted to Hokkaidō and Sakhalin, becomes briefly abundant in spring [MAB].

Latham's Snipe display by night and day over wetlands in Hokkaidō [SP].

Japan's largest natural wetland lies within the Kushiro Shitsugen National Park [MAB].

Going on a Snipe Hunt

Latham's Snipe [YaM]

While '*going on a snipe hunt*' may, in North America at least, be considered a rite of passage or a practical joke, tricking an unsuspecting youngster into attempting to catch a non-existent animal called a snipe, in Japan on the other hand, snipe-watching is a real past-time.

A number of snipe species (Common, Pin-tailed and Swinhoe's, for example) occur in Japan. All are long-billed shorebirds that forage for invertebrates in damp soil. They can be found wintering in damp and muddy rice fields throughout southern Japan, but one species is spectacularly different. Latham's Snipe winters in the Antipodes and migrates into Japan, primarily from central Honshū northwards, only for its brief summer breeding season. During May and June, it engages in an amazing aerial and clamorous display both by day and at night. In dramatic swoops and dives over its grassland or wooded wetland territory, a Latham's Snipe unleashes some of the most bizarre sounds in the bird world. It calls forcefully and repetitively, "*tsupiyaku, tsupiyaku, tsupiyaku*", as it climbs ever higher against the sky; then, like a roller-coaster reaching the peak of its ride, it swoops earthwards while emitting a strange thrumming sound, somewhat like a bullroarer, "*gwo-gwo-gwo-gwo*", by fanning its stiff tail feathers outwards, allowing them to vibrate in the wind of its rapid descent, only to swoop back skywards and repeat the process over and over again.

In winter, Solitary Snipe makes for a particular avian challenge for birders. As its name indicates, it lives alone, and it does so beside cold, rapidly flowing mountain streams, typically between banks covered with snow during much of its period in Japan, where the cold brown and grey tones of its plumage combined with its slow-motion creeping behaviour make it difficult to spot. It crouches for minutes on end, barely moving other than to rock gently back and forth on the spot.

A larger relative of both Latham's and Solitary Snipes, Eurasian Woodcock is a summer visitor to northern Japan but may be found in winter in the south. Like Latham's Snipe, it, too, has a distinctive display flight, known as roding, during which it flies on a regular circuit, beating the bounds of its territory, at dusk and dawn. While roding, it flies in straight lines, just above treetop height, giving a low grunting call followed by a sibilant "*twissick*". Wonderfully camouflaged with highly cryptic plumage, woodcock nest on the ground on the forest floor. If disturbed, they are capable of grasping their young chicks between their legs and taking flight to carry them to safety.

Common Snipe is a common winter visitor in southern Japan [TsM].

Latham's Snipe is a summer visitor to Hokkaidō from wintering grounds in Australia [MR].

The aerobatic display of Latham's Snipe is accompanied by extraordinary sounds [MR].

Solitary Snipe winters only along fast-flowing rivers [MR].

Pin-tailed Snipe is a winter visitor to southern Japan [KT].

The cryptic plumage of the Eurasian Woodcock allows it to blend into the foliage of the woodland floor [MC].

Magnificent Eagles

Each winter there is an exodus from Russia southwards to Japan. The world's largest eagle, the magnificent Steller's Eagle, and its smaller cousin, White-tailed Eagle, breed around the northern Sea of Okhotsk. As this region freezes in winter, the eagles are forced south in search of food and as many as 2,500 of them may reach Japan. The eagles migrate here from their breeding grounds between late November and early March, to spend the winter scouring the coasts, shores and rivers for fish and the inland forests for deer carcasses.

Steller's Eagles are huge. This massive, fish-hunting raptor breeds in eastern Russia around the Sea of Okhotsk and on the Kamchatka Peninsula. In winter the eagles fly south by way of Sakhalin to reach feeding grounds around Hokkaidō's Shiretoko Peninsula, across Hokkaidō and in lesser numbers south into Honshū. The eagle was named to honour Georg Wilhelm Steller, the 18th-century German zoologist and explorer. He also has species of jay, eider, sea lion and extinct sea cow to his name. Male Steller's Eagles measure 88 cm from bill to tail tip, and females over a metre. With a wingspan of 2·5 m, this species soars and glides with little effort. It is not just the size that makes it so impressive. The striking plumage and massive bill, along with its distinctive habits and behaviour, make this a powerful magnet for naturalists. Its narrow-based wings and massive white shoulders distinguish it in flight from its cousin, the White-tailed Eagle.

During the day, along the Shiretoko Peninsula, a traditional wintering area for the eagles, they fish the Nemuro Strait and use the drifting sea-ice as perches, where they are easy to observe. Adults are boldly patterned, with white shoulders, rump, diamond-shaped tail and 'trousers'; younger birds begin entirely dark brown, the white appearing only through successive moults. The forehead blaze develops last. The stiletto-like talons are ideal for grasping slippery fish prey, while the massive hatchet bill is used to rip into the thick, scaled skin of salmon, a material so tough that local humans have used it for clothing and footwear.

When blizzards strike, the eagles return to roost during the daytime in their hundreds in the shelter of the wooded slopes leading down to the sea. On fine days, they begin to return to the roost during the late afternoon, before it becomes dark.

In the heyday of pollock-fishing in east Hokkaidō in the early 1980s, almost all of Japan's wintering Steller's Eagles could be encountered along the Shiretoko Peninsula, where they benefited from the waste of the fishing fleet. Before dawn the eagles flew out to the Nemuro Strait to await the fleet. With the decline in the inshore fishery, fish pickings have become scarcer, but the ever-adaptive eagles now range farther afield and can be found in scattered concentrations at lakes and lagoons around east Hokkaidō

The Shiretoko Peninsula is synonymous with an abundance of Steller's Eagles, winter visitors from across the Sea of Okhotsk [MAB].

The wing shape and wing pattern of an adult Steller's Eagle are immediately recognizable, while its hatchet-sized bill, enormous talons, and blackish-brown and white plumage render it impressive and unmistakable [*above* MAB; *below left* JW; *below right* BOTH PP].

White-tailed Eagle is common in winter in Hokkaidō, if somewhat overshadowed by the presence of its larger cousin. It is not afraid to snatch food from Red-crowned Crane or from the larger Steller's Eagle [*left* MAB; *right* JW].

wherever fishermen cast out waste from their nets. Even today, when their wintering numbers are more dispersed throughout east Hokkaidō, drift ice still provides them with a resting platform, and gatherings of a hundred or more at a time are not uncommon. The world population is estimated variously at about 3,200–4,200 pairs, of which as many as 2,200 individuals may winter in Hokkaidō.

Steller's Eagles forage along coasts and river mouths, where they take various marine fish and salmon that are heading back into rivers. They will also scavenge at seal carcasses, and sometimes kill birds (swans, ducks and gulls). Fish, such as Walleye Pollock, falling from drift nets as they are winched aboard boats, make easy pickings for eagles. The eagles circle overhead for just such a moment, and then swoop, and snatch the prey in their huge yellow feet. Floating sea-ice drifting in from the Sea of Okhotsk makes a suitable feeding platform on which to dismember prey. The massive bill of a Steller's Eagle makes short work of a Walleye Pollock.

After the decline in the local fishing industry during the 1990s, the eagles began to disperse inland, away from coasts and rivers. There they feed on Japanese Deer shot by hunters. This 'new' food source has exposed them to an added health risk – poisoning from the lead shot used to kill the deer. The natural mortality of Japanese Deer is also highest during winter, and the eagles have learned that there is food to be found in the forests during this season. Steller's Eagles sometimes engage in piracy – stealing fish and other food from their smaller White-tailed cousins. Although a large bird itself, White-tailed Eagle is no match for Steller's and yields its catch. Now, wherever one travels in Hokkaidō during winter, there is always the chance of seeing a magnificent eagle soaring overhead.

As spring approaches, pairs of Steller's Eagles engage in aerial display flights. During these they reveal an astonishing aerobatic capacity, engaging in power dives, rolls and spins, and occasionally they even lock talons and cartwheel together.

Steller's Eagle has a very restricted breeding range, with nests increasingly at risk from disturbance. Meanwhile the easy pickings of the waste from the northeast Hokkaidō fishing fleet are no longer so easy, as the fleet fishes far less often and catches far fewer fish than in the past. Local fishermen, concerned for this situation, have begun to provide food for the eagles and in doing so have spurred a local ecotourism opportunity. Nowhere else in the world can one approach so closely to these normally nervous birds when they are feeding.

Four more large raptors share central and northern Japan with the magnificent sea eagles. On their own they are impressive, but in comparison with the sea eagles they are dwarfed. The largest of this quartet is Golden Eagle [EN], now confined largely to a narrow range in the Japan Alps. Next in size is Japanese Hawk-Eagle [EN], a solitary, very broad-winged, almost hawk-like eagle, most frequently seen soaring over montane forests in search of Mountain Hare and Hazel Grouse. Rough-legged Buzzard is a scarce winter visitor mostly to coastal areas of Hokkaidō, where its very pale appearance, white tail with black terminal band or bands, and distinctive habit of hovering over grasslands and cliffs all serve to make it noticeable. The commonest of these four, however, is Eastern Buzzard.

Eastern Buzzard is very similar in size to the ubiquitous Black-eared Kite, but differs both in its plumage and in its habits. It is a sit-and-wait vole- and mouse-hunting raptor often seen atop roadside trees and poles. At a glance, it may be mistaken for a kite, especially when perching quietly atop a roadside utility pole, but take a closer look and you will notice significant differences. The proportions and general appearance, the brown tones, are similar, but a look at the tail will reveal that the buzzard has a short, rounded tail, in contrast to the kite's long tail with a notched tip. Eastern Buzzard is smaller and darker than Rough-legged Buzzard but shares with it the pale underparts and very dark flank patches; it does, however, lack the greyness to the plumage and the white tail base of its larger relative.

In flight, a buzzard's wings are broad, and rounded at the tips, whereas the kite's are long and broad. When seen from below, the kite sports prominent pale panels across the bases of the underside of the primary wing feathers, and from above it has broad pale panels across the upperwing. At close quarters, the dark patches of feathers on each side of the head, the black 'ears' that give the kite its name, are readily visible.

Eastern Buzzard has a plainer, gentler-looking face (although that belies its nature) and dark flank patches on its otherwise pale underparts. The buzzard is a true predator, albeit of the sit-and-wait persuasion. It watches for its small-rodent prey from a prominent perch and, once it spots one, the buzzard swoops down to capture and kill its prey on the ground. While the buzzard waits for prey and then swoops, the kite quarters low over the ground, forever searching, its wingbeats slow and measured, the fingers of its primaries spread widely, as if grasping outwards for each iota of lift that it can find. It picks up a breeze and rises, drifts, and turns with a mere twist of its long tail.

Buzzard and kite: they both range so widely in Japan, appear so similar in many respects, and share much the same environment, yet they live such different lives.

Eastern Buzzard, along with Red Fox and Ural Owl, focuses on hunting voles [FT].

Black-eared Kite is Japan's most widespread scavenger [OT].

Night of the Swimming Dead – Salmonids of Japan

Corpses of exhausted salmon litter the streams and rivers of east Hokkaidō, while living fish seem oblivious to their mortal wounds [MAB].

Before the rebirth of a new year is possible, sacrifice and suffering of the old year are necessary, and for some this self-sacrificing process can be called '*the night of the swimming dead*'. This is not some new ghoul-driven festival but, rather, an event that draws to a close in major northern rivers as the year turns: it is the running of the salmon. The annual miracle of death and rebirth leads to strange ghost-like creatures inhabiting cold rivers.

In darkness, the delightful sound of running, rushing and gurgling of a northern stream is broken by brief, more deliberate splashing perpetrated by shadowy pale beings that are part of the final great sacrifice of the season. Close inspection reveals that the shallow stream bed is littered with pallid rotting corpses, the 'ghosts' of sleek marine lives past. Here and there, corpse-like, but not yet dead, are animated, partly skeletal fish, the 'swimming dead'. Their powerful instincts to return 'home' and to breed ultimately overtax their physical resources. Their bodies, in decline, literally wear away and break apart as they batter themselves against the rocks protruding from the stream. At first sight, they appear pale, with scale-less flanks and torn fins, but up close it is clear that they have worn their pectoral fins to the bone, and their dorsal fins protrude not as sleek, silvery blades, but as ragged skeletal fans.

Ugly apparitions of skeletal fish fight again and again to climb upstream, only to be overwhelmed by the water's flow, washed and battered back downstream, and cast up on some exposed rocks. The fish's thoughtless, genetically encoded drive to achieve its goal is admirable, and in stark contrast to the hampering burden of self-awareness that we humans carry and that prevent us from achieving life's fulfilment. Suddenly, an apparent fish corpse bursts briefly back to life, its great tail thrashing and sending gouts of water splashing upwards, then more, their ghastly heads rising monstrously from the stream's flow as they gulp and gasp through their last nervous impulses, triggered perhaps by a human shadow falling across their eyes, and sparking final bursts of energy from almost entirely spent muscles.

Standing in a stream surrounded by the dead and dying remains of once great salmon, returned from their lives at sea, one might easily become maudlin, even depressed by the apparent futility of life's struggle, but instead they are symbols of life to come, of rebirth. These anadromous fish, born in freshwater but adapted to living their lives in saltwater at sea, undergo a great physiological transformation to return to the freshwater river of their birth. Here they fight: the females to scrape out a redd, a shallow depression in the gravel stream bed undisturbed by other females, the males to gather females to them, to possess them in a mini-harem. The female lies on her side to release her reddish-orange eggs like precious pearls into the gravel, while the possessive male ejaculates his milt over the eggs to fertilize them before the female sweeps gravel over them with her tail. This pattern of fighting and fecundity is repeated over and over until the female has exhausted her flesh while laying several thousand eggs, and the battered and bruised male has no more fight left in him.

As the year wanes to a close, the very last of the running salmon thrash their way upstream. On the Shiretoko Peninsula, earlier autumnal spawning salmon are savoured by Brown Bears before the latter hibernate, while more widely in east Hokkaidō later-running fish are still present in rivers when Steller's Eagles arrive in early winter. Then, certain river mouths attract fascinating congregations of eagles, gulls and crows. Yet even farther south, in Honshū and Kyūshū, where neither bears nor eagles now live to take advantage of the bounty of dying salmon, their flesh is not wasted. The remains of the adult fish sink to the riverbed, and in the cold autumnal and winter water they break down slowly. They release their nutrients back into the freshwater ecosystem, providing food for riverine invertebrates, which in turn will form the food for the next generation of riverine fish.

As part of the new year's rebirth, the salmon's pearly eggs hatch into a new generation of tiny fingerling parr; these make the river their home, feeding off invertebrates that may have themselves benefited from the nutrient 'fertilizing' of adult salmon bodies, in part of nature's endless recycling system. Then, from several months to three years later, as silvery smolt salmon, they make their way to the sea, where they live off the riches of the ocean for several more years while growing into large, magnificent adult salmon.

Throughout each autumn and on into winter, the returning hordes of mature salmon gather in spectacular shoals just offshore, awaiting the physiological shift that allows them to switch back to living in freshwater after a life spent in saltwater. Their arrival upstream heralds an enormous bounty of energy and nutrients carried back from the sea; they deposit it, in the form of their bodies, in their natal rivers, and so, as part of the ceaseless cycle of the year and of life, they contribute their own living matter to the freshwater ecosystem, fuelling it with their own energy, and allowing the next generation to thrive.

Six species of salmon occur in the west Pacific waters around Japan: Pink or Humpback Salmon, Chum or Dog Salmon, Sockeye or Red Salmon, Coho or Silver Salmon, Chinook or King Salmon, Masu or Cherry Salmon, and Steelhead Trout. Thanks to a range of artificial-incubation projects, the numbers of salmon in Japan grew to staggering proportions after having been greatly reduced. In 1983, an all-time record was set when 33.3 million Chum Salmon alone returned to Hokkaidō and northern Honshū, this being more than twice the peak catch of the late 1800s. These days, catches are declining, the many adverse factors cited including overfishing at sea and, especially, changing seawater temperatures leading to fish stocks adjusting their breeding ranges.

Chum Salmon, having re-entered freshwater, have only one purpose, namely to battle until they can breed [KS].

Both Steller's and White-tailed Eagles take advantage of the bounty that awaits their arrival in Hokkaidō in late autumn [SP].

Cherry Salmon bring colour back to their birth rivers [KS].

Pink Salmon showing why its other name is Humpback [KS].

The Angels of Winter

Whooper Swan [YaM]

Whooper Swan, the most wide-ranging of all the world's swans, is a common winter visitor to Japan, particularly to northern regions of Hokkaidō and northern Honshū. The arrival of the early autumn icing, weather that can trap unfledged cygnets or even isolate them from their food supply, drives these swans south from their boreal breeding grounds in Russia. The swans pass over Sakhalin to reach Hokkaidō ahead of autumnal snow squalls. Family groups linger through the coldest, bone-numbing months of winter; they seem at home along freezing rivers and lake shores, even in ice-filled bays, often gathering where geothermal activity provides partial ice-free access. There they swim amid rising mist and appear ethereal and angel-like.

Gracefully elegant and seemingly angelic, Whooper Swans can also be boisterous and aggressive. This is mostly in the form of noisy clamouring, neck-stretching displays and jousting thrusts with their long necks and bills, although this can escalate to combat when a family's feeding patch has been breached. The males then grasp with their beaks, thrash with their wings and kick with their feet, sending water spraying and, eventually, their opponents fleeing. Male and female then turn towards each other and perform an emotional display combining success with relief. They have strengthened their own bond and successfully defended their family against intruders. In this way, there is a pecking order among families, pairs and singletons, with those with larger families usually dominant. Where food is found or given in great plenty, these same swans may put aggression aside and gather so closely that their creamy-white wings and backs are arrayed like densely arranged pillows of snow, no water showing between them. They gather in flocks when grazing on land, too, and gain safety benefits by being part of a crowd. While feeding, each swan occasionally pauses to lift its head and look around, and the randomized way in which this happens means that some flock members are always vigilant. As a consequence, they all benefit; the group vigilance makes it difficult for a predator to surprise them and it is easy for the flock to respond to any potential danger, human-related or otherwise.

Numbers of Whooper Swans in Japan have increased over recent decades, from about 7,000 in 1982 to more than 32,000 by 1999. The wintering Whooper Swans leave Japan as the snowpack melts, before true spring takes its grip, to be ready on their breeding grounds when the thaw arrives.

Although Whooper Swan is a well-studied species, there is always more to learn, as I discovered when I witnessed a superbrood of Whoopers for the first time one winter. One pair of swans was being followed by no fewer than 16 cygnets, and this seemingly was the first record of brood amalgamation

Whooper Swans return each winter in large numbers to northern Japan [MAB].

for a species that typically averages 2·5 to 3·2 cygnets per pair in winter. Judging by the plumage of the various young in the 'gang' of 16, they were probably derived from at least three separate broods, but why were they all together and how had those particular parents ended up with a crèche?

Almost exclusively a winter visitor to Japan, Whooper Swan has on occasion bred here. The isolated cases seem to involve an injured bird unable to migrate paired with a healthy partner, and young have been raised successfully.

Our knowledge relating to Whooper Swans continues to grow, for these large birds are easily watched, relatively approachable and easily studied. Nevertheless, much remains to be learned, ranging from the esoteric (do you know that many of them have blue eyes?) to the essential (which migratory resting sites are vital to them and hence important to conserve?). But the very best way to watch them is from a piping hot *onsen* bath as the swans swim through the morning mist rising from the edge of a sparkling, snow-covered frozen lake in east Hokkaidō.

Three species of swan are regular in Japan. Two, Whooper Swan and its smaller cousin Bewick's Swan, are abundant winter visitors, while Mute Swan has been introduced to city parks and castle moats throughout the country. Where Whooper and Bewick's Swans have a yellow-based black bill and a rather straight neck, Mute Swan is readily identified by its sinuous neck and black-based red-orange bill complete with a black knob. Size and bill pattern separate Whooper from its close relative, Bewick's Swan. The latter has a more rounded head, a blunter profile and a beak that is largely black with a variably shaped yellow patch at the base. Whooper Swan has a longer, more imperious profile, and the yellow of the bill base extends forwards into the black in a long wedge along the sides of the bill and past the nostril. Bewick's Swans have very variable bill patterns. Among Whoopers there is also individual bill-pattern variation, but it is far less conspicuous than among Bewick's. In general, the smaller-bodied Bewick's Swans winter farther south and west in Japan than the larger-bodied Whooper, many of which remain in Hokkaidō and northern Honshū over winter.

Bill patterns and head-staining are individually variable characters [MAB].

Bewick's Swan is also a common winter visitor, but mostly to western Japan [MAB].

Winter swan flocks, helped by local swan-feeders, are important tourist attractions and draw photographers from around the world [MAB].

Introduced Mute Swans are most likely to be encountered in city parks, but now also frequent various lakes [MAB].

Red-crowned Crane – Symbol of Happiness and Longevity

If any northern creature is symbolic of Japan it is the Red-crowned Crane. You would be forgiven for believing that it is one of Japan's commonest birds. In symbolic form it certainly is. From saké bottle labels to wedding kimono, from elevator doors to chopsticks, many different things are decorated with them, while Japan's innumerable shrines are frequently draped with thousands of origami cranes, like colourful lei. Once it was even in every wallet, etched to perfection on the reverse of the old ¥1,000 bill. Known as *Tanchō* in Japanese, it seemed the perfect choice for this as not only is it an ancient symbol of happiness, but myth tells us that it lives for a thousand years.

Tanchō is a symbol so powerful that it shouts Japan, almost as boldly as do images of sacred Mt Fuji. But there is a paradox here, for in Japan image and reality are often at odds and perhaps no more so than in the case of this sacred crane. It is common as an image, yet few Japanese people have ever seen the real thing. Well, there is good reason for that. Early last century it was believed to have become extinct.

The cranes were pushed to the verge of extinction during the late 19th and early 20th centuries through being hunted for their meat. Miraculously, a handful survived in east Hokkaidō, and once they were protected, and food provided for them during the winter to help the survival of the young cranes, their numbers began to increase slowly. The advent of a winter feeding programme in the 1950s enabled more and more to survive the harsh northern winters. Now, in 2021, the population stands at 1,500–1,750 birds, and they are spreading westwards and northwards. A few pairs have even reached the northern tip of Hokkaidō. Red-crowned Crane's current population is deemed sufficiently secure for the Ministry of the Environment to scale back the winter feeding programme, although this hallowed bird still remains endangered by threats to its wetland breeding habitats.

If any place is typical of the crane's year-round lifestyle, then it is Kushiro Shitsugen National Park. This enormous peat swamp was designated a protected wetland because of the species' presence. During summer, the cranes disperse widely, occupying traditional nesting grounds in large breeding territories amid the marshes of east Hokkaidō, such as that in Kushiro, where they will raise a single chick, or two if they are lucky. In the lush green of the summer reedbeds even these tall birds are well hidden, but come autumn, as the waters of the marshes cool, the cranes forsake their breeding grounds in search of other food along streams and marsh edges, and then on agricultural land.

Red-crowned Crane survives some of the harshest winters in Japan [WM].

Left, top Red-crowned Crane courtship displays can occur at any time during winter, but increase steadily towards spring [MH]. *Left, bottom* They invariably produce two eggs, but often raise only one chick [MAB]. *Right, top to bottom* For some years courting cranes even appeared on the reverse of the ¥1,000 note, based on an image by renowned photographer Hayashida Tsuneo [MAB]; cranes are depicted on gorgeous silk wedding kimonos as symbols of happiness and long life [JW]; and are used widely in advertising, promoting subjects ranging from airlines to saké, such as the barrel shown here [MAB].

Red-crowned Cranes roost in the Setsuri River, Tsurui Village [WaM].

As autumn turns to winter the crane families gather into ever larger flocks, sometimes of more than 100 birds. These can be watched at close quarters at three major sites north of Kushiro City, two of these in Tsurui Village and one in Akan Village. On winter days from December to March, the cranes can be seen displaying. Almost man-high, and with a 2.4-m wingspan, *Tanchō* is long-lived, slow to mature, slow to breed. The epitome of grace, its winter courtship displays in the snow are moving to watch. As the birds dance, they call, twirl and pirouette with one another in the snow. By duetting and dancing they rekindle their pair bonds and prepare for the breeding season ahead. And as they dance they resemble graceful, monochrome snow ballerinas, creating a wonderful winter spectacle on the fringe of Kushiro Marsh.

Northern Hokkaidō

The northern tip of Hokkaidō, in sight of the Russian island of Sakhalin, is remote and rich in wildlife. Not surprisingly, then, it is largely protected as the Rishiri–Rebun–Sarobetsu National Park. The coastal meadows in the Sarobetsu area, and the surroundings of the shallow lagoons in the coastal plain are not only alive with bird song, but also pretty with wildflowers during summer; the yellow-orange lilies, white cotton grass, white rhododendrons, purple irises and almost black fritillaries are all highly photogenic. Just 20 km offshore from this area the startling volcanic peak of Rishiri Fuji (1,719 m) appears dramatically, rising straight from the sea. Its partner island of Rebun is lowly (490 m) in comparison, but this, Japan's northernmost island is renowned as the 'isle of flowers'. These islands are ideal for walking, hiking and birdwatching, with coastal fishing villages, flower meadows and mountain trails, and migratory birds passing through.

Travel a little farther south and two more islands, Yagishiri and Teuri, appear off Hokkaidō's west coast. Teuri hosts an extraordinary wildlife spectacle.

Facing page At the end of each winter's day Red-crowned Cranes fly to roost at a local river [JW].

Teuri-tō – Seabird Island

Teuri Island lies off the western coast of Hokkaidō [TT].

Picture this: darkness; a strong wind blowing; fog thickening and swirling; the ceaseless crashing of waves against a rocky shore. Not the kind of night for nature observation you may think; but think again. At the southern cape of Teuri Island, known as Aka-iwa, there is a scene of mayhem and madness. Teuri-tō is a small island, barely six kilometres long and only about a kilometre wide. It is situated an hour and a half by boat west of the small harbour town of Haboro, which is just three hours north of Sapporo by car. Along its eastern side, facing the mainland, Teuri Island is low, sloping to the rocky shore, but towards the south the island rises gently towards the southern cape at Aka-iwa. From east to west the island rises, so that the treeless west coast consists of magnificent cliff scenery. From near the northern end of the island there are far-reaching views eastwards to the distant Teshio Mountains of Hokkaidō. Far to the north the beautiful, conical shape of Rishiri Fuji rises from the sea on the skyline.

Deep and ominous whirring sounds approach at speed out of the darkness. There are sinister thumping sounds, and irregular loud crashing and crunching noises. Only the sound of squealing tyres is missing for this to be a series of cars crashing off the road into the hillside. Instead, this is a 'puffin' colony! The erratic and irregular noises suggest that dodging and ducking are appropriate to avoid the threat of incoming missiles with poor guidance systems. This is what it is like at a colony of Rhinoceros Auklet on a typical night during the species' early-summer breeding season from April onwards. They may be adept at swimming (essentially underwater 'flight' as they use their wings for submarine propulsion), but in the air they have serious steerage problems. Their wings whirr almost to a blur to keep their heavy bodies in the air. For landing and take-off, they need the additional surface area that a fully spread tail and broadly webbed feet provide to give them sufficient surface area to remain airborne at low speeds.

Isolated and rugged, Teuri Island is home to one of the largest bird colonies in Japan [TT].

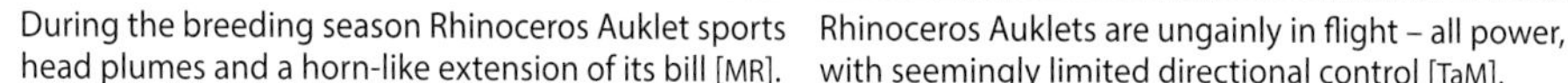

During the breeding season Rhinoceros Auklet sports head plumes and a horn-like extension of its bill [MR].

Rhinoceros Auklets are ungainly in flight – all power, with seemingly limited directional control [TaM].

Down below the viewpoint at the south end of Teuri Island, there are squadrons of Rhinoceros Auklets. These heavy-bellied, dark and dumpy shapes zoom by in lines of half a dozen to a hundred or so. They buzz along just above the waves, busily heading off to feed at sea. Their wings, honed by aeons of evolution to be effective as underwater sculls, seem only marginally airworthy, propelling these portly sea-parrots at high speed with limited manoeuvrability. During the day, they hurtle back and forth low over the sea, intent on reaching their distant fishing grounds by the shortest possible route – a straight line. At night, they return to their colony, perhaps identifying it by the island's black silhouetted shape against the night sky, or by other uncanny senses that birds use for homing. Their ability to know the location of their home colony seems undone by their poor control. They seem to bustle aimlessly in the dark, circling; then, once committed, they whirr groundwards, thumping into the turf on impact.

The auklets return from the sea flying at island height and home in fast and low and with little control just above their headland colony. They turn upwind (if they can) to use the extra lift, but woe betide anything (or any person!) in their way. Be ready for near misses and glancing blows; direct hits happen occasionally, too. With such poor control, the auklets simply crash-land, as close as they can to the area of the colony that they call home. The thumps and crashes are the sounds of their landing. There is no delicate touchdown for them as they alight, like a falcon homing in on a crag or a flycatcher on to a branch; instead the auklets rely on a hefty sternum, massively developed flight muscles attached to a solid keel (an extension of the breastbone), and a dense feather coat to lessen their crunching, chest-first impacts. After a brief bounce, they pick themselves up and stand upright on short greyish-yellow legs; stunned for a moment or two, they scrabble for purchase with their claws, and then they bounce up, running. These perky colonial clowns scurry off in a comical, hunched, semi-vertical posture, heading for their burrows. As they do so, they crash noisily through the dry remnants of last year's vegetation.

A scan across the hillside will reveal that it is dotted with dumpy, dark grey and white standing figures – those that are still stunned, and those that are readying themselves for take-off. Between them hundreds of other birds scurry at speed into the darkness, heading for their burrows while carrying food for their young. Three types of fish predominate in their diet, namely Japanese Anchovy, Japanese Sand Lance and Japan Sea Greenling, but it is the anchovy that makes up the bulk of the diet, typically over 60 percent and sometimes over 90 percent.

The auklet colony on Teuri Island, comprising as many as 400,000 pairs, is reputedly the largest in the world. The breeding birds sport two pairs of remarkable silvery facial plumes, one commencing over the eyes, the other starting at the corner of the bill. Their most distinctive feature, however, is their extraordinary beak. Although generally known as the Rhinoceros Auklet, the birds so closely recall Atlantic, Horned and Tufted Puffins in so many ways that the old alternative name of Horn-billed Puffin is far more appropriate. While many bird species have distinctive changes of plumage between breeding and winter coats, the puffins have a fascinating characteristic of changing their bill, and Horn-billed Puffin is no exception. All breeding puffins develop a brightly coloured outer bill

The hillsides at the south end of Teuri Island are riddled with auklets' nesting burrows [MAB].

Puffin-like, Rhinoceros Auklets return to their burrows carrying multiple prey [SuT].

sheath, or rhamphotheca, which falls off after breeding, leaving a smaller, duller-looking beak. During summer the base of the orange-yellow bill develops a grey-white protrusion that juts forward bluntly, just like a rhinoceros's horn. It is this horn that is shed outside the breeding season. The true function of this adornment remains a mystery, although it seems likely that it may serve some purpose during displays to attract females, or in competition between males. My own particular speculation is that it may serve as some kind of aqua-dynamic device. These birds dive for their food, and during one trip are able to catch several fish, the heads and tails of which dangle from each side of the beak. The protruding portions of the fish must cause considerable drag under water and make their underwater 'flight' less efficient. Could that protrusion act like the bulb beneath the bow of Japanese fishing boats? Could it somehow cause a broader bow wave and reduce turbulence around the jutting fish heads and tails?

The experience of visiting a Horn-billed Puffin's colony differs in only one respect from that of visiting any of the other puffins' colonies. Other puffins visit their colonies during the day, whereas the Horn-billed Puffin visits its breeding grounds at night, when all terrestrial activities occur.

Like all seabirds, the Rhinoceros Auklet (to revert to its 'official' name) faces a single massive constraint on its lifestyle every year. All its needs, bar one, are met at sea. There it can forage underwater, feed on the surface, and rest while bobbing on the waves. Thanks to its possession of an extraordinary gland, it is even able to drink seawater. This gland is the auklet's own mobile desalination plant, making life at sea possible. The only thing it cannot do at sea is breed. Like other seabirds, Rhinoceros Auklets are forced to return to the land of their birth each year in order to nest. Only coastal sites, and usually only small offshore islands, supply their twin needs of suitably accessible real estate and a safe neighbourhood. With many different seabirds needing to return to land at the same time, and with such a limited supply of suitable locations, there is considerable pressure on nesting space. The situation is made worse by the fact that they must all breed at much the same time of year. Fierce competition between species has led to the division of nesting sites into distinct neighbourhoods, and the Rhinoceros Auklet is the species that has staked its claim on the upper slopes of the cliffs. There, where soil has accumulated to a sufficient depth, they excavate their burrows.

Puffins are burrow nesters, and this 'puffin' is no exception. The southern cape and the southern half of the western side of Teuri Island provide stunning scenery to the human eye, but to a burrow-nesting seabird they offer the important additional ingredients of thick vegetation and deep soil. The birds dig tunnels up to 2 m into the clifftop soil and excavate a nesting chamber at the end, so that between them the hundreds of thousands of birds move an inordinate amount of soil around each year.

The earliest auklets may return to their breeding grounds in February to stake their claim on a burrow and begin refurbishing work. At the end of the burrow a broader scrape of earth may be lined with local vegetation, and here the single egg is laid. The parents make a massive investment in their solitary egg, the incubating of which takes 40–50 days, and another couple of months are required for the chick to fledge. By mid-August, the colonies are becoming deserted again.

Spectacled Guillemot is a close relative of Black Guillemot and Pigeon Guillemot, but with a much smaller range than either: it is restricted largely to the Sea of Okhotsk; its red legs, red lining to the bill and white face patches are in stark contrast to the dark plumage [*left* SP; *right* TT].

Once abundant, Common Guillemot is now a rare breeder in Japan; decoys such as these (*right*) are placed on abandoned nesting cliffs in the hope of tempting birds to return to breed [*left* TaM; *right* MAB].

Teuri-tō is situated off the west coast of Hokkaidō about halfway between Sapporo and Wakkanai, but the birds are said to feed at sea as far away as Shakotan to the southwest and beyond Cape Sōya to the northeast, giving them daily one-way commuting distances of perhaps 150–200 km. In a few weeks their eggs hatch; from then onwards the adults return, their beaks filled with tiny fish, in a frenzy of feeding visits to help their chicks to survive and grow during their first weeks of life. Eventually the chicks become large enough to be tempted out to the entrance of their burrows. While this facilitates parental feeding, it also places the chicks at greater risk from the predators of the colony, the gulls, which are eager to grab a beakful of fish from the hard-working parents, and to seize a chick when the opportunity arises.

Auklets, like many other seabirds, are under severe pressure. We humans have introduced to their nesting islands alien land predators, such as rats and cats, which eat their eggs and their young, and sometimes even the adults themselves. We have also stepped up competition for their food, as fishing fleets mine the oceans of their fish stocks. Nets set in the ocean are dangerous barriers for diving seabirds. Fishing nets worldwide are implicated in the drowning of large numbers of diving birds and marine mammals. Alcids and other seabirds have become our 'miner's canaries' for the health of the oceans.

Common Guillemot, or Common Murre, once a common breeding species around Hokkaidō and on Teuri Island, has declined almost to zero, despite great conservation efforts. Thankfully, another species, Spectacled Guillemot, still survives in good numbers, and on calm days their high-pitched thin whistling cries can be heard below the cliffs. These smart sooty-black seabirds, with white eye-patches and brilliantly red feet, have declined around much of Hokkaidō. Teuri-tō is something of a stronghold for them, although they, too, are another indicator of the health of our now plastic-laden oceans.

Southwest Hokkaidō

Situated to the southwest of the island's capital of Sapporo is the two-section Shikotsu–Tōya National Park, named after its two spectacular volcanic caldera lakes, and offering some of the most dramatic scenery in the country. The surroundings of Lake Shikotsu are dominated by the rugged peak of Eniwa-dake (1,320 m) to the north, and the recumbent forms of Fuppushi (1,103 m) and Tarumae-san's (1,038 m) cinder cone to the south. Forty kilometres farther southwest is Lake Tōya, which has a large central island. Nearby stands Japan's youngest volcano, the Shōwa Shinzan volcanic lava dome, which grew between 28 December 1943 and September 1945, situated beside active Mt Usu. Meanwhile the dominating feature on the western horizon is the magnificent 1,893-m high Mt Yōtei, or Ezo Fuji, which translates as Hokkaidō Fuji. This graceful, symmetrical stratovolcano with a 700-m wide summit crater and deeply eroded flanks last erupted 5,000–6,000 years ago. The whole region of the national park is like an open-air classroom of vulcanology.

Hokkaidō's southernmost park is the Ō-numa Quasi National Park. Located on the Oshima Peninsula of southwest Hokkaidō, the park encompasses the Komagatake volcano and lakes Ō-numa, Konuma and Junsai-numa. The scenic landscape of the three lakes, home to more than 120 islets, forms much of the Ō-numa Quasi National Park. They provide welcome relief to the eye after the unzoned sprawl of Hakodate and Sapporo cities. The diverse mixed forests surrounding the lake are home to Hokkaidō's typical large mammals, including Japanese Deer and Red Fox, and the lakes support a range of waterfowl species, particularly during the spring and autumn migration periods. In summer, the forest floor and lake shore are carpeted with wildflowers, rhododendrons are in bloom, and the songs of cuckoos, thrushes, flycatchers and warblers can all be heard, while overhead Latham's Snipe perform their bizarre display flights.

The forested surroundings of Ō-numa provide a beautiful setting to the lakes year-round, and are outdone only by the graceful form of Mt Komagatake (600 m), which provides a stunning backdrop to the north. There is an easily followed trail to the upper mountain, which is like a natural alpine flower garden, providing views across the landscape of southwestern Hokkaidō from Ezo Fuji to the north and Mt Hakodate to the south. The view to the south from Mt Hakodate is across the Tsugaru Strait to the northern capes of Honshū, Japan's main and most populated island.

Lake Shikotsu (*distant right*) in one of two large calderas of the Shikotsu–Tōya National Park [MAB].

Alpine flowers, such as Clubmoss Mountain Heather, can be found near the tree line [MAB].

Once the ownership of burrows has been settled, the order shifts, and when it is time to hibernate the adult females begin first, usually at some point in September or October, depending on the altitude and the weather. While the onset date may vary from year to year, the order is fixed. Only after adult females have entered and sealed their hibernation burrows will the adult males enter theirs, followed by the juvenile females and, finally, the juvenile males. It is unusual to see chipmunks above ground from November onwards, and once they have entered and plugged their burrows they will not emerge until spring, up to six months later. In Hokkaidō, the males emerge during the first half of April, up to 20 days before the females. On emerging from hibernation in spring, they are often to be seen dashing across the snow layer from tree well to tree well in search of food around the exposed bases of the tree trunks. There are good reasons why the orders of selecting burrows and of actually hibernating are different, and sex is at the heart of it. In spring, the males emerge first, to be ready to mate with females as they emerge during April. Breeding may occur once or twice during spring and summer, the females producing litters of 3–7 young (average 5), and these young become independent in the season of their birth.

As with the cavities and holes used by nesting birds, there is competition for burrows among chipmunks, and adult chipmunks are better able to find and occupy burrows than are youngsters. Between selecting their burrows for the winter and actually hibernating, each chipmunk must store up food and nesting material to last at least six months underground. To gather in sufficient stores, the animals carry supplies to their burrows up to 40 times a day. In Hokkaidō, one of their favourite foods for hoarding is the acorn. The large size of these oak seeds means that the chipmunks have to make fewer trips to build up a good store. The males, because they emerge early, before there is much fresh vegetation available in the forest, must also hoard food in other places, to be eaten in the spring. Where they have been introduced in parts of southern Japan, it is likely that the warmer climate allows their period of hibernation to be shorter, and their period of activity each year to be longer.

At home from the tundra-like alpine zone of Hokkaidō's highest mountains to coastal dune forest, Siberian Chipmunk is widespread but nowhere common. Asleep for more than half the year, it spends much of its waking time in gathering and storing food, carrying food in its cheek pouches to cache it in an underground burrow [*middle* MR; *others* WaM].

There is good reason why the hibernating males go to sleep late and rise early. Males live in large home ranges that overlap those of up to ten females. In autumn, each male keeps watch on each of the females in his range, and only after he has confirmed the location of the burrows of 'his' females will he himself hibernate. Then, in spring, the males emerge early so that they can check again on the females' burrows and be there when they emerge. Each female will mate only on one day each year, about ten days after emerging from hibernation. So, with the chances of mating limited to such a very brief period, it is important for males to know where their females are.

If you think that sleeping for up to six months each year is a wasteful approach to life, consider the fact that early relatives of today's chipmunks first roamed the earth 25 Mya. Sleeping in is a strategy that works!

The Cutest Creature in the Woods – The Siberian Flying Squirrel

The forest dusk descends slowly like a steadily thickening curtain of anti-light, creating a bewildering interface between day and night. Creatures of the anti-light slip through the crack in our dimensions, emerging like an undefined colour in our limited spectrum. These creatures are crepuscular, their light-loving cousins are diurnal and their light-shunning relatives are nocturnal. Their activities bookend both day and night; they favour the times of changing light levels, choosing then to emerge and explore their own realms, to forage and to lay claim to their territories.

Colour and texture fade from trunk and branch, from limb and twig, as light levels fall towards darkness. A perceptible dimming of the bulb of day is seen first as a softening of all that surrounds us; the blue dome of the sky turns grey, the forest, so deep and multi-layered, begins as a multi-textured brown, grey and green, but shifts imperceptibly towards a flat pen-and-ink line-drawing of itself in silhouette. Then, once all colours have gone, only faint outlines linger, strongest in our peripheral vision, and best seen against moon or starlight. Turn to look at something and it seems magically to dematerialize; the trick is to glance sideways, as things on the periphery of our vision linger longest. Or is it just that they are easier to imagine?

The chuckling song of a river muffles the impact of dusk. As twilight spreads, sounds seem to shift and change; for a while they seem softer, like muted voices in a cathedral. With a final brief chatter from a disgruntled-sounding Eurasian Wren as it heads to its roost in a tangle of branches beneath a fallen tree, the day shift falls into a hushed silence. The night shift has yet to emerge on to the darkened forest stage. Only the river breaks silence – as if gently laughing at itself. After a prolonged hush, an owl hoots sonorously; its mate responds. Off in the distance, a dog fox yaps: at once a piercing warning to other males, and a welcoming to any nearby vixen. The dwarf bamboo, or *sasa*, that covers the forest floor is an unwelcome sound barrier which crackles and snaps when a human pushes through it, yet the fox seems to shift its shape ethereally around the *sasa* and achieves a silent passage. Perhaps the same cries from owl and fox cast fear into the minds of voles and mice, causing their hearts to race and forcing them to pause in dread of the silent descent of talons, claws and teeth. The river mocks on.

A tree trunk, stark against the slight moonlight, seems suddenly to undergo a slight rippling, presaging movement, but not breeze-driven. There it is again, a slight brief deformity in the smoothness of the trunk's silhouette. An almost subliminal flicker of action occurs in that still rift between night and day. It is so hard to resist turning to focus on it, yet so hard actually to locate it. No sounds carry over the chortling noise of the river as it forces its way over and between rocks, yet it seems that a small shape has skittered and scurried skywards up the trunk; in an instant it is gone, leaving only the impression of the scritch-scratching of its claws on the bark at the entrance gate of night. Glimpsing the creatures of the anti-light before human eyes are overwhelmed by darkness seems a matter of good fortune.

By night, the forest in central Hokkaidō is mysterious, a black bewildering void, filled with unidentified sounds; at dusk and dawn the forest comes alive with mammals on the move and birds beginning to call and forage; by day the forest is a labyrinth of interlocking branches, tangles of fallen trunks. Drifts of sound now flood the early-morning forest, in waves of familiarity: here a Eurasian Nuthatch, there an

Siberian Flying Squirrels emerge from cracks, crevices and woodpecker holes soon after dusk and like a flickering will-o'-the-wisp, they glide through the dusk [*left* MAB; *right* IiM].

Oriental Greenfinch, and in the distance the soft buzzing call of a Japanese Pygmy Woodpecker. The day shift is awake now and out in full force. Searching by day for the creators of those 'subliminal flickers of movement' is a task that I have set myself; it is a challenge, but one with tremendous rewards, not least being a collapse in my vocabulary, as revealed in the following paragraph.

Finally, as a scurrying, pale grey shape emerges suddenly from a tree hole, I glimpse a wisp of its tail. Then, I lock on to enormous limpid eyes. Like mysterious, shining domes of night, the huge eyes resemble rounded mirrors, reflecting the whole scene. All of the forest is somehow held there in its eyes: the sky, the tree on which it rests, the woods, and me on the forest floor. It is as if this one creature is an interface between dimensions. Its short pink nose and long whiskers, its tiny hands and claws, the wrinkled roll of fur along its sides and its broad tail arched upwards and flattened along its back, all conspire to defeat the scientist in me, overwhelm the naturalist in me, and connive to conjure up just one adjective in my mind – *Kawaii* – Cute! I have lost count of the number of articles I have written over 40 years, but you could probably count on one hand the number of times I have used that word in print before; it is not a normal word in my vocabulary. Now I am a convert. Any field guide that does not describe Siberian Flying Squirrel, or the closely related Japanese Flying Squirrel, as 'cute' is missing a primary attribute.

The scratchy sound of a Siberian Flying Squirrel as it leaves its roosting cavity and scurries up towards the treetops is distinctive. Dashing out along a tree limb, it launches itself into oblivion. Reaching out to all points of the compass with its arms and legs, it stretched taut its patagia, a kind of wing membrane, and sky-sailed! What had appeared earlier as a wrinkled fold of fur along its sides became a smooth gliding surface, grey above and white beneath, and now I understood why its tail was also so flat – it, too, was part of the gliding surface. The distance over which the flying squirrel flew astonished me. I had expected a precipitate drop to a nearby tree, but instead it flew a long, almost level glide, drifting slowly downwards towards another distant stem. There it sat awhile, surveying its forest territory from a burl on the trunk; then, in brief scurrying movements, it made its way towards a crack in the bole. It disappeared into the crack head first, but soon popped out again from another hole to survey its home patch from this new vantage point. As it turned and disappeared for a final time into its resting hole, I could not help but comment to myself: 'Definitely cute!'

This small, grey-with-a-hint-of-brown, flying squirrel is a nocturnal and arboreal indigenous inhabitant of native broadleaf deciduous and coniferous forests from the lowlands to the montane zone across Hokkaidō. Its underparts, from chin and throat to lower belly, are creamy white. It has a rounded face, with distinctively large, bulbous eyes appearing disproportionately large for its head size, and has short rounded ears. The tail is long, flattened and about three-quarters of the body length. When settled, it appears to have a ruffled two-tone flap of skin and fur along its flanks. This is the retracted flying membrane, which extends between fore and hind limbs. The flying squirrel weighs just 81–120 g and measures a mere 15–16 cm long, with an additional 10–12 cm of tail. These animals emerge from their roosting sites at dusk, and climb higher into the tree; then, they launch themselves off, using their flying membrane to glide down to another tree and so off around their territory in search of edible leaves, buds, flowers, seeds and nuts of deciduous broadleaf trees.

The species occurs across northern Eurasia from Scandinavia eastwards to the Korean Peninsula and northeast Russia. In Japan, it lives only in Hokkaidō, though nowhere is it common. During the breeding season, males have large home ranges (up to 2 ha) and females[1] have small ones (of about 1 ha). Siberian Flying Squirrels occupy natural tree cavities or disused woodpecker holes as breeding sites and daytime roost sites. They breed twice a year, the females giving birth to litters of 2–6 young in spring and summer. These young animals may then survive for up to five years in captivity, but probably live for only three in the wild.

Siberian Flying Squirrel's bulging eyes seem to reflect the whole of their woodland world [WaM].

Siberian Flying Squirrels occasionally emerge in daylight, perhaps for a little solar warming [IiM].

Japanese Pygmy Woodpecker is not merely Japan's smallest woodpecker but also an important primary cavity-producer [ImM].

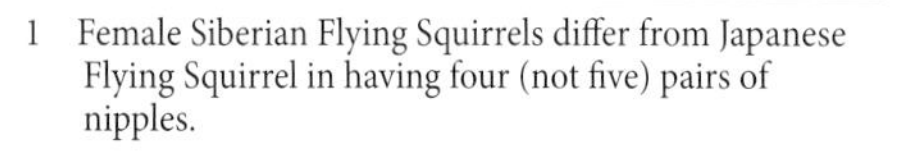

1 Female Siberian Flying Squirrels differ from Japanese Flying Squirrel in having four (not five) pairs of nipples.

Japan's most voracious predators

Japanese Marten [YaM]

There is movement in the snow – the snow layer bulges, bursts, and out pops a creamy-faced creature with rounded black eyes like tiny beads and a stare that seems to say: "I can kill". Powder snow dusts its head as it stares around. Then it has gone, like a periscope submerging into the snow, the face disappearing in an instant. A few metres away it pops up again through the snow, first the face, then the head and neck, and then, finally, the whole creature emerges. Lithe and long with short stout legs and a long bushy tail, it has as soft and as gentle a face as you can imagine, one that belies its true nature. Its tiny black button nose, black beady eyes, short rounded ears and tapered face give it a fox-like facial appearance, but that horribly overworked word 'cute' rises unbidden to describe it.

'Cute' is not the half of it. This creature is a Sable, and it belongs to the mammal family Mustelidae, the largest family in the Order Carnivora. Yes, this sweet-faced creature is a predator, and its relatives include various species of otter, badger, weasel, stoat and mink. Its diminutive relative, Least Weasel, has been described as the most ferocious predator of the British Isles, and the stoats, martens, minks and sable rank alongside as highly active pursuit predators of the undergrowth. Like high-speed heat-seeking missiles they home in on hot-blooded prey such as mice and voles; woe betide any '*Wee, sleekit, cow'rin, tim'rous beastie*' that pauses even for a moment in their sights.

Japan is well blessed with ten species of mustelid, although not all of them are native. Japanese Marten ranges widely in Kyūshū, Shikoku and Honshū. Sable lives in Hokkaidō. Japanese Weasel is a little smaller and lives in the same region. Siberian Weasel, a continental species that used to occur in Japan only on Tsushima, has been introduced and spread through Kyūshū, Shikoku and southwest Honshū. Least Weasel, the same species as that which terrorizes British hedgerows and woodlands, is found in Japan only in northern Honshū and Hokkaidō. Stoat or Ermine occurs from highland central Honshū north to Hokkaidō, and, like Least Weasel, it is present also west to Europe. It is best known for its white winter coat and contrasting black tail tip. Then there is American Mink, a voracious predator introduced to Hokkaidō – and one can only wonder how many Red-crowned Crane chicks have been lost to this wetland-loving species. Because of its semi-aquatic habits and size, the mink is occasionally

Sable is found only in Hokkaidō [WaM].

mistaken for Eurasian Otter, but that species has been extinct in Hokkaidō for perhaps a century. Like Eurasian Otter, Sea Otter was exterminated by hunters for the fur trade. There is, however, an expanding population in the Kuril Islands which has begun to extend its range southwards, and now Sea Otters are regularly being sighted off the Shiretoko and Nemuro Peninsulas once more. Japanese Badger is the final member of the mustelid family in Japan. Like Japanese Marten and Japanese Weasel, it occurs only in Kyūshū, Shikoku and Honshū, but not Hokkaidō. Unlike them it is not a long, lithe predatory creature but a stout, more heavy-set omnivore, which hibernates during the winter.

Japanese Marten and Sable are counterparts in Japan; whereas the marten lives in Japan's southern three main islands, Sable lives only in Hokkaidō. To hibernate or not seems to be a question that has not bothered the majority of the Mustelidae, with the exception of the badger. For the predatory species it simply is not an option; they burn so many calories in just staying alive that they need to top-up regularly with fuel – hence the snow-burrowing antics of Sable in winter in search of prey. Its thick winter coat has been its undoing in the past, and its willingness to enter holes, nooks and crannies in search of its rodent prey has meant that it has fallen prey itself to human hunters, ever ready with noose, trap and snare.

Sable hunts in a very particular fashion, for hunting rodents beneath snow is a specialist art. It can burrow, almost snake-like, twisting and turning through the snow in pursuit of rodents living in the subnivean layer – the interface between the ground and the bottom of the snow layer. Then again, above the snow layer, it will even chase down animals larger than itself, such as Mountain Hare, although the hares have a good chance of escape because, with their large hind feet and densely furred toes, giving them broad surfaces like mini-snowshoes, they can skim fleet-footed over the snow.

Scent is a vitally important cue for predators such as the marten and Sable when seeking food, which is why they are occasionally attracted to birdtables or to carrion. Their speed, agility and antics are entrancing. Sable chitter and mewl to each other, dashing back and forth, and chase each other

Japanese Marten, of Honshū, is a close relative of Sable [MM].

Japanese Weasel is endemic to Japan's main islands south of Blakiston's Line [FT].

Least Weasel, one of ten mustelid species in Japan, becomes fully white in winter [IiM].

Japanese Badger is a Japanese endemic found only in Kyūshū, Shikoku and Honshū [IiM].

across the snow, as likely to disappear up into a tree as down a hole into the snow. Aloft, they are nimble like squirrels, running along branches and leaping across great gaps between tree trunks. On the ground, they halt their headlong rush abruptly to stand up on their hind legs and peer about with bright, shiny eyes. These are creatures able to pursue and tackle prey, ranging from small rodents to heavy hares, while also sufficiently fast to chase down native Eurasian Red Squirrels as they dash about the forest canopy, and Siberian Chipmunks in the undergrowth. Their cousins farther south are similarly capable.

Where Japanese Marten is a golden-brown animal with a black face, black legs and a brown tail, the arboreal and terrestrial Sable is a pale, creamy-yellow colour, almost white on the face. It has a somewhat triangular-shaped head, broad across the skull, with short, erect slightly pointed ears, and a pointed muzzle, with prominent black eyes and a black nose tip. Its limbs are short but strong, the feet well furred, often black, and with strong, white claws. A Sable's tail is long and bushy. Males and females differ marginally in size, with males slightly larger than females; differences in proportions also make females appear slighter, and gentler. These two mustelid species are similar in size, the marten measuring 60–65 cm from nose tip to tail tip and weighing 1·1–1·5 kg, while Sable measures 67 cm in length and weighs in at 1·5 kg. It is Sable that smacks of wilderness and conjures up images of the immense Russian taiga forests that make up the majority of its natural range. Because of persecution throughout its continental range, the very best place to see this creature is in Japan, where it lives solitarily or semi-socially in extended families in mixed deciduous and coniferous forests from the lowlands to the tree line. It is not entirely carnivorous; although an aggressive rapid-pursuit hunter, it also eats invertebrates during summer, fruits and nuts in autumn, and seeds in winter.

The smallest relative of Sable in Japan is Least Weasel, which occurs only in Hokkaidō and northernmost Honshū. This species is dark chocolate-brown above, from head to tail, and white below from chin to vent. Its very short tail (2 cm) is often not noticed. Males (18 cm) are slightly longer than females (15 cm), but much heavier, weighing 80–100 g against the female's 50 g. Unlike the larger Sable, this tiny weasel is terrestrial and solitary, ranging from lowlands to mountains in a wide range of habitats from grasslands and pastures to montane forests, where it typically hunts voles, mice, birds and insects.

The inquisitive Sable often scans its surrounding in a standing posture and climbs trees to predate nests, even those of Blakiston's Fish Owl. Keen eyes, high speed and sharp claws allow it to prey on fast-moving shrews, voles and mice [*top* WaM; *others* MAB].

Shrews – Spirits of Darkness

Japanese White-toothed Shrew [YaM]

As winter approaches, Japan's two largest mammals, Brown Bear and Asiatic Black Bear (see *p. 162* and *p. 260*), seek to avoid the cold weather by retreating into torpor in a den. At the other end of the scale, Japan's tiniest mammals, the shrews (12 species), cannot afford to rest since they would freeze to death if they lay down to sleep for very long.

These superficially mouse-like mammals are unrelated to the rodents but, rather, they are insectivorous and belong in the family Soricidae. Considered to be among the most primitive living placental mammals, the shrews live right at the limits of survival for vertebrates. Among their number is the world's smallest mammal, and some record-breaking anatomy and physiology. These tiny creatures live life in the fast lane. Compared with us humans, they breathe at a blur, taking up to 800 breaths per minute (whereas we begin life taking 30–60 breaths per minute, although this declines to 12–20 breaths per minute in human adults). Shrews are right at the physical limits in terms of their heart rate, too: no leisurely 60–80 beats per minute for these creatures; at rest, a shrew's heart purrs along, beating an astonishing 1,000 times a minute. Although perhaps not surprisingly at that speed, they usually do not live for more than one or two years.

Dentition is a fair indication of diet, and in the case of the shrews their teeth indicate that they are predators. Shrews, like moles, have for long been considered to be in the taxonomic order Insectivora[1] and, as that name implies, their diet consists largely of invertebrates – a wide range of insects, such as beetles, grasshoppers, butterflies and moths, but also spiders, snails, slugs and earthworms. They are not averse to eating small birds, rodents, reptiles and even other shrews, and some species also consume seeds, nuts and roots. The dark pigmentation on the tips of the teeth is a deposition of iron in the enamel, which may make the teeth stronger and resistant to wear; presumably this is an advantage in a tiny creature that needs to eat every few hours and to consume as much as three times its own body weight in food each day. Because of their small size, shrews have a proportionally high surface-to-volume ratio and lose body heat rapidly, so they must refuel almost constantly to stay warm and alive.

The world's largest shrew, House Shrew of tropical Asia, is 15 cm long and weighs 45–80 g, and it occurs in southern Japan in the Yaeyama and Okinawa Islands. The smallest shrew is Eurasian Least Shrew, one subspecies of which, measuring only 4–5 cm long and weighing less than 2 g, has been found in northernmost Hokkaidō. And not quite a shrew, nor a mole, is the strange Japanese Shrew Mole.

1 The latest mammalian taxonomies based on molecular analysis have discarded 'Insectivora' as an order, as it proved to be an unnatural group. Currently the shrews and moles are considered to belong to the new and unfamiliar Order Eulipotyphyla.

The tropical Asian House Shrew is found only in the far southwest of Japan [IiM].

Japan's smallest mammal is the Eurasian Least Shrew [IiM].

Long-clawed Shrews only occasionally appear during daylight [MAB].

The endemic Shinto Shrew is found only on the islands of Honshū and Shikoku [IiM].

Not quite a shrew, nor really a mole, this is the endemic Japanese Shrew Mole. It is found across much of Japan, but not in Hokkaidō [IiM].

Japan's Seals and Sea Lions

Kuril Seal [YaM]

Considerable numbers of Northern Fur Seal, an abundant eared seal, migrate south into Japanese waters following the cold Oyashio current during the winter and spring months, from December to May. They can be encountered during those months well out to sea off the shore of the Sea of Okhotsk and southward along the coastlines of the Pacific and Sea of Japan almost to the latitude of Tōkyō and Sado. They migrate south from their breeding location on a surprisingly small islet off eastern Sakhalin, just to the north of Japan. Today's breeding population numbers 55,000–65,000 individuals, a population that is slowly recovering from historic overhunting. They also breed farther north in the Kuril Islands, the Aleutian Islands and around the Bering Sea. At home in cold waters, they adopt an odd jug-handle posture when resting at the sea surface, gripping a hind foot with a fore foot to create an arc as they rest at the surface and thermoregulate.

Northern Fur Seal is a medium-sized sexually dimorphic seal with a silvery-grey to dark, blackish-brown coat. The adult males weigh four to five times the weight of adult females, and have different proportions: the males have a massive neck and forequarters, while those of females are more slender. The male may weigh up to 210 kg and grow to 2·0 m in length, and is dark, blackish-brown, with brown tips to the guard hairs of its neck and head. The female may weigh up to 44 kg and grow to 1·3 m in length, and is paler, greyer above. Their pinnate limbs are hairless.

Northern Fur Seal has a broad head with a sharply pointed snout. The eyes and muzzle are black, with extremely long stiff bristles on the muzzle, while the ears are short and rolled, and, although visible, they lie flat against the sides of the head. The forelimbs are long, broad and modified as flippers, and very mobile, making the seals extremely agile in water. They are found within the cold current flowing south past northern Japan, or in areas where cold and warm currents mix and where food is abundant; there, they feed on various species of fish and squid.

From the mid-18th century onwards, the species was commercially harvested for its fur, causing many populations to be dramatically depleted. The fur is especially fine and dense and is covered with long guard hairs.

Although male Northern Fur Seals may reach 210 kg as adults, they are dwarfed by the mighty and impressive Steller's Sea Lion, males of which may weigh 1,000 kg. Steller's Sea Lion also reaches Japan in winter.

The winter sea-ice on the Sea of Okhotsk is the southernmost in the Northern Hemisphere. At its greatest extent, it reaches the shore east of Abashiri, the Shiretoko Peninsula, and flows erratically around

Seals and sea lions are best looked for around offshore islands, off rocky coasts and headlands, and in sandy bays [TaM].

The long, pointed snout, long whiskers and short rolled ears distinguish the Northern Fur Seal from the true seals, and from sea lions. Large numbers breed to the north of Japan and migrate into Japanese waters following the cold Oyashio current during winter and spring [BOTH NB].

Impressive mobs of Steller's Sea Lions can sometimes be found off the Shiretoko Peninsula in winter [KS].

Although typically nervous and shy on land, Steller's Sea Lion is curious when in the water [YM].

the tip of that peninsula and down into the Nemuro Strait between east Hokkaidō and Kunashiri Island. Sea-ice is a harbinger of deep winter; it contributes to the further chilling of Hokkaidō and brings with it a web of life ranging from macro-planktonic Sea Angels to marine mammals, including sea lions and seals. As the ice approaches day by day, birds, such as Harlequin and Black Scoter, seem to move ahead of it, and with it come the sea lions.

The tawny-brown Steller's Sea Lions may appear like a pack of eager dogs, rushing pell-mell in one direction before charging off excitedly in another. They can turn in an instant and dash and splash back again in a moment; they are adept at diving, swift and lithe in their movements, and frequently lunge forwards and up, craning their heads and necks out of the water to peer about. Then, they resemble enormous inquisitive otters. Watching them off the Shiretoko Peninsula World Heritage Site (WHS) is thrilling, but there are more than nature-watchers after them.

The unmistakable loud, sharp crack of rifle fire is not often encountered in justifiably gun-cautious Japan, but it reverberates across the waves, not unlike that of brittle sea-ice shattering. Rifle fire at sea is unusual, but you may trace it to a hunter in his harness, standing in the bow of a small hunting boat, braced against the kick of the waves and bundled against the chill of the biting wind. Hunting happens because the seals are considered pests for disturbing the fisheries, and the meat finds its way into restaurants and cans to be sold at souvenir stalls.

Large male Steller's Sea Lions, the largest eared seals in the world, reach lengths of 2·87 m. Males are massive, weighing up to 1,000 kg, and are mid to dark brown, while females are smaller, weighing up to 300 kg and measuring up to 2·27 m, and are a paler sandy-brown. Their pinnate limbs are hairless. Like the fur seals, this species exhibits extreme sexual dimorphism, adult males far outweighing the more slightly built females.

Steller's Sea Lion has a broad, flat-crowned head with a broad, blunt snout. The eyes and muzzle are black, the latter with extremely long stiff bristles. The ears are small, short and rolled, and although visible they lie flat against the sides of the head. The fore limbs are long, broad and modified as flippers, and are extremely mobile. The similarly modified hind limbs can be drawn beneath the body, allowing the animal to reach a lumbering run on shore.

In Japan, sea lions are found most frequently off north and east Hokkaidō during winter and spring, from October to May (occasionally later). Elsewhere in their range they go ashore in summer to breed on isolated islands, the nearest of which to Japan is Tyulenii Island, off Sakhalin, in the Sea of Okhotsk.

Considerable declines in populations of Steller's Sea Lion have occurred in the Sea of Okhotsk, and on certain of the Kuril and Aleutian Islands and down the Pacific coast of northern North America. These declines are most likely the result of depletion of their food resources caused by extensive fishing by humans.

In the water, sea lions are both extremely agile and generally fearless and inquisitive. They approach intruders to find out what they are and what they are up to, and this can be their undoing when there is a hunter abroad. While swift enough in their turns, and clever enough in their tactics to elude hunters repeatedly, like all marine mammals they need to come up for air, and each time they break the surface to breathe they give away their new position. Once tired, or just too curious, they break away and lunge from the surface to watch; it is then that they become vulnerable to the human marksman. A bloody trail in the water marks an injured sea lion, which is easily followed and killed. Considered an endangered species in parts of its North Pacific range, and protected as such, in Japan the species is considered a pest of the fishing industry. Its meat may make its way eventually into local restaurants, or even cans for sale in souvenir shops for tourists.

It is an extraordinary irony that the very purpose of listing the Shiretoko Peninsula as a World Heritage Site can only have been to promote it for further nature tourism – after all, it was already afforded Japan's highest protected status as a national park – yet tourists attracted by Shiretoko's natural environment and wildlife may witness a shocking event. It is a further irony given that Steller's Sea Lion is listed internationally as an endangered species because of its precipitous decline in the Bering Sea and

Largha Seal is the common species around Hokkaidō's coasts. Its rounded muzzle and large, gentle-seeming eyes distinguish it from the Northern Fur Seal [*left* SuT; *right* MAB].

along the Aleutian Islands. It is also a threatened or endangered species under the USA's Endangered Species Act, and will soon be on Russia's Red Data List. So killing this sea lion is strictly prohibited in the USA, Canada and Russia. In Japan things are different; some are still shot each year, ostensibly to protect local fisheries. And in Japan this endangered marine mammal falls within the jurisdiction of the Fisheries Agency, not the Ministry of the Environment as might be expected. It is particularly disappointing that, despite hard work to keep it safe, a revision of The Wildlife Protection and Hunting Law in 2002 failed to list it as a species requiring protection.

As a result of decades of promoting Japan as a destination for nature tourism, in recent years there has been an upsurge in visitor numbers. Consequently, Japan has a golden opportunity to promote itself as a destination for special-interest tourism in whale-, dolphin- and seal-watching, not to mention all of its other fascinating wildlife. Where marine mammals are concerned, however, Japan appears hard tracked towards persecution and human consumption of these species.

The most numerous 'true' or earless seal occurring around Japan's northern coasts is Largha Seal (or Spotted Seal). Like its near relatives, Largha Seal is a creature of cold coastal waters that ranges from the frigid Bering Sea south to the northwest Pacific Ocean. It is a fairly common winter visitor to northern Japan, and especially Hokkaidō, where it can be found northwards from the Shakotan Peninsula in the west, around northern and eastern coasts and along the Pacific coast west to Cape Erimo. Most arrive in Hokkaidō with the sea-ice drifting down the Sea of Okhotsk, and return northwards as the ice retreats. Largha Seal feeds on various benthic species, including a wide range of fish, and octopus. These blunt-nosed creatures float upright in the water, their bobbing heads pointing upwards like broad-based floating bottles. When they haul out at low tide on exposed rocks they appear as lethargic campers enveloped in thick sleeping bags. Sometimes they are indelicately described as 'blubber-slugs'! The larger males may reach 168 cm in length and weigh 90 kg, less than half the weight of the more agile Northern Fur Seal, and are only slightly larger than the females, which weigh up to 75 kg and measure up to 162 cm.

A medium-sized pale brown and spotted littoral seal, Largha Seal has a long, rotund body with a relatively small head. The entire body, including the limbs, is furred. The dorsal surface is dark grey with an irregular scattering of many small dark brown or blackish-brown oval spots varying in size and number, but the spots on the body resemble sesame seeds, from which its Japanese name of *Gomafu Azarashi* is derived. The ventral surface is paler, grey or greyish-white with fewer, smaller spots. In water, it appears dark grey, and after hauling out for some time it becomes a lighter, sandy-grey colour as its fur dries.

Largha Seal's head is broad across the crown and tapers to a rounded muzzle. The eyes are large and dark brown; the muzzle is furred and has extensive long stiff whiskers, and the nostrils can be completely closed. There are no external ears. The fore limbs are short, broad, and modified as paddles. The heavily modified hind limbs are largely hidden within the body; the large, paddle-like feet extend backwards and are held 'palms' together, like praying hands. The hind limbs cannot be drawn beneath the body, hence locomotion on land is limited to an ungainly wriggling motion. In water, where they are agile, flexing of the spine and the hind limbs provides propulsion, while the fore limbs provide additional control.

Whereas playful dolphins and whales splash, blow, spy-hop, breach, roll and fluke at the surface, seals are sedate. Thus, seal-watching is not the finest spectator sport and does not come close to whale- or dolphin-watching. Not only does by far the bulk of seal behaviour take place beneath the surface, but watching what little takes place at the surface is about as exciting as watching someone relax on a sofa. To understand fully what a seal is like when it bobs vertically in the sea, one needs to have at some time worn a neoprene wetsuit and walked out into deep water to gain an inkling of what a seal's subcutaneous fat layer offers – both buoyancy and warmth.

Emerging slowly, seals briefly open a tight-shut nostril, sniff at the wind, squeeze their nostrils shut again and sink from sight. They have the advantage over whales of being able to emerge from water on to land, although their ungainly manner of doing so leaves them in a class of their own. Imagine, if you can, donning the aforementioned wet suit – nothing less than a thick one will do – then climbing into a thickly padded sleeping bag. For a true seal-like experience on shore you would in fact need to cut two slits at the narrow end of the sleeping bag, just large enough for your feet (but no part of your legs) to protrude through. Just below the broad end, which will be closely clinched around your neck, you need two more slits from which your hands, but not your arms, protrude. Now lie down on a beach, and attempt to 'run' a fifty-metre dash. If the best you can approximate is a five-metre wriggle like a caterpillar or a maggot, then you are doing quite well in the seal stakes.

Thus evolutionary adaptation to their aquatic environment, combined with the necessity of emerging on shore or on ice to give birth and raise their young, has left the modern true seals with a flexible, propulsive spine, short trailing hind limbs suitable for steering in water, but of no value for propulsion on land, short fore limbs with strong nails for sculling in water and scuttling or scrambling on land or ice, and no external ears or external genitalia. Their descent from an otter-like ancestor some 25 Mya is indicated by the lateral movement of the lower spine when swimming, and scientists currently think that seals evolved in the region of the North Atlantic or Europe. While at the surface, or on the shore, they may seem as ungainly as CS Lewis's Narnian Dufflepuds, but give them deeper water and they come into their own. It is to that epipelagic realm that they are supremely adapted, with a physiology that allows them to dive down 100–200 metres below the surface and remain underwater for an hour or more.

Kuril Seals breed at just a few sites along the coast of southeastern Hokkaidō [YM].

Bearded Seal is a scarce visitor to northern Japanese waters [IiM].

With frigid seas, washed by cold currents from the north, the coastline of Hokkaidō seems just right for Largha Seals in winter. There they are sometimes joined by their larger and endangered relative the Kuril Seal, also known as West Pacific Harbour Seal and sometimes considered a subspecies of Harbour Seal. Males of this species reach 190 cm in length and can weigh up to 170 kg. They have a much more restricted range, being found only around the headlands of southeast Hokkaidō, especially Nemuro, and south to Cape Erimo at the southern tip of Hokkaidō.

From the 1940s to the early 1970s the population of Kuril Seal fell dramatically to just a few hundred individuals, from as many as 4,800, as a consequence of commercial harvesting, as well as bycatch in the autumn set-net salmon fishery, and other coastal fisheries. Although commercial harvesting ended in the late 1980s, hunting probably had an effect on population trends until the 1990s, and at least two haul-out sites from which the species disappeared during the early 1980s have not been recolonized, while bycatch rates in the 2000s remained similar to or higher than those in the 1980s. The seals have been observed recently at just nine haul-out sites during the pupping/moulting season along the coast of southeastern Hokkaidō, with approximately 70 percent at just two of them: Cape Erimo (up to 500 and the largest group in Japan) and Daikoku Island and Akkeshi (up to 250).

Other species that the naturalist might see include Ringed and Ribbon Seals, rare winter migrants to east Hokkaidō shores, and Bearded Seal and Walrus, which are very rare, accidental visitors.

Why individual seals should occasionally stray not just a little farther south, but a thousand kilometres south, to appear in rivers in the Tōkyō and Yokohama areas, is a strange quirk of natural history. There are those among us, however, who thank them for so placidly raising the heat on issues of identity, residency, citizenship and nationality here in Japan. During its appearances in the Tama, Tsurumi, Katabira and Naka rivers during 2002 and 2003, the non-taxpaying, non-human Bearded Seal, known locally as *Tama-chan*, not only was fêted by locals and given celebrity status by national television but, astonishingly, was also awarded honorary *juminhyo* (residency registration). Likewise *Ara-chan*, the Largha Seal, which appeared more recently, in autumn 2011, in Saitama Prefecture's Ara River. The diplomatic visits of these silent 'spokes-seals' subtly exposed the Japanese government's radical variant on evolutionary thinking, or should that be biased intelligent design, legitimizing a hierarchical taxonomic tree wherein Japanese nationals by descent are placed above temporarily resident seals (not requiring re-entry visas or registration cards). Those resident seals are in turn placed above foreigners (forever dubbed aliens not entitled to *juminhyo*, requiring re-entry visas and required to carry registration cards at all times). The further question, that of where foreigners rank vis-à-vis off-world aliens, remains unanswered, and the spokes-seals have remained ominously silent on this bewildering issue.

Ribbon Seals may be encountered occasionally on the ice drifting off the Shiretoko Peninsula [*left* WaM; *right* FT].

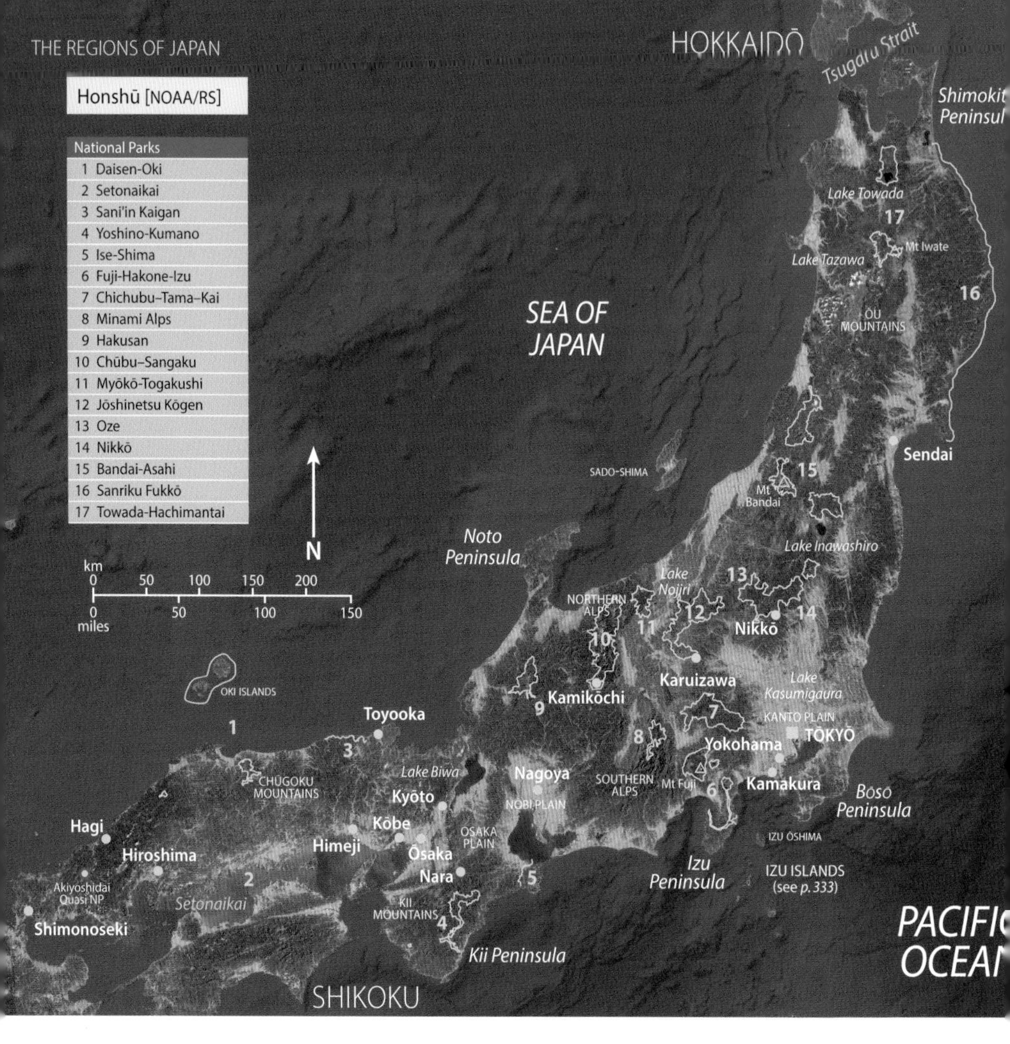

Honshū, the main island of Japan, is largely mountainous and heavily forested [MAB].

Honshū – The Heart of Japan

Shintō is Japan's own animistic religion, and shrines can be found throughout the country [MAB].

Honshū is the largest of Japan's islands and is generally considered to be the *mainland* of Japan. Honshū straddles a number of different climatic zones, producing great natural diversity from north to south and from east to west, while its mountainous spine divides the Sea of Japan coast from the Pacific coast and generates local weather effects. These regional and local climatic patterns in turn affect the distribution of various types of forest and so also the distribution of a range of other organisms.

The mountains held sacred and home to many deities of the Shintō pantheon, were taboo until barely 150 years ago. Even today, religious taboos combined with long-established culture prevent development on the mountain slopes, with the exception of forestry. While visitors may regard this as a missed opportunity, these taboos are grounded in common sense. Today, typhoon-related flooding and erosion, earthquakes and landslides often lay waste to housing situated at the base of hillsides. If occupied, the mountain slopes would be the scene of great annual devastation and disaster. One consequence of this attitude is the immense crowding of human settlement, as dwellings, industrial buildings and transport infrastructure, including rural roads, highways, railways and elevated bullet-train tracks, are restricted to the lowlands. The impact of the large human population is at its greatest here in Honshū, and the rapid pace of development since 1945 has been at its most extreme in the scarce lowland areas.

Much of Japan's human population is concentrated in Honshū. Although it is only the seventh largest island in the world, it is the second most populous large island after Java. Of Japan's nearly 127 million people, 103 million (in 2005), more than 81 percent, live in Honshū. Tōkyō, Japan's capital, houses

Historically, Japan's mountains, such as these in the Shirakami Sanchi World Heritage Site, were both sacred and taboo [PUDQ].

more than 13 million people, while the greater metropolitan area spreads across three prefectures and is estimated to be home to more than 36 million people. Recent development has overwhelmed the coastal lowlands that were once used almost exclusively for agriculture. Farming, people and industry are now all crowded into the same narrow space spreading from the foot of the mountains to the sea. Inland, much of the rail-and-road network burrows and tunnels its way through the mountains from coast to coast, while settlements and agriculture are crowded into narrow valleys. Thus, the scene is one of contrasts: small coastal fishing communities cheek-by-jowl with major shipping ports, oil refineries and industrial complexes; small rural communities at odds with vast urban sprawl; and narrow tracts of farmland among steep-sided, densely forested mountains. At first sight there seems little room for nature.

Honshū's northern region, Tōhoku, is extensive, occupying 18 percent of the island, yet has a mere 8 percent of the human population. It is extremely rugged, with a backbone formed by the longest mountain chain in Japan, the Ōu Mountains, and includes Mt Iwate as its highest peak at 2,038 m. Tōhoku experiences severe winters, as the seasonal winds from the Asian mainland bring heavy snow to the Sea of Japan coast and the mountains. In contrast, in the rain shadow farther east along the Pacific Coast, Sanriku is cold and dry.

The Shirakami Mountains, in northwest Tōhoku, are home to extensive native beech forest, a host of azaleas and other flowering shrubs, along with Asiatic Black Bears and, perhaps, the last Black Woodpeckers in Honshū. Farther inland, the crater lakes of Towada, Tazawa and Inawashiro, and the highland regions of Hachimantai and Urabandai are scenic highlights. To the north of the region's capital of Sendai, the wetlands of Izu-numa and Uchi-numa are not to be missed. These two lakes host one of the most spectacular gatherings of wintering geese one could hope to see in Asia. These are predominantly Greater White-fronted Geese but lesser numbers of other species occur, sometimes including rarities. Together with an array of other waterfowl, including swans and both dabbling and diving ducks, each winter the lakes host many tens of thousands of birds, making the cacophony and sight as they depart at dawn and arrive at dusk truly sensational.

Mountains and forests provide vital water catchment, make up the largest areas of wild habitats in the country, and support tremendous biodiversity [*left* MAB; *right* Silver Dragon Plant KuM].

Lake Towada and the Oirase River in northern Honshū are reminders of Japan's natural bounty [*top right* AOTA; *others* MOEN].

Izu-numa and Uchi-numa wetlands host a spectacular gathering of wintering waterfowl [YM].

Farther south, the Kantō Region, while the smallest of Japan's seven main regions and occupying just 8 percent of the country, is home to the nation's capital, Tōkyō, and many other cities, too, so that as many as 40 million people, close to a third of the nation's total, live there. Enormous Tōkyō Bay was once home to vast numbers of migratory waterfowl and shorebirds, but today only tiny fragments of semi-natural wetland survive and only small numbers of birds pass through. However, the capital's wetland reserves (such as Kasai Marine Park, also known as Kasai Rinkai Kōen) and its urban parks (such as Meiji Shrine) are renowned among birdwatchers. To find more wildlife, though, it is necessary to travel to the western outskirts of the city. Yet, within sight of the capital are regions where Japanese Badger, Japanese Giant Flying Squirrel and the endemic Copper Pheasant can all be found. Meanwhile, to the east, in the reedbeds along the Tone River, are Green Pheasants and rare species such as Marsh Grassbird and Japanese Reed Bunting.

To the south of the Tōhoku and west of the Kantō regions, the Chūbu Region spans both coasts. It is home to Mt Fuji and the Japan Alps (see *p. 242*) and hence is the source of much of the region's water, as many major rivers arise here, some flowing west to the Sea of Japan, others east to the Pacific. It is also home to Sado, the island home of Crested Ibis; the Noto Peninsula and, for nature photographers, the single most famous site of all, Jigokudani in Nagano Prefecture, the winter home of the world-famous Snow Monkeys (see *p. 253*). While Chūbu's Sea of Japan coast is known as Snow Country and the Japan Alps have what are among the heaviest snowfalls in the world, the Pacific Ocean coast experiences mild winters and hot summers. It is not surprising, therefore, that the highland regions of Karuizawa, Tateshina and the Fuji Five Lakes are popular summer retreats for residents of the Kantō Region and perennially popular with birdwatchers and wildlife photographers. Even this region of Japan has endemic species of plants and insects, for example the delightful Japanese Luehdorfia, or Gifu Butterfly.

To the west of the Chūbu Region lies the Kansai Region, home to the famous industrial cities of Ōsaka and Kōbe, the second largest metropolitan area in Japan, and the cultural capitals of Kyōto and Nara.

Urban parks, such as here at Meiji Shrine in Tōkyō, are havens of nature for city dwellers [PUDQ].

Many endemic and near-endemic species, such as Marsh Grassbird (*left*) and Japanese Reed Bunting (*right*), survive even within sight of the nation's capital [BOTH KT].

Crested Ibis forage in natural wetlands and wet rice fields. Their grey seasonal breeding plumage is unique among ibises. In flight they reveal the uniquely named (*toki-iro*) colour of their flight feathers, named after the Japanese name for the bird [BOTH MAB].

Sado is the largest island in the Sea of Japan and home to the reintroduced population of the Crested Ibis [MAB].

There are mountain ranges to the north, the Chūgoku Mountains and Tanba Highlands, and to the south the Kii Mountains, separated by a broad lowland zone which to the west faces the sheltered Setonaikai, Japan's Inland Sea, with its rather Mediterranean Climate. The Kansai Region is also home to one of Japan's three most beautiful views, that of the pine-clad coastal spit at Amanohashidate in Miyazu Bay, in northern Kyōto Prefecture.

From space, one feature above all others in Japan is striking – enormous Lake Biwa. Not only is it Japan's largest lake, it is also one of the oldest surviving lakes on earth, dating back some four million years. Lake Biwa's long uninterrupted history of isolation has allowed the evolution of numerous endemic species of fish in the lake. There are also numerous endemic freshwater snails and bivalves, alongside many other species that are more widespread natives. In winter, the lake attracts many thousands of migratory waterfowl. Unfortunately, the lake's biodiversity has suffered through the introduction of alien and invasive fish species, including Largemouth Bass and Bluegill, which prey heavily on native species.

Farther west from Lake Biwa, where the Maruyama River meanders its way north of the city of Toyooka and the community of Kinosaki into the Sea of Japan, is the home of Oriental Stork. Here, brave efforts have been made to breed in captivity and then release this endangered species that was once exterminated as a breeding bird in Japan. This almost crane-sized stork is a dramatic sight, and it can now be seen easily in captivity and in the wild at the Hyōgo Park of the Oriental White Stork, and often, along with many other species, at the Hachigorō Toshima Wetlands.

The westernmost region of Honshū is known as Chūgoku. To its south lies Shikoku and to its west Kyūshū. Here the Chūgoku Mountains are aligned east–west, as are those of Shikoku; between the two lies the island-studded Inland Sea. Since ancient times the Inland Sea has been a crucial route of communication; religion, culture and goods moved this way into the country via what was essentially a marine section at the eastern end of the ancient Silk Road. Before the more recent development of Japan's railway network, the only efficient means of transporting goods through the region was via this inland sea. The distinctive forests of this region support both Japanese Black Pine and Japanese Red Pine and the many islands protected in the Setonaikai National Park have an attractive flora which includes numerous species of azalea. The marine environment here supports both the strange, living fossil known as Japanese Horseshoe Crab and Indo-Pacific Finless Porpoise. Human communities here are centred on major river mouths or harbours, restricted by the general lack of flat land.

Key natural features of the Chūgoku region are the extraordinary Tottori sand dunes facing the Sea of Japan. These dunes are wind-driven marine sands. In a country of otherwise lush greenery, this strangely desert-like landscape is home to some very specialized plants, such as Asiatic Sand Sedge, Beach Silvertop and Beach Vitex. Protected as part of the San'in Kaigan (Coast) National Park, and also recognized as the San'in Kaigan Geopark, the area includes diverse, distinctive and beautiful coastal scenery and geology, incorporating coastal crags and dunes, caves, extensive areas of columnar basaltic lava and inland mountains spanning 120 km from east to west and covering an area of 2,458 km^2 – one of the largest geoparks in Japan. The area has both aesthetic and scientific value, particularly because the land forms here tell the story of the early development of the Japanese Islands with formations dating from the period of the Sea of Japan's creation.

Farther along the Sea of Japan coast to the west the twin lakes of Shinji and Nakaumi are important sites for wintering waterfowl, such as the highly vocal Bewick's Swan, and the dapper Tufted Duck. To the east of the lakes is the dramatic Mt Daisen massif, presenting steep walls when viewed from the south or north, with diverse flora ranging from beech groves and Japanese Yew to alpine flowers. Offshore to the north are the numerous Oki Islands, washed by the warm Tsushima Current, which have a mild climate and even warm-water corals. The forests of Chūgoku are home to a wide range of Japan's resident and summer-visiting avian species, while the cold montane rivers host the endemic Japanese Giant Salamander (*p. 267*).

Hiroshima, the largest city of the region, has been entirely reconstructed since it was devastated by an atomic bomb on 6 August 1945. It has now, with its peace park and memorial museum, become a major tourist destination, as has nearby Miyajima, with its freestanding Torii gate out in the bay,

Amanohashidate, in Miyazu Bay, is considered one of the three most beautiful classic views of the country along with Matsushima and Itsukushima [MAB].

The strange Japanese Horseshoe Crab is considered a kind of living fossil [KHCM].

The desert-like dunes along the Sea of Japan coast of Tottori Prefecture are a rarity in Japan [MAB].

Enormous Lake Biwa is Japan's largest and oldest lake [MAB].

Bewick's Swan migrates to spend the winter in western Japan [MAB].

Endemic Japanese Giant Salamanders live in cold montane rivers [FR].

The enormous Torii gate marks the entrance to the beautiful Itsukushima Shrine [MAB].

The karst landscape of Akiyoshidai Quasi National Park is the largest in Japan [MAB].

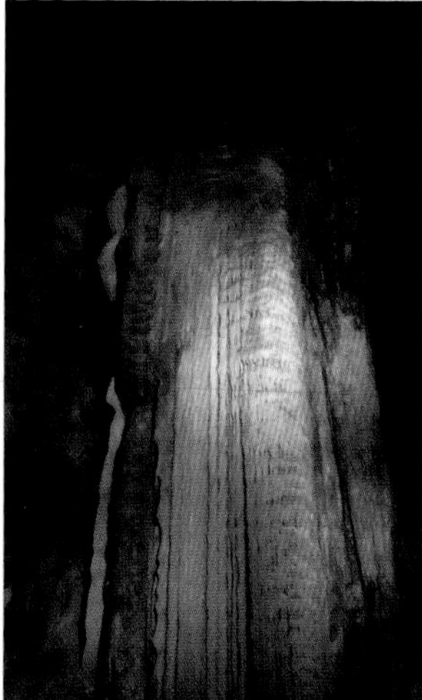

The large cave complex of Akiyoshidō lies hidden beneath the Akiyoshidai landscape [ALL MAB].

and Itsukushima Shrine that are partly inundated at high tide. Inland from Hiroshima lies the Nishi-Chūgoku Sanchi Quasi National Park with its not-to-be-missed Sandan-kyō (Three Level Gorge), a national special place of scenic beauty. The area abounds with hiking trails, and the beautifully forested river gorge can be explored on foot to admire the various waterfalls and the Siebold's Beech, Japanese Zelkova and Japanese Horse Chestnut groves. Here you can watch Brown Dipper, Common Kingfisher and other birds beside the water, and can explore the area by kayak and by boat.

In far western Chūgoku, in Yamaguchi Prefecture, between the delightfully sleepy, historical town of Hagi and the bustling port city of Shimonoseki, lies the extraordinary Akiyoshidai Quasi National Park. In Hagi, once capital of the Mori Clan – one of the most powerful clans during Japan's feudal age – one can explore the well-preserved former samurai district, delightful Tōkōji and Daishōin temples and its distinctive Hagiyaki pottery. Shimonoseki, the prefectural capital, is nicknamed locally as the *Fugu* (Pufferfish) Capital because it is the largest market for species such as Tiger Pufferfish in Japan.

Akiyoshidai provides a window into some fascinating aspects of non-volcanic geology in Japan. Akiyoshidai has the largest karst landscape in the country, stretching 17 km from east to west and about 8 km from north to south. Its limestone plateau with greenstone basement was formed about 350 Mya and is the remnant of extensive coral reefs. The uplifted landscape, formed from permo-carboniferous limestone and now exposed to air and rainfall, has been eroded[1], resulting in the unique complex karst landscape, with rugged limestone pinnacles and numerous dolines[2], or sinkholes. Where water has penetrated the ground through these sinkholes, it has carved out runnels and hidden streams, and eventually an immense cave complex, Akiyoshidō. This cave complex, the largest in East Asia, comes complete with an underground river and several waterfalls.

The famous Three Level Gorge or Sandan-kyō
[BOTH HIPR]

1 Limestone, a carbonate sedimentary rock consisting of the skeletal remains of marine organisms, gradually dissolves in water, especially if it is slightly acidic.
2 A doline is a funnel-shaped basin in a karst region.

The intricate structures in the cave include numerous stalactites and stalagmites, columns like organ pipes, a 15-m lace curtain of rock known as *Kogane Bashira*, and the extraordinary *Hyakumaizara*, the sinter-terrace-like disc-shaped formations or rim-stones resembling miniature flooded rice paddies. Everything has been formed over eons, as water has first eroded the limestone, then deposited it again, drip by drip, in wonderful forms and shapes. The limestone here has yielded many fossils, including those of corals, brachiopods and bryozoans. Various invertebrates have evolved in these dark, damp conditions, giving the caves a unique fauna. In some sections of the caves several thousand of as many as six species of bat roost. While the cave complex extends over more than 10 km, a readily accessible and easily walked 1-km trail traverses an engrossing section and, with a stable underground temperature of 17°C, the caves are not only an enjoyable diversion throughout the year, but also one of the most fascinating geological excursions in Japan.

Japan's Southern Squirrels

Blakiston's Line, the great biogeographical divide in northern Japan, separates a multitude of species from one another, including Japan's various squirrels. North of the line, across the Tsugaru Strait in Hokkaidō, occur Eurasian Red Squirrel, Siberian Flying Squirrel and Siberian Chipmunk (as described on *p. 188*). South of the line, all are replaced by different endemic species and joined by a well-established introduced species.

The squirrels share an elongate, slightly pointed face with protruding black eyes and long, prominent whiskers. They have long fore limbs and particularly dexterous fingers with which they can grasp and manipulate their food. They are mostly arboreal, and strong limbs and sharp claws mean that they are as comfortable climbing or descending head down as when moving head up.

In Central Japan, occupying essentially the same niches as in Hokkaidō, we have Japanese Squirrel and Japanese Flying Squirrel, but larger than all others there is also Japanese Giant Flying Squirrel. The interloper here is Pallas's Squirrel, introduced from Taiwan and designated an Invasive Alien Species by law in 2005, and occasionally you may bump into a chipmunk introduced from peninsular Korea.

Japanese Squirrel is a sister species to Eurasian Red Squirrel and is endemic to Central Japan. It is wide-ranging within Japan, but may have become extinct in Kyūshū. It favours mixed woodlands with plenty of deciduous trees and it occurs also in parks and large gardens. Smaller and lighter than its more northerly relative, Japanese Squirrel weighs 150–310 g and measures just 16–22 cm in head-and-body length, with a tail 13–17 cm long. Its coat is reddish-brown in summer, turning greyish-brown in winter, and it has a pure white belly and a white tip to its tail. Its short ears grow tufts in winter, but never so long as those of Eurasian Red Squirrel.

The introduced Pallas's Squirrel, at 20–22 cm long plus a tail of 17–20 cm, is similar in size to Japan's endemic squirrel, but is heavier at about 360 g. It has dark grey-brown fur on its upperparts, with golden or grey tips to the hairs lending a brindled appearance. Its underside is greyish-brown, not white as in the sympatric Japanese Squirrel. It has short, untufted ears and a long (almost as long as the body) very bushy tail, usually held aligned with its body. In its natural range it occurs from northeast India to Taiwan, from where it was introduced to Japan in the 1930s; it has thrived locally in Japan since its first release.

So far, it remains confined to pockets of forested habitat in the warmer southern parts of central and western Honshū and Kyūshū – including parts of Kamakura, on Izu Ōshima, in Ōsaka, or in Himeji. It can be locally common in dark, shady broadleaf evergreen forests, where it feeds on a range of buds, flowers, leaves, bark, fruits, nuts and insects. Although usually a solitary species foraging mostly in the trees, it will visit feeders, but it appears not to occur in those habitats favoured by the endemic Japanese Squirrel or to compete with it, nor is it regarded as a pest.

Male Pallas's Squirrels occupy non-exclusive home ranges of up to 3 ha; these overlap with those of several females (each with a much smaller range of approximately 0·5 ha). Breeding occurs once a year in autumn, when they use tree cavities in which to nest, or construct a large spherical drey of twigs in the fork of a tree, where the females produce litters of one or two young.

The endemic Japanese Giant Flying Squirrel (*left*) and Japanese Flying Squirrel (*right*) are both confined to the forests of the main islands south of Hokkaidō [BOTH MM].

Whereas the Eurasian Red Squirrel occurs in Hokkaidō, the endemic Japanese Squirrel is found only south of Blakiston's Line [MM].

The introduced Pallas's Squirrel is currently restricted to warmer forested regions in Honshū and Kyūshū [MM].

With a somewhat similar geographical range to that of Japanese Squirrel, Japanese Flying Squirrel is also confined to Honshū, Shikoku and Kyūshū but, instead of occurring in lowland broadleaf forests, it favours montane and subalpine forests. This small, pale grey-brown nocturnal flying squirrel is endemic to Japan. In winter it is largely pale grey on the upperparts but with a pale brown back in summer. The underparts, from the chin and throat to the lower belly, are creamy white. It has a rounded face, with distinctively enormous, bulbous black eyes appearing disproportionately large for the head size, and has short rounded ears. The tail is long, flattened and about two-thirds of the body length. When settled, the species appears to have a ruffled flap of skin or fur along its flanks; this is the retracted flying membrane, which extends between fore and hind limbs. It weighs 150–220 g and measures 14–20 cm in length, with a tail 10–14 cm long.

Occurring in native forests from the mid montane zone to the subalpine zone, Japanese Flying Squirrel uses natural tree cavities, including disused woodpecker holes, as breeding sites and daytime roost sites. At dusk, the squirrels emerge from their roosting sites, climb up the trunk of the tree and then launch into a glide, using their flying membrane, down to another tree and so off around their territory in search of their food, which consists of leaves, buds, seeds and fruits of trees and some fungi. They breed once or twice each year, the females giving give birth to 3–5 young in each litter.

Dwarfing Japanese Flying Squirrel is Japanese Giant Flying Squirrel. One of a number of giant flying squirrels across Asia, this one is endemic to the lowland and montane forests of Honshū, Shikoku and Kyūshū. This large, cat-sized flying squirrel is brown on the upperparts (pale brown in the north, dark brown in the south) and pale greyish-white on the underparts. It has an oval face, more pointed and hence more squirrel-like than those of the two smaller flying squirrel species in Japan, with smaller eyes and slightly longer, more pointed ears. The face pattern is distinctive, being slightly paler on the forehead between the eyes, and with pale crescents extending from the crown, in front of each ear, onto the cheeks. The tail is long, flattened and about three-quarters of the body length. During gliding, the flying membrane is extended between the fore and hind limbs, as with the two smaller species, but also between the neck and the fore limbs and between the hind limbs and the tail. It weighs a hefty 700–1,000 g, and measures 34–48 cm in length with an additional 28–41 cm of tail.

Nocturnal and arboreal, this is a mostly solitary creature, although you may occasionally see it in family groups. It ranges from the lowlands to the montane and even subalpine zones in native primary mixed deciduous forests and mature secondary forest, and occurs also in conifer plantations. Japanese Giant Flying Squirrel uses natural tree cavities or nestboxes as breeding sites and daytime roost sites. Some of its finest habitats are the large old trees left standing around hillside temples or shrines. There the trees are left to mature and the wildlife is left in peace.

At dusk, as with other flying squirrels, individuals emerge from their roosting sites, climb up the trunk of the tree and then launch themselves, stretching out their arms and legs, and glide off to explore their territory in search of food. This giant flying squirrel's diet includes the leaves, buds, flowers, seeds and nuts of deciduous broadleaf trees, and because it occasionally eats tree bark it can cause damage, especially in conifer plantations. Cat-sized and silent as a cat burglar, it is usually visible only as a dark shadow flitting between the trees. When 'flying' at night, its pale belly and the white undersides of its 'wing' flaps make it appear quite ghostly and ethereal as it passes silently among the treetops.

Japanese Giant Flying Squirrels are territorial: males have larger home ranges (up to 3 ha) than those of females (up to 1–1·5 ha). They breed twice a year, mating in winter and spring, the females giving birth to two litters of 1–4 young in spring and autumn after a gestation period of about 75 days. They survive for up to 14 years in captivity, and probably to about ten years in the wild.

Cormorants and *Ukai*

From top to bottom
Great Cormorants occur widely in Japan [YaM]; Temminck's Cormorant is essentially a bird of the coast, where it nests on crags and cliffs. [JW]; Pelagic Cormorant in northern Japan [MC]; the range of Red-faced Cormorant just reaches Japan in southeast Hokkaidō [KT].

With Japan's coastline measuring 29,751 km (greater even than that of Australia: 25,760 km) and with so many islands, it should come as no surprise that Japan hosts several species of cormorant. Great Cormorant breeds throughout the main islands; Temminck's Cormorant is resident around northern coasts and winters around much of the rest of the country; Pelagic Cormorant breeds locally around Hokkaidō and winters widely farther south; and rarest of the four is Red-faced Cormorant, which, at the southwesternmost limit of its North Pacific range, occurs locally in southeast Hokkaidō.

The coastal cormorant of Japan is Temminck's Cormorant, which resembles closely Great Cormorant of freshwater areas, and greatly dwarfs the much more slender Pelagic Cormorant. These cormorants dive for their prey and kick their enormous webbed feet for propulsion, rather like an underwater paddle steamer.

It is not the cormorants themselves that are surprising but what people do with them! During the summer, on several rivers in Honshū, fishermen can be seen out in pursuit of Sweetfish, not with rod and reel but with cormorant, leash, boat and brazier.

The practice, known as *Ukai*, has certainly continued for more than 1,300 years on Gifu Prefecture's Nagara River. There, in both Gifu and Seki cities, the master fishers are entitled – Imperial Fishermen of the Royal Household Agency. The practice occurs also on more than ten other rivers in western Japan, including on the Uji River south of Kyōto. The fishermen go out at night not because cormorants are nocturnal but because, by using a burning brazier on their fishing boats, they can draw fish up to the surface, where the leashed cormorants are able to catch them easily. Rings around the necks of the cormorants prevent them from swallowing the fish completely, so that, once a cormorant has caught a Sweetfish, the master fisherman reins it in, removes the fish from its gullet and sends the bird back to work, giving it a break every now and then and a reward. Most surprising of all is that the fishermen do not use the local Great Cormorants, which are so widespread and common in Honshū, but instead they use Temminck's Cormorants from the coast.

Left Temminck's Cormorants are used for *Ukai* – catching Sweetfish – at night [GPTF].

Pheasants – Copper and Green

Copper Pheasant [YaM]

Both of Japan's pheasants are endemic, and one, Green Pheasant, is the national bird known as *Kiji.* This somewhat secretive metallic-green bird favours woodland and farmland edge habitats and may occasionally be seen wandering out in fields, although it prefers to remain hidden in tall grasses, revealing itself by its rather hoarse, loud "*ko-kyok*" call. It can be found in the lowlands of all three main islands of Japan, but not Hokkaidō.

The second species, Copper Pheasant, is one of the most sought-after Japanese endemics among visiting birdwatchers. In contrast to the Green, the Copper is a bird of montane forests. There are grisly folk tales from northern Honshū of these birds, known as *Yamadori*[1] in Japanese, playing pranks on newcomers to the mountains by creating terrific crashing sounds and mimicking severed heads rolling down from the mountains over the snow. In reality the most likely sign of a Copper Pheasant is the occasional cast feather on the forest floor, or the whirring sound of rapid wingbeats as a bird flees deep cover. Should you be fortunate enough to find a long, drooping coppery-brown tail feather, you will have found a real jewel. Not only are those feathers exquisite, but they are also said to be endowed with magical qualities. These beautiful feathers have transverse bands or nodes and special properties. A feather of 13 nodes is said to be proof against evil spirits, while folk tales say that a pheasant with such a feather may trick people. And at night *Yamadori* feathers are said to phosphoresce and frighten people away. The human residents of Hida (now in Gifu Prefecture) believed that very old *Yamadori* emitted light when they flew, making them visible on moonless nights.

Folk tales apart, the *Yamadori* is an extremely attractive bird unique to Japan, and, as its English name suggests, it is brightly copper-coloured. If you are willing to ignore the ghost stories, you can look for this shy, secretive and elusive bird in broadleaf and mixed coniferous-and-broadleaf forests in montane regions of Japan up to the subalpine zone in Honshū, Shikoku and Kyūshū.

There is an aura about Copper Pheasant and it has nothing to do with the supposedly magical and mystical nodes of its tail. Naturalists who visit Japan invariably rank this among the top five species they dream of seeing.

It is a bird of consummate camouflage, as copper as the leaves, as silent in its movements as the first tendrils of spring, and rarely heard calling. Like other pheasants, Copper Pheasant is not overly fond of

1 In the avian context Yamadori means mountain bird. In a botanical context, however, the same word means the taking of small trees or plants from the mountains and is, in essence, the origin of bonsai.

The endemic Green Pheasant is Japan's national bird [AK].

Male Copper Pheasants drum in display and show off their long tail plumes [TK].

The elusive Chinese Bamboo Partridge is far more frequently heard than seen [TK].

flying unless it can glide downhill. So, look for it as it runs uphill to escape between the trees. If it should take off, it will do so with an explosive burst of speed, leaving the undergrowth like a springing deer, and clattering through the trees, only then to glide downslope and away from its pursuer.

Copper Pheasant is best detected when the males are drumming during late winter and spring. As a Copper Pheasant strolls its favourite forest paths, its long tail tips fall slightly apart and, sometimes, gently brush the ground. To drum, the male pauses, stands tall and still atop a fallen log, or rock, or on the forest floor, and beats his short, broad wings so rapidly that they emit a strange muffled thrumming like a bull-roarer. As he drums, he shows off the full length of his 125-cm long tail drooping down behind him. With his raspberry red eye-wattle engorged, he is a magnificent bird, especially when, with a blurring whirr of his wings, he produces that loud "*phhhrrrrp*" again.

The whirring *phhhrrrrp* sound is made by the rush of air over the stiff wing quills, and forms an important part of spring courtship. Although it is quite a strong sound and far-carrying, it can easily be overlooked because it is so low, much like muffled thunder, or a distant truck tipping its load of earth. Count yourself fortunate if you hear one, let alone see one.

Once very common, even abundant, with nearly a million shot each year, this sedentary and generally solitary bird has declined under the onslaught. Although the shooting of females has been illegal since 1976, this highly sought-after bird is still a difficult one to find.

The highly vocal Chinese Bamboo Partridge, a distant relative of the pheasants, is widespread in the southern half of the country. Despite its loud and frequent calls, however, it is typically elusive in dense undergrowth.

There are several subspecies of Copper Pheasant, including a white-rumped subspecies in Kyūshū [TK].

Wintering Waterfowl

While sea-ducks (see *p. 118*) are at home in the surf, in harbours and along rocky coasts, another 25 or more species of dabbling and diving ducks, geese and swans make the freshwater and brackish habitats of Japan amazing magnets for wintering waterfowl. These habitats include lakes, marshes, lagoons, rivers and river estuaries. The waterfowl arrive each autumn from Siberia to spend the winter in the relative warmth of Japan's temperate climate, feeding, and forming pairs, in readiness for their return northwards in spring to breed.

Piping flocks of Northern Pintail occur in their thousands at many rivers and lakes around Japan, and these are joined by many other species, including Eurasian Wigeon, Mallard and Eurasian Teal. For waterfowl aficionados it will be the distinctively Asian Falcated Duck and Baikal Teal [VU] that will be the most enticing.

Skeins of Greater White-fronted Goose and Tundra Bean Goose [VU] migrate, by way of the Kuril Islands or via Sakhalin, first to Hokkaidō and then to Honshū. Their numbers exceed 100,000 birds, and they spend several months each winter in commuting between traditional night-time roosting sites. Major roosting sites for them include the Izu-numa and Uchi-numa lakes, in Miyagi Prefecture, and Miyajima-numa, in west Hokkaidō, from where they commute to their feeding grounds on fields, where they glean the waste left over from the grain harvest or graze the grasses along the strips between the fields. Rarer geese, especially Cackling Goose [CR], Snow Goose and Lesser White-fronted Goose, are sometimes among them.

Less numerous than the ducks, and even larger than the geese, are the 'Angels of Winter'. On a blizzard of wings, Bewick's and Whooper Swans migrate south across the Sea of Okhotsk, alighting with much calling and whooping, as they find their resting spots at lagoons along the coast of Hokkaidō. Family members call and greet one another as they settle in with their flock to feed and recover from their long flight. Some may move on. All of the Bewick's Swans will pass on to milder regions of Honshū, especially in the north and at sites along the Sea of Japan coast, but many of the larger-bodied Whooper Swans will remain in Hokkaidō. Cruising amid the spiralling steam from thermal springs, they appear ethereal, and mystical, true Angels of Winter (see also *p. 176*).

Northern Pintail [YaM]

Wetland habitats draw enormous numbers of wintering waterfowl from northeast Asia [MAB].

Greater White-fronted Geese extend their autumn migration over several weeks, pausing along the way at important feeding and roosting sites such as at Miyajima-numa, in western Hokkaidō [MAB].

Migrating Whooper Swans arriving in Hokkaido for the winter [WaM].

The Asian Snow Goose population is recovering and many now winter in northern Honshū [MAB].

Once rare in Japan, Cackling Goose numbers are recovering rapidly [WaM].

Baikal Teal is among the prettiest of Asian duck species [TK].

Male Falcated Duck sports full breeding plumage during winter in Japan [MAB].

Mandarin – the Hooded Love Bird of the East

Mandarin Duck [YaM]

Appearance

Almost like phantoms, or moving shadows, silent shapes jump into the air, take wing and flee along a densely wooded river valley and are soon out of sight upstream. Mandarin Duck is a shy shade-loving bird, but catch a clear glimpse of one and its colours shout – *Oshidori*.

Secretive and difficult to observe in their secluded summer habitat along mountain streams and at wooded lakesides, it is from autumn onwards, when Japan's forest colours are brilliant enough to match even the male Mandarin's gorgeous plumage, that they become much easier to watch. At this time Mandarins gather in increasingly conspicuous numbers at secluded lakes, even on moats and shaded city-park ponds in the lowlands of Japan. These wintering congregations are flighty, full of activity and prone to rising *en masse* to form a whirling airborne flock that flashes orange. It is then, too, that avian predators, such as Northern Goshawk, drive Mandarin Duck flocks into the air and pick off their prey.

Only when the males sail out into full sunlight do their colours and patterns come into their own, and then they are splendid. Some consider them the most gorgeously plumed of all waterfowl. Each aspect of the Mandarin's plumage has been honed by competitive attention to detail. The elongate, creamy eye stripes, like designer spectacles, contrast with the dark forehead, the long orange and green crest and the orange ruff. At first sight, the stubby pink bill with its yellowish-white tip seems out of place – until it is shown in vivid contrast when it is tucked back against the dark maroon of the breast, as the male rocks back on his orange feet and thrusts out his chest in display. A double necklace of contrasting narrow white and broad black bands dramatically divides the maroon neck and chest from the bird's finely vermiculated orange flanks. The mantle and wings are resplendent in purplish-black, white and orange, while the lower chest and belly are of the purest white.

The Mandarin male appears as a duck in designer clothing. In the shady woodland habitat, even the seemingly gaudy colours of the male are surprisingly camouflaged, and the dappled greys and browns of the females virtually disappear against the browns of the woodland floor and in the rippled shadows on the water.

Male (*right*) and female (*left*) Mandarin Duck [FT]

For all the intricate details of the Mandarin's colours and its pattern, it is a singular, unique feather that defines this bird. None of the more than 10,000 other avian species on earth shares such an oddity. The drake Mandarin is the only one to carry such an instantly recognizable and highly modified wing feather, and it serves a singular purpose. This immensely asymmetric feather, a single tertial on each wing, has a greatly enlarged vane which stands proud on its strongly arched rachis, while the vane below is narrow. The upper portion of the vane stands conspicuously tall, like a raised sail, and is brightly orange, whereas the portion of the vane below is mostly hidden and is as deep and as iridescent a blue as a Mallard's speculum.

Behaviour

The twin biological requisites of survival and reproduction have, through sexual and natural selection, driven avian mating systems that coincidentally enchant and delight the human eye and ear. That male birds sing, display, posture and carry great adornments, to show off to their mates, is widely known and commonplace from the diminutive wren to the enormous peacock, and to the Mandarin's tertial feather. The popular press, meanwhile, may mistakenly describe these males as being showy, flashy or gaudy, as if they are revealing some kind of terribly chauvinistic character flaw that they should grow out of, but, in reality, the males are being driven by a biological imperative: mate choice by the female has driven the development of these plumage characteristics and behaviour. Likewise, the same press may anthropomorphize further and describe the females as drab or dull and as hoodwinked into maternal care by their lazy mates, all the while ignoring the evolutionary selective pressures that have honed their very different and very successful sexual strategies over many millions of years.

If this bird seems to wear designer clothing, then its movements, too, seem to have been choreographed for the catwalk, mingling swagger and swish with bravado and allure. Meanwhile, his suitress is clad in a cryptic blend of subtle greys, browns and cream, with a delicate pair of whitish spectacles. She has all the understated elegance borne of the confidence of being in charge.

Male Mandarins pursue the subtly plumed females and engage in a range of behaviours, including chest-puffing courtship [FT].

Mandarin Ducks prefer the shadiest areas of rivers, ponds and lakes. They like those places where overhanging trees cast shade on the water below, where they can stand on a low branch over the water or on a stump at the water's edge, or where they can forage safely on the ground, looking for seeds and nuts, especially acorns. On the water, the Mandarin males swim leisurely back and forth, each holding its orange 'sails' high above its back like a samurai's pennant. Their capes and ruffs are spread for the greatest effect, and they puff out their chests as they try to outdo each other and, at the same time, attract the attention of the trim grey-plumed females nearby. It is she, not he, who chooses, while her potential mates must endlessly display and joust for her attentions.

Each male Mandarin Duck spends many hours during the course of the winter and spring in showing off his finery in the hope that a female may select him as her mate. The males gather and posture, raising their crests, fluffing out their ruffs, puffing up their chests. They swim together, suddenly surging through the water, showing off their strengths; meanwhile, the cryptic females are watching and listening, spending their time in selecting and choosing.

Courtship and Breeding

By spring, when the wintering flocks are breaking up, the females will have made their choices, the pairs will have formed and mating will have started. Eventually, the pairs fly off to their chosen rivers and begin carving out their territories. Each pair will seek out a mature deciduous tree in which there is a large cavity, which perhaps began life as a woodpecker's hole or as the crack left by a storm-smashed limb. If the cavity suits, and once the pair has mated several times, the female will lay her clutch of up to 14 eggs in the seclusion of the tree hole. For two or three more weeks during April or May, the male remains nearby, perhaps to come out of seclusion to protect his mate should a predator appear, but soon he must leave or risk drawing unnecessary attention to the breeding site.

The tiny yellowish-brown chicks emerge as mere wisps of animated down, and soon they must learn the avian equivalent of base-jumping. Long before they have matured, the mother calls her chicks forth from the safety of their nest cavity, and there is only one way for them to leave. They are clad in just their natal down and have yet to develop their feathers, so they completely lack the plumes which they need for flight. They must therefore scrabble to the lip of their high-rise home, sometimes as much as 18 m above the ground, then take the plunge and jump from the nest down to the woodland floor. Like tiny parachuting fluff-balls, legs and webs spread and tiny wings fluttering, they fall to bounce and land in the leaf litter far below. Then, they must make their way hurriedly to their mother and the safety of water, which may be more than a kilometre away from the nest tree. They rely on their own cryptic plumage, and their mother's continual state of alert, for their protection during the weeks when they grow. If startled or disturbed by a predator when they are very young, they dive and flee to hide beneath the riverbank, and when older they skitter and skim away across the water surface following their mother to safety. When young they are at risk from predatory fish and from mammals such as martens and Tanuki, and as adults they are at risk from attack by forest-dwelling birds of prey such as Northern Goshawk and Japanese Hawk-Eagle.

Meanwhile, the brightly coloured drakes have left their mates, their breeding territories and their offspring behind, in order to find seclusion where they can moult. The males now drop their bright plumes in favour of a more cryptic eclipse plumage. As do many other waterfowl, they drop all their primary wing feathers at once, and so they become temporarily flightless. While the males are moulting, the females are tending to their broods. Once autumn returns, the males emerge from their eclipse phase into their bright nuptial plumage and, once more, it is time for them to compete for the opportunity to mate.

Mythology

The English name 'Mandarin' for this delightful duck is perhaps derived from the fact that its distinctive orange ruff and orange sail-like wing feathers brought to mind the orange robes of the high officials in the former Chinese imperial civil service. Its scientific generic name, *Aix*, is derived from an Ancient

Greek word, and its use attributed to Aristotle for an unknown duck, while its scientific specific name, *galericulata*, means 'the hooded one'. The hooded one's Japanese name, *Oshidori*, written in ancient pictograms, carries within it the meaning of love bird, and in numerous folk tales this species epitomizes both conjugal love and marital fidelity.

The long-entrenched belief that male and female *Oshidori* consort together throughout the year, much as do cranes, has led to the Mandarin having a powerful symbolic role in the art and literature of Japan. Mandarin is revered here as a symbol of conjugal love and affection, and of marital bliss and fidelity, and its image appears in a myriad of forms. Happily married couples in Japan are known as *Oshidori hūhu*, meaning *Oshidori* couple.

Folk tales are told of the remarkable bond between the male and female, and one of these was transcribed into English by that great interpreter of things Japanese, Lafcadio Hearn. Hearn relates, at some length, the heartbreaking tale of a falconer and hunter by the name of Sonjō, who lived long ago in the northern province of Mutsu, and who destroyed the bond that ties a male and a female Mandarin together.

But not all tales told of the Mandarin are sad, and another tells a story of affection and gratitude. A drake Mandarin, captured alive and kept in a cage by a wealthy man, was pining and refusing to eat. A kind-hearted maidservant, working in the wealthy man's household, persuaded the servant responsible for its care that a bird so devoted to its mate would pine until death if kept alone. Out of compassion the servant took her advice and released it. When the owner discovered the Mandarin's absence, however, he immediately suspected the servant and began treating him cruelly. Distressed by the cruelty she saw him suffering, and feeling partly responsible having given him the advice, the maid was drawn to him and, in time, the couple fell in love. With the power wielded by the wealthy in those days, the master sentenced the young lovers to death, without trial, for causing the loss of his valued pet. As luck would have it, and before the sentence could be carried out, two messengers arrived from the provincial governor carrying the news that the sentenced offenders were to be taken before the governor.

The journey to the provincial capital was long, and the young lovers, overcome by fatigue, became increasingly unable to keep up with the messengers until eventually they realized that they had been left behind and were lost. Eventually, they came upon a rustic hut and there the exhausted lovers spent the night. To both of them, in their dreams, the messengers of the day before returned and explained that they were, in fact, the Mandarin which they had released and his mate. Knowing of their fate at the hands of the master, they had returned to rescue them. When the lovers awoke the next morning the messengers had gone, but instead, before the hut, was a pair of Mandarin Ducks, who bowed to them and then flew away into the forest. The young couple fled, found employment in another province and were married. They had saved the magical union of the *Oshidori* and, in return, the avian messengers had made their union possible; it was a union that was to become as strong and enduring as that of the birds.

A female Mandarin Duck emerges with her brood after breeding in a large tree cavity [FT].

The Intelligence of Crows

Japanese Crow [YaM]

Crows in Japan are ubiquitous, and, before you ask, that crow you are watching, however big you think it is, is not a Northern Raven, although it is often mistaken for one. Two species are abundant, the more slender-billed Oriental Crow and the massive-beaked Japanese Crow (known also as Large-billed Crow and before that, the Jungle Crow). Both are widespread and both can be found wherever there are people – including in rural villages, major conurbations, coastal fishing communities and narrow-valley agricultural areas in the mountains. They differ in their size and behaviour, but most noticeably in their head and bill shapes.

One question arises repeatedly: 'What can we do about the crow problem?' The answer is to rethink the question. Crows are not the problem; it is what humans do that has precipitated the problems.

Crows, birds in the Corvidae family, occur around the world. There are various branches of the family, some wearing the familiar mournful black raiment of Northern Raven, Oriental Crow and Japanese Crow. Others are less obviously crows, examples being Eurasian Jay with its dazzling blue wing patch, Spotted Nutcracker with its spangled plumage, and the various tropical jays of Central and South America, which tend to be more specialized in their habitat and food requirements, too. Some of these family members are at home only in forest, while others, more flexible in their lifestyle, are comfortable close to human habitation. A small and select group of crows, generally the black ones of the genus *Corvus*, are out-and-out opportunists, finding food, nesting material and nesting sites wherever they can. Opportunists are like that. It is no good complaining about them; the only answer is not to put temptation their way.

Opportunism takes many forms, but, if a bird has that nature, then consider its options. Put yourself in a crow's feathers. Would you, given the choice, fly a long way from your night-time roost, and spend all day searching for food on the off chance of finding enough when all the time you know that just around the corner someone is putting out a heap of food? Of course you wouldn't; you'd think of an extra hour asleep, nip around the corner, grab the grub and go back to a cosy resting site. I am unashamedly anthropomorphizing, but that is pretty much what certain members of the crow tribe will do. I do not recall any country other than perhaps India that has so many crows as Japan.

In most countries crows tend to be cheeky, pugnacious and quick to take advantage, but generally they are not very common. They gather at roost sites to spend the night in the safety of large numbers but,

Japanese Crow is both scavenger and predator. Its massive bill gives it access to garbage and tree cavities, allowing it to feast on urban waste and nesting birds [KT].

Crows, such as Oriental Crow, are among the most widespread of birds in Japan [MM].

Northern Raven, a rare winter visitor mostly to eastern Hokkaidō, is known to indulge in play behaviour [MaH].

come the morning, they disperse in search of food. Crows have found agriculture to their liking. After all, wherever there are domestic animals, there are fields with lots of droppings and these serve to attract insects, earthworms and the like – all food for hungry crows, which are just continuing their tradition of following herds of wild ruminants. Many crows are happy with carrion, too, so wherever animals die, or where larger animals waste their food, they are quick to take advantage and sneak in for a free treat. They gather also where nature provides a bounty. For example, a salmon river dotted with dying fish after the spawning will attract crows. Where a storm has tossed up flotsam from the sea, crows will scour the tideline in search of a free meal of dead fish or the like.

We tend to forget, with our natural predisposition for anthropomorphizing animals, that other species do not do things because they are good or bad, or out of spite, they do them simply because they are motivated by natural instinctive goals – food, a resting place, a breeding place; so far as we know, they make no value judgements.

Crows are not attracted to cities in themselves. What attracts them is a supply of good food, good nesting sites and good roosting sites. Anywhere that supplies all three is going to attract crows, and enable them to breed and produce more crows looking to meet the same needs. From a crow's perspective, modern cities provide valuable resources. In the main, cities are more important in winter, especially in temperate parts of the world. Cities are major energy-consumers and thus major waste-producers. Much of that waste is in the form of heat, so cities now serve as significant heat islands. Not surprisingly, birds are aware of this, and crows in particular take advantage of winter roosting sites that are a few degrees warmer than elsewhere. While human commuters are returning to their suburban homes, the crows are literally flocking into our cities to sleep.

Crows are renowned for taking advantage of the urban landscape, we leave an almost unlimited amount of food available for them [MaH].

In other countries the crows, as morning arrives, fly out of the city in search of food, but one thing distinguishes Japan from many countries that have crows and that is the extraordinary, and messy, means of garbage disposal – piling up garbage in plastic bags out in the street for collection. Garbage bags are flimsy and a person does not need to be armed with anything more than a determined finger to be able to poke a hole in one. Any creature armed with teeth and claws, such as a rat, or armed with a meat-cleaver of a beak, such as Japanese Crow, must just think that it has died and gone to heaven. All that garbage, much of it containing edible waste, is literally dumped on the streets and often overnight! Talk about temptation. In Hokkaidō, some citizens feel proud that they carefully cover the heap of plastic bags with a mesh net but, while this may stop garbage from rolling or blowing away, it certainly cannot stop a determined crow wielding a powerful beak. Temptation is lying there in a heap, and a crow has a degree of intelligence that it cannot fail to use in such circumstances.

The so-called crow problem is not really a *crow* problem, it is a people problem. Step one: dispose of organic waste and non-organic waste separately (preferably composting the former). Step two: change the method of depositing garbage so that flimsy plastic bags of rubbish do not lie out in the street. Other countries use permanent bins that are animal-proof and, not surprisingly, where they are used crows are rarely seen foraging in the streets. Crows appear to be at an extremely and unnaturally high level of abundance in Japan because of the enormous volume of garbage produced in both urban and rural areas by people. There are so many open garbage dumps that crows benefit greatly from our wasteful way of life. Until we change our ways, crows will remain abundant.

Crows are not merely abundant, they are intelligent, and their capacity to solve problems is widely known. Their ability to take advantage of our infrastructure and waste is less well known, but crows in Japan are known to gather wire coat-hangers (the type used by dry cleaners) with which to build their nests. They are also known to use passing traffic to crush hard nuts for them![1] Japanese crows, like gulls, are not averse to flying up with shellfish, sea-urchins and other hard items, and then dropping them from a height on to a hard surface so as to break them open for the food inside. In several Japanese cities, Oriental Crows are aware of traffic lights and traffic movement and take advantage of the traffic to crush nuts placed or dropped on to the road, and then make use of the traffic lights and the pedestrian crossing to allow them to retrieve the nuts without being run over![2]

So intelligent are birds that some have time for behaviour that can be described only as play[3]. The animals involved seem to engage in the behaviour for the fun of it, and among the birds it is the corvids that are considered to exhibit the most complex play behaviour of all.

Northern Raven is the largest corvid that can be found in Japan, and is a rare winter visitor, mainly in the eastern third of Hokkaidō. More common than it once was, it can today be found wintering very locally, most often in the coldest part of the winter during January and February, although sometimes as early as November and until as late as early May. Two areas where ravens may be found in winter are the well-forested, mountainous Akan–Mashū National Park, in central eastern Hokkaidō, and the equally mountainous Shiretoko Peninsula of extreme northeast Hokkaidō. In east Hokkaidō, I have observed ravens engaging in various elements of play behaviour such as I have never seen with Japan's more commonplace crows, including 'snow-romping'[4]. I have watched both individuals and pairs engage in aerial chases, displays, and calling activity, both while flying and while perched among the rocks on the upper slopes of various mountains, and very rarely they also snow-romp, this sometimes even involving several pairs at the same time.

On one particular occasion on 8 February 2000, I watched as three pairs engaged in play. One raven,

1 A simple internet search for Japanese crows and nuts will reveal numerous videos of this amazing behaviour.
2 It seems that they retrieve only their own nuts. Andrew Davis, who has watched them for hours doing this in Sapporo, has not observed them squabbling with each other, in the middle of the road, over the most recently squashed nut.
3 Play in animals is generally interpreted as any behaviour that seems not to have any direct adaptive advantage.
4 A term coined by Derek Ratcliffe.

after landing in the snow, lay on its breast and slid head-forward downhill, apparently 'sledging'[5]. Its partner, who was nearby, began by lying sideways to the slope and then rolled over and over downhill. The pair continued sledging and rolling downhill for more than ten metres before flying back upslope to repeat this behaviour. Just minutes later I noted another raven fluttering in deep powder snow as if bathing, while its partner was rolling sideways down the slope nearby. As the latter rolled over repeatedly on to its back, I was able to see its legs in the air, its wings flicking in the snow each time, before it rolled over upright again. A further pair nearby also engaged in similar play behaviour; one member rolled sideways down the slope alone in the snow, and was quickly joined by its partner, which landed on its mate, sat down, and remained there for at least a minute. Meanwhile, one member of the first pair was still sliding head-first down the slope. The three pairs were engaged almost simultaneously in play behaviour in the snow. Rolling ravens disturb powder snow in an erratic manner; 'sledging' does, however, leave clear linear tracks downslope in the snow.

Eurasian Jay is a member of the crow family, and in Japan is most frequently found in woodland or forest with distinctive subspecies in the main islands [*top*, Honshū KT] and in Hokkaidō [*bottom* MR].

On the Shiretoko Peninsula, at a roost gathering of ravens, I observed a pair preening each other[6] while a single individual nearby began pecking at the branch on which it stood. After a few moments, it slipped into a hanging position beneath the branch, holding on with both feet. While upside-down beneath the branch, it let go with one foot, holding on with the other. It then grasped the branch in its beak and let go with both feet so that it hung beneath the branch holding on only with its bill. Finally, after a few tugging motions, as if attempting to break the branch by using its weight, while hanging by its bill, it flew off.

Spotted Nutcracker is a crow-family member found high in the mountains [BOTH MR].

Not far away, in the Akan–Mashū National Park, I watched one of a group of ravens land on the snow slope on the inner rim of the Mashū Caldera. It proceeded to peck out a large chunk of snow crust (larger than its head), and flew off with it in its bill. It was immediately chased by another individual and, after circling for a while, it returned to the same part of the slope and dropped the chunk of snow. Then the raven repeated the process of pecking out snow crust, carrying it into the air, being chased for it, and then dropping the snow crust down on to the slope. On another occasion, I watched a pair among a group of birds that was engaged in aerial pursuits, paired flights, swoops, stalls, and rolling displays. The members of the pair flew towards each other, grasped each other by the beak and descended slowly with their wings and tails spread like two black spiralling parachutes. After several seconds, they disengaged and flew separately. Such apparently playful behaviours are not confined to Northern Raven in Hokkaidō. Oriental Crows in Honshū have been observed to slide down solar panels, and down a children's slide. Such behaviour may in fact be a means of 'showing off' to other individuals and serve as status-enhancing displays of critical importance when establishing pair bonds.

5 Carrion Crows are known to indulge in similar sledging behaviour in Europe.
6 This behaviour, known as allopreening, involving two birds preening each other, often involves them in preening each other about the head and neck – areas that are difficult for a bird to preen by itself.

Cuckoos and their Hosts

Eleven species of cuckoo have occurred in Japan, four of which are regular summer visitors: Northern Hawk-Cuckoo, Lesser Cuckoo, Oriental Cuckoo and Common Cuckoo. They breed in wide areas of the country, from Okinawa to Hokkaidō. It is the last species' call that lends its onomatopoeic name to the whole family. They are all insectivorous birds, feeding largely on the larvae of the Lepidoptera, and they have another, extraordinary feature in common – cuckoldry.

A wide range of native species, including Azure-winged Magpie, Bull-headed Shrike, Oriental Reed Warbler, Eastern Crowned Warbler, Blue-and-white Flycatcher, Japanese Bush Warbler, Asian Stubtail and Olive-backed Pipit, all inadvertently play host to the cuckoos' eggs. Cuckoos do not spend time on building a nest for themselves or in rearing their own young. Instead they enlist other species to do the hard work, for they are brood parasites. They lay eggs the shells of which are patterned to mimic those of their hosts, hosts which vary among cuckoo species and among individuals. Cuckoo eggs hatch quickly, after only about 12 days of incubation, ahead of the eggs of the host. This gives the young cuckoo a head start for its grisly business of ejecting the host's eggs or young from the nest. As the young cuckoo grows it outstrips its hosts, so that, by the time the young cuckoos fledge (after about 17–19 days), the small host parents will be feeding a monstrous chick much larger than themselves.

These parasitic cuckoos provide clear evidence of the instinctive nature of migration in passerines. Cuckoo parents are not involved in raising their young and migrate weeks before their offspring. The young cuckoos find their own way to their wintering quarters in Southeast Asia without guidance (other than from their own genetically provided aptitude).

An Oriental Cuckoo fledgling dwarfs its Siberian Rubythroat host [WaM].

Northern Hawk-cuckoos mimic sparrowhawks in shape and plumage [KT].

The smallest cuckoo species in Japan is Lesser Cuckoo [KT].

Common Cuckoos favour woodland edge and grassland habitats [SP].

Oriental Cuckoos closely resemble Common Cuckoos, although their vocalizations are remarkably different [WaM].

Occasional brown-morph females occur [ImM].

The endemic Japanese Macque's main range is in Honshū and Kyūshū [MAB].

The alpine mountain ranges of Honshū are dissected by deep valleys [MAB].

The Sky Islands at the Heart of Honshū

Brown-eared Bulbul is an important disperser of forest-tree seeds [ImM].

Wild Honshū

The rugged mountains of Honshū are isolated from one another, from other ranges in Japan such as those in Hokkaidō and Kyūshū, and from other mountain ranges in Asia. They are effectively another archipelago of isolated islands – sky islands of high mountain habitat, divided by deep valleys and surrounded by lowland habitats. They experience high annual precipitation as rain and snow, and they have steep flanks incised by frequent stormflow events, giving the rivers here steep profiles.

With the greatest altitude range of any of Japan's islands, Honshū also supports the greatest range of habitats, and, as with islands of other types elsewhere, Honshū's mountains are home to a range of very special species. At the island's most northern extent, the mixed forests of deciduous trees and conifers of these mountains resemble those farther north in Hokkaidō. Meanwhile, in the far south, the evergreen laurel forests are similar to those in the subtropics. Winter snows are deep and heavy along the western (Sea of Japan) side of Honshū's spine of mountains, while during winter a dry Pacific climate dominates much of the rest of the island, which lies in the rain shadow of the mountains. In the south and west in particular, long hot summers are preceded by heavy rains and followed by the powerful typhoon season. This combination of Japan's seasonal and regional climatic patterns dictates the distributions of plants and animals, making travelling around Honshū particularly interesting.

Amid the thousands of square kilometres of urban and ribbon development, only the hardiest and most flexible species survive in a landscape dominated by man. Brown-eared Bulbuls and Japanese Crows seem ever-present, even in the heart of Tōkyō and Ōsaka, but leave the cities and you will find a fascinating diversity of natural wildlife. Honshū shares many links to the south with Kyūshū and Shikoku, and rather fewer with Hokkaidō to the north, but its mountain ranges have been isolated long enough for endemic species to have evolved here, too.

Honshū contains the main part of the range of the famed Japanese Macaque. A winter visit to the mountains of Nagano Prefecture provides a unique opportunity to watch them bathing in hot-spring waters. The goat- and antelope-like Japanese Serow is an animal of forests where snows are deep in winter and, in the same region, in fast-flowing, cold rivers there lives one of Japan's most extraordinary

The endemic East Japanese Toad is widespread in Japan's main islands south of Hokkaidō [HT].

The endemic Japanese Stream Toad is restricted to cold montane streams of Honshū [FR].

The Japan Alps offer the finest scenery in Honshū [MAB].

The spectacular Bandai–Asahi National Park is located in northern Honshū [SHUT].

creatures – Japanese Giant Salamander. This, the second largest of the world's amphibians, is a lie-and-wait predator in a habitat that elsewhere would be occupied by river otters. Japan has a number of endemic amphibians, such as East Japanese Toad, but even more restricted in both its range and its habitat is the unusual Japanese Stream Toad. Unlike most toads, which live in still water, it is found only in fast-flowing, cold, montane torrents, to which it is specially adapted. Its range is a narrow transect of the mountains that stretch across central Honshū between the Sea of Japan and the Pacific Ocean.

High in the Japan Alps there are breathtaking views combined with a wonderful array of alpine flowers and butterflies at or above the tree line. Butterflies that may range more widely and at lower elevations elsewhere in East Asia, such as Eastern Blackvein, which occurs in Tibet, Mongolia and Amurland, and Orange Tip, which ranges from western Europe to eastern Asia, are found in Japan only in the high mountains of central Honshū. Here, too, is another unexpected relict from the last ice age – a small population of Rock Ptarmigan.

In north-central Honshū, in the snowy Tōhoku Region, is an array of peaks within the Bandai–Asahi National Park. Here, Dewa Sanzan (the Three Mountains of Dewa) are revered as sacred for mountain worship, while the Asahi and Iide Mountain Ranges, spanning Yamagata and Niigata Prefectures, are known as the Tōhoku Alps. The Bandai plateau, with its numerous ponds and lakes, was formed following the last great eruption of Mt Bandai, in 1888. This is the second largest national park in Japan, home to a lush alpine flora on the higher ridges and extensive beech forests on the mountain flanks.

Farther to the south lie the national parks of Nikkō and Oze. The former is home to the Nasu group of volcanoes, with many peaks over 2,000 m and sacred Mt Nantai. Nikkō National Park, in Tochigi Prefecture, is famed for its lovely lakes, waterfalls and autumn foliage, and for the marshlands of Senjōgahara, with its colourful Japanese Azaleas and delicate sedges known as Hare's-tail Cottongrass. This national park presents a unique combination of stunning natural scenery and indigenous and endemic wildlife (including Asiatic Black Bear, Japanese Serow and Japanese Macaque), along with the World Heritage Site shrines and temples of Nikkō, the most famous of which is Tōshōgu.

The national parks of Honshū would repay a lifetime of exploration, offering dramatic and spectacular montane scenery, delightful seasonal changes and a very rich flora and fauna. Among the many it is possible to mention only a few, such as Oze National Park, spanning parts of Fukushima, Tochigi, Gunma and Niigata Prefectures; Jōshinetsu Kōgen National Park, spanning parts of Gunma, Niigata and Nagano Prefectures; Myōkō–Togakushi Renzan National Park, spanning parts of Niigata and Nagano Prefectures; and Chichibu–Tama–Kai National Park, spanning Saitama, Tōkyō, Yamanashi and Nagano Prefectures.

Oze National Park offers not only stunning scenery and a plethora of wildflowers but also its own endemic genus and species, the Oze Japanolirion, which is restricted to serpentine soils [KREO].

While all are sensational and deserving of far more attention, one that cannot be glossed over is the Fuji–Hakone–Izu National Park, home to iconic Mt Fuji and the hearts of the Japanese people. For long the subject of Japanese art, this impressive peak is a geological anchor point for the Japanese psyche, providing both an identity and a sense of place. The mountain itself is listed as a World Cultural Heritage Site as a 'Sacred Place and Source of Artistic Inspiration'. The disjunct structure of this national park is not unusual in Japan, where several non-contiguous features are recognized and combined into one protected entity. In this case, the non-contiguous areas include the Hakone district, the coast of the Izu Peninsula and, much farther to the south, the Izu Islands (see *p. 333*).

The cherry blossom front that spreads up the country each spring is widely known, but is rivalled by an array of native and endemic azaleas [MAB].

Scenery within Nikkō National Park [MENR]

Sacred Mt Ontake (3,067 m) on the borders of Gifu and Nagano Prefectures is Japan's second-highest volcano [MAB].

The Nature of Mt Fuji

Mt Fuji's symmetrical profile is recognized worldwide. The isolated cone of Japan's archetypal stratovolcano so dominates the scenery of lowland central Honshū that it is difficult to achieve a true perspective of its scale and natural significance. Climb to over 3,000 m in the Japan Alps, however, and look southwards and you will see the mountain for what it is: a distant, isolated cinder cone on the horizon, like a small island amid a sea of cloud.

Thinking of it as an island, we begin to understand the nature of this mountain. Any island is by definition isolated, making it difficult for species to colonize, and their ultimate success in their new home depends on the qualities of the island's habitat. Its isolation means that the natural history of Mt Fuji is somewhat less diverse than that in the Alps, although it is nevertheless fascinating.

Surrounded by temperate forest on its lower flanks, the natural flora higher up approaches that of an alpine zone, with an abrupt transition from lush green forest to the dark battleship-grey of loose scoria[1]. It is difficult for plants to take hold here because strong winds and winter snows undermine any attempts by plants to take root in the friable substrate. From a natural-history perspective the sacred peak itself is of little significance, except perhaps as a place from which to watch migrating birds, and it is the middle and lower levels of the mountain that are of most interest.

1 Scattered rough clinker-like masses, the result of rapid cooling of material ejected violently into the air during a volcanic eruption.

Red-flanked Bluetail [*left* MM], Narcissus Flycatcher [*centre* TK] and Blue-and-white Flycatcher [*right* MC] all occur in the mixed forests on the flanks of Mt Fuji and elsewhere in central and northern Japan.

Mt Fuji seen from the Japan Alps [CC].

Mt Fuji stands in splendid isolation and on clear winter days is easily visible from the Tōkyō area [UNSP].

The endemic Japanese Marten is a significant predator in the forests of Honshū [IiM].

Japanese Serow have only one or two young each year [HT].

The impoverished alpine flora above the tree line quickly gives way to lower-elevation mixed forest, and these habitats are best explored early on a summer's morning. By then the nocturnal nightjars have just ceased their nightly churring and hunting of insects, and Red-flanked Bluetail and Japanese Accentor will be singing from the forest edge. No fewer than four species of cuckoo (see *p. 234*) may be heard in the forests here, and lower down brightly coloured flycatchers, particularly Narcissus and Blue-and-white, are common. Early-rising explorers may be rewarded by sightings of a Red Fox, a Tanuki, a Japanese Marten, or perhaps even a Japanese Serow.

The Japan Alps

Mountains occupy around 80 percent of Japan, many peaks rising above 3,000 m. Foremost among them, the steep-sided *Nihon Arupusu* (Japan Alps) extend for 120 km across the breadth of central Honshū in a broken three-part rugged range split by deep, water-cut, gorge-like valleys. Formed mostly from ancient Palaeozoic (542–251 Mya) and early Mesozoic (252–66 Mya) sedimentary rock layers, igneous rocks including Cretaceous (145–66 Mya) granites have intruded and metamorphosed the sedimentary rocks. The Japan Alps consist of the steep Hida, Kiso and Akaishi mountains (Northern, Central and Southern Alps) spanning mainly the prefectures of Nagano and Gifu, extending into Toyama in the north, Aichi and Shizuoka in the south and Yamanashi in the east. Kita-dake is the highest alpine peak, at 3,192 m above sea level.

The dominant landscape feature of Honshū consists of its alpine mountain ranges [MAB].

Shintō Fortune Tellers of the Avian Kind

From top to bottom Tōshōgū Shintō Shrine, at Nikkō, where trained Varied Tits once performed religious rites [NCTA]; the widespread Japanese Tit [MC]; Marsh Tit is found only in Hokkaidō [MR]; the colourful endemic Varied Tit [MR]; the Izu Island endemic Owston's Tit [MC].

The small group of woodland birds known as tits or chickadees is well represented in Japan, more so now that advanced research techniques allow the recognition of isolated populations as full species. The common and widespread Japanese Tit recalls Great Tit of Eurasia with its neat black 'neck-tie', but it lacks that species' bright yellow underparts and green upperparts. Marsh Tit and Willow Tit are both wide-ranging Eurasian species, essentially grey with a black cap, that occur also in Japan. The most colourful member of the group is the East Asian endemic Varied Tit. For long known for its extraordinary fortune-telling behaviour, it is now recognized not just as one species, but as a complex of four, one of which is confined to Taiwan and the other three of which occur in Japan.

Varied Tit is an attractive chestnut-bodied tit with white cheeks, a black cap and grey wings. Widespread and familiar throughout Japan, it is quite fearless where it is fed, and is easily tamed. For several hundred years[1], it has also performed a cultural and religious service known as *omikujihiki*.

As a First World nation, Japan is unique in having its own, home-grown, animistic religion – Shintō. More than 80,000 Shintō shrines have been dedicated throughout the country, many of them in the mountains which were traditionally held sacred, and each is marked by a distinctive entrance gate known as a *torii*, symbolically marking the transition from the mundane and profane to the sacred realm. The etymology is obscure and contested, but the literal meaning of *torii* is 'bird's perch'. Shrines, with their sacred precincts and ancient trees, are often excellent places to look for wildlife, such as Japanese Giant Flying Squirrels and a wide range of birds, although it is rare to find a bird perching on the *torii*. Nevertheless, trained birds can occasionally be found at shrines, performing the religious rite of *omikujihiki*.

Where I have witnessed *omikujihiki*, at Tōshōgū Shrine in Nikkō, each human visitor first offered a Varied Tit a ¥100 coin. The bird carried the coin to a small model Shintō shrine, deposited the coin into an offering box, rang a small bell by striking it with its beak, and then opened the doors of the shrine, from which it drew out a fortune-telling slip[2] and gave it to the visitor, something which the visitor typically purchases directly from a shrine.

Varied Tit's newly recognized relatives live isolated lives and differ in their breeding biology from the mainland species, having smaller clutches. The larger and more deeply chestnut-coloured Owston's Tit is endemic to the southern Izu Islands, while the smaller and paler orange Orii's Tit is found only on the remote Yaeyama Islands, in the far south of Japan's Nansei Shotō.

1 Said to date back to the Heian Period (8th to 14th centuries).
2 *Omikuji* in Japanese.

Kita-dake, in the Southern Alps, stands 3,192 m high [KREO].

The 3,180 m high peak of Yarigatake (the spear) is dramatic [CC].

From Kamikōchi the Azusa River valley draws the eye towards the Hotaka Mountain Range [MAB].

Above 2,500 m the alpine zone, which resembles the Arctic tundra, is home to species such as Komakusa [CC].

The northern Japanese Alps fall within the Chūbu Sangaku National Park, which is truly representative of the mountainous nature of Japan's main island, and a mecca for Japanese mountain hikers. Several peaks here rise above 3,000 m, the more famous being Shirouma-dake, Tateyama, Yarigatake, Hotaka-dake and Norikura-dake. The whole region is photogenic, but the approach along the valley of the Azusa River at Kamikōchi, with a view towards the Hotaka Mountain Range, is perhaps one of the best-known and best-loved.

Hakusan National Park protects another wonderful scenic montane area rich in alpine flowers and lying just to the west of the delightful World Heritage Site villages of Shirakawa-go and Gokayama. To the south the Minami Alps National Park, spanning parts of Yamanashi, Nagano and Shizuoka Prefectures, contains no fewer than ten 3,000 m peaks[1] listed among the 100 famous mountains of Japan. The mountains here support such rich forests that they are known locally as the Green Mountains. Broadleaf forests below 1,500 m include beeches and Kousa Dogwood, while just above them are subalpine conifer forests of endemic Maries' Fir and endemic Northern Japanese Hemlock. Finally, above 2,500 m, there is the alpine zone with numerous rare alpine plants, including several that are endemic such as Japanese Callianthemum.

The clearest signs of glacial activity in Japan are to be found in these high Alps. Many cirques can be seen, especially in the Hida Mountains. Many narrow snowy valleys and ravines contain perpetual snow banks, and here and there tiny remnants of glaciers have been found. The forests generally extend up to about 2,500 m, above which is the alpine zone of creeping pine and specialized alpine plants.

1 Kaikomagatake, Senjogatake, Noutori-dake, Kita-dake, Aino-dake, Shiomi-dake, Warusawa-dake, Akaishi-dake, Hijiri-dake and Tekari-dake.

The delicate-beauty of the Kousa Dogwood hides the fact that the white 'petals' are actually modified leaves [NZUE].

The Weston memorial at Kamikōchi commemorating Reverend Walter Weston, one of the pioneers of alpinism in Japan [MAB].

The Alps are relicts of an ancient past, when Japan was still connected to the Asian mainland as a peninsula. They are home to relict birds, plants and insects living in an environment with only a short summer growing season, especially at higher altitudes, with slow growth and low productivity. Human impact here is multiplied by the slow rate of recovery of this extremely fragile, high-altitude ecosystem. Until just over a century ago many of these mountains were taboo and unvisited, and human impact virtually non-existent, all that has now changed.

In ancient times, certain peaks in Japan were regarded with awe and considered the inviolate dwelling places of the gods and ancestors. Many peaks are revered as sacred even today. To break their sanctity was to risk divine wrath. With the development of mountain mysticism in the 700s, however, pilgrimages to the sacred mountains began. In one form or another pilgrimages to the mountains have continued, boosted, just over a millennium later, by an interest in alpinism for its own sake. This pursuit was engendered during the latter part of the 19th century by British mountaineer William Gowland in his *Japan Guide* (1881) and by Reverend Walter Weston in his *Mountaineering and Exploration in the Japanese Alps* (1896); it was the latter who coined the term 'The Japan Alps', and who popularized hiking there. Today, the numbers of people visiting the mountains run into millions, and their impact is of serious concern.

The mountains these days suffer the multi-faceted threats of mankind, including the tramping boots of fervent mountain-respecting hikers, the cutting edges of myriad skis of the enormous ski-bunny

Subalpine forests of endemic Maries' Fir and endemic Northern Japanese Hemlock [MAB]

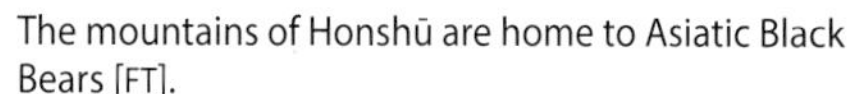

The mountains of Honshū are home to Asiatic Black Bears [FT].

A small population of Golden Eagles continues to survive in the Japan Alps [ImM].

class, and the avid golfer. Deforestation, whether for ski runs, golf courses or industry, leads to a net loss of natural montane forest and reduces the habitats of the special wildlife of the Japanese Alps. Where forestry has been attempted on particularly steep slopes, ugly erosion results and some areas have lost so much of their fragile soil that native trees can no longer grow there. The mountains are extremely vulnerable to erosion because of their steep slopes, heavy rains, and snow melt. Erosion cascades silts and gravels into watercourses that are themselves sensitive and vulnerable.

Symbols of the once natural splendour of the region are a mammal, Asiatic Black Bear, and a bird, Golden Eagle, both the largest of the region. They are both at the peaks of their food webs and potent images of the health of all they roam, of all they survey, and of all on which they depend. They are two of a kind, living where montane habitats provide sufficient for their daily and seasonal needs. At high altitudes, where productivity is low and growth slow, each requires an enormous range to support it. Symbols of security and environmental health, they are also a legacy from the past, relicts of ranges once much vaster when Japan's Asian connection was dry land. Thus the bear, the eagle and other specialists, such as Rock Ptarmigan, linger in Japan only because solely in the mountains have their habitats survived.

High in the alpine zone, above the tree line, and thus up beyond the ranges of bear and serow, one bird has particular significance. The grouse-like Rock Ptarmigan, known locally as the Thunder Bird, is the symbol of the high Alps. Because they feed unconcernedly, even at the height of the summer's electric storms, they are revered, carved and painted as a talisman to protect lodges and hikers from lightning strikes. They are, however, powerless to protect the mountains, and the wildlife that depends on them, from humankind.

Asiatic Black Bears are secretive and now rare because their habitat is shrinking and because they continue to be hunted. Golden Eagle, too, is in decline, and its breeding success was falling so steeply that, by the mid-1990s, the population of just 300 individuals was considered to be unsustainable. Soon, Japan will lose another species, and another fragment will be gone from the nation's biodiversity mosaic.

Only Japanese Macaque in the lower forests is still relatively common, although it, too, is threatened in some areas. The other significant mammal of the uplands is the extraordinary Japanese Serow, which wanders like a forest ghost through the mountains.

A hike up into the Japanese Alps takes the visitor not only through altitudinal bands of varied botanical communities, but also back through time, penetrating to the ancient backbone of the Japanese archipelago.

Relicts of Ice Age Japan

The average Japanese hillside or mountainside consists of forested slopes on dark soils, with bare rock exposed here and there. Yet occasionally an observant and well-travelled hiker may notice loose rocks,

Exposed rock runs in Hokkaidō [*left* MAB] are the habitat of the Northern Pika [*right* WaM].

not rocks recently shattered by winter's freeze–thaw processes, nor formed by crashing rock falls, but rounded rocks, weathered and worn, although not so smooth as if they had spent millennia in running water. Such an observer might also notice that they occur in discrete patches, oddly like a dry-stone garden, amid clearings on forested hillsides, like short, broad rivers of rocks. The rocks themselves tend to be lichen- and moss-covered; here and there shrubs and young trees are gaining a toehold in the thin soils accumulating among the mosses. The surrounding forest will eventually claim them and overwhelm them, as they have no doubt overwhelmed many other such rock patches that now lie hidden beneath deeper soils. Such rock or boulder patches form an increasingly rare habitat, one that is disappearing steadily as soils and forests spread ever upwards.

A jumbled pile of somewhat rounded rocks on a mountainside may seem at first sight to be a very odd place to look for wildlife. In fact, in most of Japan, even if you could find such odd places, you would not find special mammals associated with them, but in certain parts of Hokkaidō, in the mountains of the centre of the island, in Daisetsu and Hidaka, not only do such places still survive, but in some of them a strange creature is at home. Spend a while; take a seat, watch and listen. A sharp half-cry or half-whistle is a hint that some creature may have emerged. The animal blends in so well with the weathered rocks and their patina of fine vegetation that it is difficult to spot at first, then suddenly your eyes adjust and you realize that a small rock has moved. Not rolled, not fallen – moved. It's a Northern Pika.

This compact rabbit-like creature, with short, rounded ears, short limbs and an almost non-existent tail, pauses, sits still again, raises its head, opens its mouth and releases a high-pitched half-whistle.

Rock runs (*arrowed*) remain visible at some locations in the mountains of Honshū [CC].

Itaji Chinkapin [TsM]

Both males and females vocalize, males giving multiple high-pitched calls and females single whistled calls. This species' Japanese name is delightful – *Naki-usagi*, a singing or calling rabbit. In fact, it is not a true rabbit, but a distant relative in its own family, the Ochotonidae.

A small stocky creature, weighing just 150 g and measuring 13–19 cm in length, with a very short tail of 5–12 mm, Northern Pika has short (15–20 mm) but prominent rounded ears, and large hind feet (24–27 mm). It resembles a small rabbit with shorter ears and less prominent hind limbs. It inhabits rocky areas in the alpine zone, particularly exposed and stable boulder-scree slopes, where it occupies cavities between the rocks. It feeds on the leaves and stems of a wide range of plant species, as well as ferns, mosses and mushrooms, but only those growing in close proximity to its den site. It gathers and stores food, collecting actively during autumn to provide itself with a supply of food for the winter. Pikas make hay when the sun shines – literally. They gather bundles of grasses or autumnal-coloured leaves and lay them out on rocky ledges in the sun to dry, before dragging them beneath the rocks into crevices below. Their remarkable appeal lies not only in their appearance and their behaviour, but also in the habitat on which they depend and what that tells of the past here.

Active by day and by night, and year-round, Northern Pika lives partly above ground from spring to autumn. In summer it is a somewhat reddish-brown, while in winter it is dark greyish-brown although few people see them at that season because they then have a largely subnivean existence. Most of their very particular habitat survives only above 800 m above sea level, and at that elevation deep snow blankets the landscape in winter. Seemingly, pika do not hibernate but remain active beneath the snow, feeding on snow-buried vegetation and on food that they have gathered and stored during summer and autumn. They breed just once annually, females producing a single litter usually of 2–4 young.

In the Hidaka Mountains of southern Hokkaidō, Northern Pika is at the very southern limit of its range in East Asia. These mammals are capable of surviving the harshest of the winters that Hokkaidō can throw at them, and they are tough enough to withstand even harsher conditions farther north in northeast Russia, as far north as northeast Chukotka, close to the Bering Strait.

These pikas are survivors in more senses than one, not only in terms of the seasons but also in terms of their habitat. Those weathered rock piles upon which they depend, and which form isolated 'islands' of suitable habitat for them, are far more visible at higher altitudes and latitudes. The piles of rounded rocks are periglacial features, relating to the past presence of ice and frozen ground. Northern Pikas live only where permafrost or frozen snow beneath the rocks cools the air around the rocks, creating the cool microclimate critical for their survival[1]. They are a very visible reminder of a period 18,000–24,000 years ago, when the world was last in the grip of a glacial maximum. At that time sea levels were considerably lower, and the Japanese archipelago had a very different form with several land connections to Asia, including one from Hokkaidō, via Sakhalin, to what is now northeast Russia.

Pollen and whole-plant fossil evidence has revealed that at the time of the last glacial maximum, when the climate of the entire archipelago was very much cooler than it is today, ranging from 5°C to 9°C in August, the flora of Japan was vastly different. The Japanese archipelago was almost entirely covered by coniferous forests at that time. Northern Hokkaidō supported coniferous parkland and tundra vegetation, while southern Hokkaidō and northernmost Honshū were covered by northern boreal

1 The wind holes between the rocks ensure that the temperature there remains below 10°C even in summer. Northern Pikas are unable to survive if the temperature rises above 25°C.

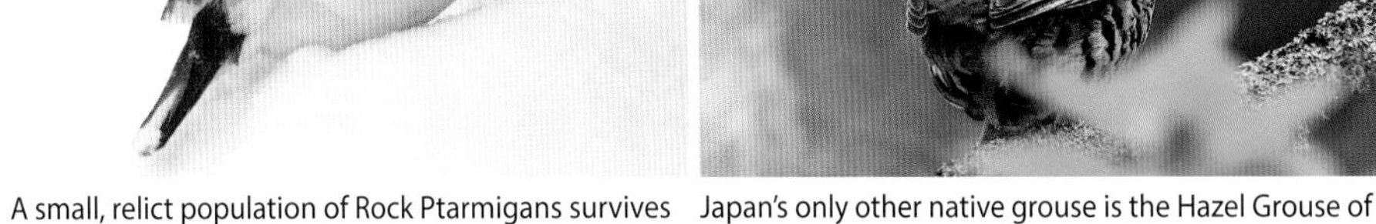

A small, relict population of Rock Ptarmigans survives only in the tundra-like alpine zone [ImM].

Japan's only other native grouse is the Hazel Grouse of Hokkaidō [WaM].

coniferous forests. Central Honshū and montane southwestern Japan supported subalpine coniferous forest, a habitat now restricted to the central mountains. Where there are now mixed temperate deciduous broadleaf and temperate conifer forests, then boreal conifer forests covered extensive areas. Where there are now warm-temperate broadleaf evergreen forests there were then temperate coniferous forests, and those broadleaf evergreens survived in a limited area of the palaeo-Yaku Peninsula.

Climatic changes and glacial oscillations have crucially affected the evolutionary histories and distributions of many species in Japan, but one tree that has been well studied is Itaji Chinkapin. Research into the genetic diversity of the species has shown that several major genetic lineages survived and evolved in isolation in multiple refugia (along the western Sea of Japan, in southern Kyūshū, and in the Ryūkyū Archipelago) during the last glacial maximum, and expanded their ranges as Japan warmed during the post-glacial period.

During past cold periods, the summer snow line in Japan's mountains has been much lower, and the boreal habitats of northern, central and even western Japan have been far more like those of parts of eastern Siberia today. Species which now are found only much farther north once extended their ranges

Spring-flowering Dogtooth Violets grow on exposed sunlit banks [MAB].

farther south and west; the forests there at that time consisted more of a mixture of needle-leaved and deciduous trees, lacking the broadleaf evergreen species that dominate central and western Honshū today.

In areas where forest could not yet reach, and in rock piles then exposed, pikas ranged freely across Hokkaidō. As climate has warmed, as broadleaf evergreens have spread northwards through Honshū, different forest types have also spread in Hokkaidō, overwhelming much pika habitat in a process that continues today. As winters become milder and shorter, plants are better able to colonize poorer soils and creep up mountain sides, and pikas are losing ground to advancing vegetation.

Northern Pika of Japan is a relict; it once had a much wider distribution, and today survives only in certain safe refugia. But it is not the only Ice Age relict here. Another Japanese relict is Rock Ptarmigan, known here as *Raichō*. Once more widely spread and occurring at lower altitudes in Honshū, its range has contracted and shrunk as it has retreated into the small confines of the only climatic and vegetation zone left that suits it – the alpine zone of the high mountains of central Honshū. Today one must hike high into the Japanese Alps to locate these birds, yet it is a species that occurs right down at sea level farther north in northeastern Russia. What these two widely separate areas share is a similar environment. The high alpine zone of the Japan Alps is cold enough throughout the year to resemble almost exactly the tundra of the Arctic. So this elevated zone, in fact, represents Japan's mini Arctic high in the clouds, and it is that tundra-like habitat in which ptarmigan thrive – so far.

Just like the pika isolated in Hokkaidō, the ptarmigan of Honshū has no further habitat to which it can retreat; the mountains simply go no higher and are not connected to those farther north. The way north is now blocked by inhospitable lowlands, agricultural development, and a marine channel. As summers and winters warm, different vegetation colonizes areas higher into the mountain ranges, different birds and mammals follow, and slowly the tiny islands of tundra-like habitat are disappearing – just as the rare rock piles of Hokkaidō are becoming overwhelmed. In recent, warmer summers, Japanese Macaques have ascended high into the Japan Alps and, for the first time, they have been observed preying on ptarmigan chicks. One day, in the not too distant future, both the ptarmigan and the pika will not be merely relicts in Japan, but lost relicts.

During the last glacial maximum, 18,000–24,000 years ago, the whole of Hokkaidō lay within the discontinuous permafrost zone. The continuous alpine permafrost belt descended to about 700–800 m in elevation in Hokkaidō and down to about 2,200 m in the Japan Alps. For decades, it was thought that Japan *once* had montane glaciers among the high peaks of Honshū, but that they had all disappeared during the post-glacial maximum warming. In the last decade, however, a team of scientists has confirmed the amazing existence of seven flowing glaciers: Sannomado and Komado on Mt Tsurugi (Tsurugi-dake) and Gozenzawa on Mt Tate (Tateyama), in the Hida Mountain Range (Northern Alps), Toyama Prefecture. These three were the first to be recognized in Japan and are the most southerly in East Asia. Subsequently, three more have been confirmed: Kakunezato on Mt Kashima-yari (Kashima-yarigatake), in the Ushiro-Tateyama range on the border of Toyama and Nagano Prefectures; and Ikenotan on Mt Tsurugi, and Kuranosuke on Mt Tateyama, both in the Tateyama range in the same general area. Most recently, in 2019, Karamatsuzawa on Mt Karamatsu, on the border of Toyama and Nagano Prefectures, was also confirmed.

Excavations around Japan since 1946, including perhaps the most famous at Lake Nojiri, in Nagano Prefecture, have revealed that the remnants of a once rich but now extinct Pleistocene megafauna, Naumann's Elephant[1], equivalent to the mammoth, along with Steppe Bison, Yabe's Giant Deer and Ancient Japanese Deer, survived until relatively recently, and were hunted by early Palaeolithic peoples of the Japanese archipelago. It seems that there are more relicts of the ice age in Japan than anyone could have imagined.

1 This species seems to have inhabited Japan from about 500,000 years ago until about 15,000 years ago.

Semi-permanent snowfields linger in the Daisetsu Mountains of central Hokkaidō [*left* MAB], but the final remnants of Japan's glaciers survive only in the Japan Alps [*right* TCSM].

Japan's once rich Pleistocene megafauna included wide-ranging Naumann's Elephant [*left* TOTF] and Yabe's Giant Deer [*right* NNEM].

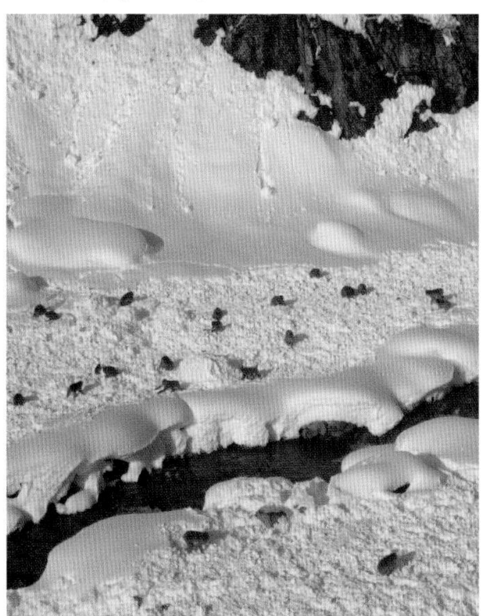

Endemic Japanese Macaques range as far north as the tip of Honshū, but are most famously seen at Jigokudani in the mountains of Nagano Prefecture [*left* MAB; *right* FT].

Japanese Macaque faces are expressive and individually variable [*top left, top right, bottom left* MAB; *others* HT].

Snow Monkey or Hot Spring Monkey?

In the valley of hell, known in Japanese as *Jigokudani*, in the mountains of Nagano Prefecture, in Honshū, lives an internationally famous and frequently televised troupe of extraordinary monkeys exhibiting unique and unconventional behaviour (for a monkey). These are Japanese Macaques and they are endemic to Japan.

A largely terrestrial and partly arboreal primate varying in coat colour from grey to dark grey-brown, this macaque is somewhat paler from the chest to the belly and on the inner surfaces of its limbs. Its face is mostly bare with some sparse dark hairs, and its buttocks, too, are naked. The skin of the face and buttocks is red, the intensity of which varies individually, seasonally, and with the animal's emotional and reproductive states.

Give a moment's thought to most of the primates you know, and you will realize that the majority occur in tropical or subtropical areas with mostly stable temperatures but often highly seasonal precipitation. Some Japanese Macaques are like that, too, as they range from isolated almost subtropical Yaku-shima, an island south of Kyūshū, through Kyūshū itself to the main island of Japan, living in a mostly warm-temperate environment with heavy rainfall. On Yaku-shima a unique endemic subspecies (*Macaca fuscata yakui*) occurs which is smaller, stockier and more vocal, and has darker grey-brown fur than those living farther north (*M. f. fuscata*). Japanese Macaque, however, is also something of a primate anomaly, as it ranges through the main islands of Japan as far north as the very northern tip of Honshū at the Tsugaru Strait, living in a cool-temperate region that is decidedly frigid in winter and historically prone to heavy snowfall.

Not only does Japanese Macaque live relatively far north, but it lives there in a region with a distinctly seasonal climate. The northernmost of Japan's macaques are also said to be the northernmost of all non-human primates. They are the amazingly hardy ones which, by growing long shaggy winter coats, are able to survive the severe snowy winters of the Shimokita Peninsula, at Honshū's northern tip. These are truly Snow Monkeys. Those in the mountains of central Honshū are wimps in comparison, with food provided and a hot bath in which to soak. These would be more appropriately named Hot Spring Monkeys, not Snow Monkeys.

Family members huddle together for warmth in winter [HT].

Throughout the long winters the macaques huddle together in their night-time roosts, waking at dawn to wander in the deep snows of the northern forests, eking out a meagre existence. Food is scarce, and they are often reduced to gnawing at tree bark and tightly closed buds for food, as there is so little else for them to eat during this season. Once the snow begins melting in the spring, as tree buds and spring flowers begin to swell, these omnivorous primates find plenty to eat again in the forest, including bark, buds, leaves, fruits, nuts, invertebrates and small vertebrates, although plant matter does form the bulk of the diet.

They range from the lowlands to mid-elevations in the mountains and are closely associated with broadleaf forests (both deciduous and evergreen), where they are diurnal, and both terrestrial and arboreal. They will forage also out into agricultural land, where they take a range of crops, and they have become agricultural pests in some areas.

Japanese Macaques are social, clannish creatures living in a relatively ordered society for most of the year. Each of their groups, which range in size from ten or so to more than 100 individuals, consists of multi-male *and* multi-female hierarchies. Each troupe is ruled over by an alpha male, with several subsidiary males, a hierarchy of females (dominated by an alpha female), and their offspring. Group sizes and home ranges vary considerably, from 1 km^2 to 25 km^2, depending on the productivity of their habitat; larger groups have smaller range areas in the richer broadleaf evergreen forest than in broadleaf deciduous forests.

Males are larger and heavier than females, being 53–60 cm long and weighing 10–18 kg, females having a length of 47–55 cm and a weight of 8–16 kg. Whereas many species of macaque have a long tail, Japanese Macaques have a short, often barely noticeable tail measuring just 7–12 cm. They may live for up to 25 years and reach sexual maturity at five (females) or six years of age (males). They have a distinct seasonally driven breeding cycle. From October to December both male and female Japanese Macaques develop bright skin, which is particularly noticeable on their faces and buttocks. During the autumn and winter breeding season mating mayhem can be disruptive, with matings frequent and involving multiple partners. Younger males hang out around the periphery of the group, ever hopeful of their amorous advances succeeding in drawing a female away from her group for – just long enough! Gestation is approximately 170 days in duration, most births occurring during spring and autumn. After reaching sexual maturity, females typically give birth to a single youngster once every two or three years, rarely producing twins. Young females will stay with their mother's group, but young males eventually disperse from their natal group and wander in small all-male parties.

With spring come the cherry blossoms and the majority of the infant macaques, and then, by late summer and autumn, the young are being weaned. Autumn is a season of rich nut and berry harvests for them, but soon the temperature drops and snow falls again, and they become snow monkeys once more.

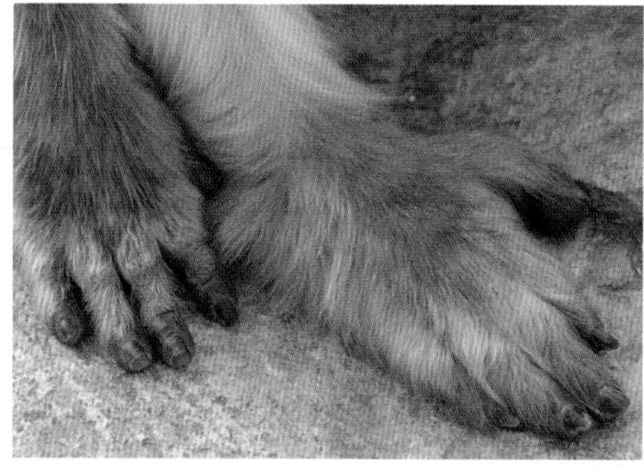

Macaques are adept at picking and plucking thanks to their opposing thumbs, and are clever climbers thanks to the additional grip of their opposing big toes [HT].

On cold winter days young monkeys focus on foraging, whereas when the sun shines they play in the warmth [HT].

Japanese Macaques are inquisitive and playful, but for many of those at Jigokudani the hot spring pool is a major attraction [*left* MAB; *right* HT].

The macaques' diet varies with the season, from twigs, buds and bark in winter to flowers, leaves and fruits from spring to autumn [BOTH HT].

Relationships within the clan are both patrilineal and matrilineal. Each member knows its own status within the group [HT].

Their strong social and family bonds have led to various cultural elements appearing and persisting in different regions. In the south, some macaques have learned to wash their food; others have learned that, if they gather food from the beach and throw it on the water, the sand falls to the bottom while the food floats. In the mountains of central Honshū, the macaques have made another discovery, the pleasure of hot-spring bathing. Their bathing pleasure is transparent. Their coats, where they are exposed, become frosted, or coated with snow, yet they appear warm and content until they walk off to the forest again, when for a while they look bedraggled and scrawny.

Hot-spring bathing among monkeys first began during the late 1960s with a macaque known as *Tokiwa*. Her bathing habits spread to many of the animals in her troop, and to other troops that shared her unusual valley. But how did she learn? It is possible that she watched human visitors to nearby hot pools, and whatever the origin of the behaviour it has remained unique to Hell Valley. Nowhere else in Japan do wild monkeys soak in hot water. At Jigokudani, the monkeys have become globally famous for indulging in their spa treatments, but in fact they are drawn down into the valley by the provision of food, given so that researchers may study their social behaviour. The macaques spend just part of each winter's day in the narrow valley close to a pool that was built specifically for them. Many of the females and their offspring spend time in the pool itself, where observation and photography at close range are uniquely possible.

The hot spring is more than just a place to warm up, as it is also a popular play area for macaque youngsters. Like human children they often break decorum by running along the side, jumping in on top of other users, and harassing those relaxing outside. Play occupies much time for the youngsters, especially on warm days, and may involve playmates, and animate and inanimate objects in the environment. Their play may be boisterous – using branches and shrubs as trampolines – and frequently involves chases that can range across the hillside slopes of their range.

Having learned to wash and salt their food, to bathe and to dive, all in the last 30 years, one wonders what Japanese Macaques will learn next.

Wild Japanese Macaques have been trapped and exported to zoos and research facilities overseas and, conversely, other macaques have been imported into Japan. Both Taiwan Macaque and Rhesus Macaque have been introduced locally in Japan, either intentionally or as escapes from captivity. Taiwan Macaque is smaller but longer-tailed, and has blacker limbs, and can now be found on the Kii Peninsula, Wakayama Prefecture, and on Izu Ōshima. Rhesus Macaque is smaller and longer-tailed, with redder body fur, and occurs on the southern Bōsō Peninsula, Chiba Prefecture, Honshū. These two species are

Streams and small rivers are no barrier to the movement of the macaques [ImM].

Hares and their relatives

The endemic Japanese Hare [YaM]

After a fresh fall of snow, the footprints left by Japan's hares reveal that they are common and widespread. Their prints are distinctive: their small front paws leave two prints close together, usually in a line pointing in the direction of travel, and their large, broad hind feet leave large widely spaced prints to each side of and just ahead of the smaller prints. While it sounds strange, the back-foot prints are at the front, and the front-paw prints at the back of a hare's track.

Japan's two species of hare, the endemic Japanese Hare of Central Japan, and the widespread Palaearctic Mountain Hare of Hokkaidō, are distant relatives of Amami Rabbit (see *p. 312*) and Northern Pika (see *p. 246*). They are also a further example of the significance of the deep marine channels separating the Japanese islands in defining the distribution of species. Mountain Hare occurs widely across Eurasia, the Russian Far East, Sakhalin, Hokkaidō and the Kuril Islands, whereas Japanese Hare is an endemic restricted to Honshū, Shikoku and Kyūshū and associated smaller islands nearby.

The widespread Palaearctic Mountain Hare is found here only in Hokkaidō [YaM].

Both hares are mostly nocturnal and extremely elusive, possibly because they have been very heavily hunted as game species in the past. Making them even harder to spot is the fact that in winter they become pure white with black ear tips. Their white coats not only serve as camouflage against a snowy background, but also provide warmth. The white hairs lack pigment and thus have a hollow medulla, leaving a central core air space that provides extra insulation.

European Rabbit was introduced on a number of offshore islands from Okinawa to southwest Hokkaidō, but thankfully not on any of the main islands. Although in the same family (Leporidae) as the hares, rabbits belong in a distinctly different genus. Rabbits are rapid breeders; their habitats soon become overpopulated, and they cause severe soil erosion by consuming most of the surface vegetation. They differ from hares in bearing altricial young in a burrow system, whereas the hares bear precocial young in a surface nest.

European Rabbits have been introduced on several offshore islands [OT].

closely related to Japanese Macaque, and there is evidence of interbreeding, especially between the very closely related Japanese Macaque and Taiwan Macaque. When sighting macaques unexpectedly in rural Japan, the observer should always look closely at tail length and coat colour to ensure proper identification.

A charming folk tradition surrounds the life of the wild Tanuki [MAB].

Wild and Mythical Tanuki

Tanuki, the partly mythical and partly genuine indigenous creature, is ubiquitous throughout mainland Japan, especially in mountainside forests. In ceramic form it is more widespread, and probably far more numerous, than the wild creature itself, appearing outside taverns and eateries, and in homes throughout the land. The wild creature, on the other hand, is native only to Japan's four main islands and on Sado Island. Elsewhere it ranges across northeast China, eastern Russia and the Korean Peninsula, and it has been introduced in northern Europe.

This rather stout, short-legged creature with a short, bushy tail is an atypical member of the larger dog family, or Canidae, with a distinctive facial mask. Whereas its relatives may opt for more open areas, Tanuki occur in heavily wooded habitat, often close to water. They are also more omnivorous than most other canids, and this is reflected in their teeth. Their grinding molars are enlarged, enabling them to tackle plenty of plant food, whereas their bone-shearing carnassial teeth are rather small. And you will not be kept awake at night by Tanuki for, unlike the true dogs, they do not bark.

Tanuki is an unusual member of the dog family in many ways, not just in being barkless and having a broad omnivorous diet, but also because it almost hibernates. I say almost because it does not actually sleep the winter through, although it does put on weight in autumn and then, from November until about April, retreats to its burrow, where it becomes torpid much like a bear. Tanuki rest throughout winter but may emerge at times to feed, and in the warmer parts of their range they may hardly sleep at all. Like most temperate mammals, they mate and rear their young in the spring and, as a result, the young have plenty of time to learn their way around their home forest in preparation for their first autumn and winter. They are most active from soon after sunset throughout the evening, and then again during the early hours of the morning, during which time they may wander 10–20 km in search of food.

Tanuki is a stocky, medium-sized animal, similar in size to a Red Fox, and superficially fox-like, with a similarly long muzzle and erect ears. The Hokkaidō subspecies (*N. p. albus*) weighs 5–9 kg and measures 54–68 cm in length, plus a tail of 15–19 cm; while the Central Japan subspecies (*N. p. viverrinus*) is smaller and lighter, weighing 3–5 kg and having a head and body length of 50–60 cm, plus a tail of 13–19 cm.

Tanuki differs from the fox, however, in having a more rounded face, very much shorter legs, a shorter tail, and a rather variable brindled coat of pale and dark browns, grey, tan and black. Molecular phylogenetic analysis has revealed, surprisingly, that it is more closely related to African Bat-eared Fox than to any other species. The muzzle and face, particularly around the eyes and ear rims, are black, as are the lower legs. The fur includes long guard hairs of white and black that lend Tanuki a grizzled appearance. The fur is also very dense and so the animal is much sought after by hunters, hence its introduction into Europe for the fur trade. The tail also is heavily furred, hanging almost to the ground, and is very broad at the tip. Its summer coat is shorter than its long, dense winter coat, and there is considerable regional variation in fur coloration and density. The northern subspecies of Hokkaidō has heavier, paler fur than the southern subspecies of Central Japan.

Tanuki [*left* FT] is native to all four main islands of Japan. It is largely nocturnal and at home in forests but ventures also into gardens and pastures; it is unusual in returning repeatedly to use a fixed latrine [*right* MM].

Tanuki [*left* WaM] is frequently mistranslated as 'badger', yet there is a true and endemic Japanese Badger [*right* MM].

Tanuki range from the lowlands to mid-montane elevations and, although closely associated with forest and forest edge, they occur also in agricultural areas and even in suburban gardens. They are entirely terrestrial, foraging only on the ground for fallen fruits and berries, and they will also catch small mammals, such as mice and voles, and occasionally take birds. A high proportion of their diet consists of insects (including beetles) and soil invertebrates, especially earthworms. Unusually among the Canidae, they use fixed latrines, either on the forest floor or in the open, such as on infrequently used paths and tracks – they visit these regularly to defecate, making it easy to study their dietary habits.

Wild Tanuki are rather shy and typically solitary, although pairs, families and even extended families may be encountered. Litters of 3–5 young born during spring are typical, and families remain together until autumn. Culturally, Tanuki is a very different creature and a prominent subject of innumerable amusing and intriguing myths and folk tales. Whereas Red Fox is perceived as a mischievous, slightly evil shapeshifter in many Japanese stories, the mythological Tanuki is mischievous in a more bumbling and humorous way, and according to an old saying it is an even better shapeshifter than the fox. Tanuki is often linked with tea-kettles, because of its similar pot-bellied appearance and its capacity to shapeshift into a kettle! This harmless friendly character is a heavy drinking carouser and womaniser and is believed to entice passers-by into drinking establishments. Consequently, ceramic Tanuki, in a range of shapes and sizes, are frequently placed outside bars and restaurants. These figurines have some features in common: a large hat, a bottle of saké in one hand, a notebook to record purchases in the other, a swollen belly and a greatly enlarged scrotum.

The omnivorous Asiatic Black Bear is Japan's second largest terrestrial mammal [*left* FT; *right* IiM].

Crescent Moon Bear

Japan's two largest terrestrial mammals are both bears, but their ranges neither meet nor overlap in Japan. In the north of the country, in Hokkaidō, beyond the Tsugaru Strait and simultaneously north of that important biogeographical delineator, Blakiston's Line, lives Brown Bear. South of that same strait and line is the domain of Asiatic Black Bear, Japan's second largest land mammal. Historically, the range of Asiatic Black Bear stretched through much of Japan from the Shimokita Peninsula, in northernmost Honshū, south all the way to Shikoku and Kyūshū.

Just as Brown Bears live within sight of Sapporo, Hokkaidō's capital (sometimes even entering the city), some Asiatic Black Bears still live in the Chichibu–Tama–Kai National Park (an area of 1,250 km^2, where Saitama, Yamanashi, Nagano and Tōkyō Prefectures intersect, containing several peaks of over 2,000 m), just 40 km west of the seat of the Tōkyō metropolitan government in central Shinjuku. As a result of human activities, hunting, habitat destruction and forestry in particular, this range has, however, become increasingly fragmented as mostly a montane patchwork. Asiatic Black Bear's Kyūshū and Shikoku populations have already been largely or completely exterminated, and other isolated populations in the mountainous areas of western and central Honshū are seriously threatened. Further isolation is ongoing in many areas, even in northern and central Honshū, and this species is thus increasingly vulnerable to human activity.

Asiatic Black Bear has a blackish pelt which is broken only by a white, crescent-moon-shaped pattern across its upper chest, hence its alternative English name of 'moon bear'. It is much smaller than Hokkaidō's Brown Bear, closer in size to a large, heavy dog, and a different species entirely from the North American American Black Bear.

Honshū's Asiatic Black Bears are omnivorous. In spring, they prefer fresh green vegetation and newly grown tree leaves; in summer, they are partial to berries, cherries, and animal material such as bees and ants; then, in autumn, they forage for beech nuts, acorns, chestnuts and wild grapes. They will also eat forest crabs, earthworms and beetles and will grub up ant and bee nests, particularly the latter, which they raid for honey. They wander widely in search of food, travelling up to four kilometres a day, and move over a range measured in many thousands of hectares. Compared with those in the Himalayan part of their range, male Asiatic Black Bears in Japan are about half the weight, weighing around 70 kg, and they range over more than 100 km^2, females having a smaller range of around 38 km^2.

The food species taken by the bears vary in tune with the relative proportions of those crops. In some areas, the acorns of Mongolian Oak are their main autumnal food supply, but there is dramatic annual variation in acorn production. Such variation in food availability affects bear behaviour, and there is greater human–bear conflict in years when acorn crops are low. In autumn, the bears are driven by instinct to lay down body fat to see them through the winter, which they spend in a state of torpor.

In autumns when food is plentiful, or with mild weather, the bears may not become torpid until late December, although in some years they become lethargic during mid-November. They usually remain hidden until mid-April or early May, but females with newborn cubs may remain in their dens until late June, choosing to emerge once the summer vegetation is tall enough to hide their cubs.

These bears favour broadleaf forests that are rich in food, and they are less numerous in areas where conifers have been planted. Their basic requirements for survival include suitably secure den sites in an area with an adequate food supply. They require a plentiful supply of den sites, because they like to change their den sites from year to year. They often choose sites in gullies or ravines, selecting holes under tree roots or under fallen trees. Not surprisingly, they are more likely to select dens on south-facing slopes, as the winter sun helps to keep these warmer.

By all accounts, the nature of the Asiatic Black Bear is to be shy and elusive. Although individuals of this species are essentially denizens of remote wild mountainous areas, however, their ranges frequently include some human habitation, or they may even live close to large urban areas. Some individuals have learned the benefits of visiting garbage dumps for the tempting and irresistible titbits available there, and others have found that persimmon trees near villages offer a tasty food source.

Of Japan's total forest cover, more than 40 percent is now conifer plantation, meaning that the bears of Honshū have lost a great deal of their energy-rich habitat to energy-poor human forestry. As a consequence, bears sometimes cause damage to trees by debarking Hinoki (or Japanese) Cypress. Such feeding does not endear them to foresters. Very occasionally bears have injured people. As usual in such human–animal interactions, the role of humans as the aggressor and interloper is casually brushed aside.

In summer, Asiatic Black Bears may wander high into the mountains, gradually working upwards as the season progresses and eventually grazing and browsing at high elevations on fresh, lush growth. There, they may be found in high meadows between groves of Japanese Stone Pine. In early winter, when snow lies deeply, and when food is hard to find at higher elevations, the bears are pushed back down to lower altitudes and are forced to eat lower-quality forage, including tree bark. For much of the winter, the bears den up in a cave or under the roots of a tree, passing time in torpor. In cold winters with heavy snow they tend to remain at a single denning site, while in mild winters they tend to wander between several denning sites, sometimes visiting six or more during the course of one winter. They probably suffer most not from the direct impact of hunting but from hunting's indirect effects. The main hunting season in Japan is during winter, from mid-November to mid-February. During this period, because of the density of the dwarf bamboo and steep topography of Japan, many dogs are used by both boar- and deer-hunters. Noisy hunters and their dogs disturb bears away from some suitable dens and foraging areas, thus affecting their survival during winter.

The Asiatic Black Bear is highly dependent on annual tree-seed production; when food is scarce it feeds on tree bark [FT].

Furthermore, young and bold bears are easily shot. In the late 1970s and early 1980s, more than 2,000 were being killed each year, just over 40 percent during the hunting season and the rest under the pretext of pest control. There is no doubt that bears do cause damage to forests, but it is too easy to claim this so as to hunt illegally outside the prescribed season.

Bears are popular among hunters both for their meat and for various organs used medicinally. The prices paid for bears are now enormous and fuel international trade, and illegal trade, too, to supply restaurants and 'herbalists'. Trade in their paws, their meat, and their gall bladders is wiping out local bear populations, and Japan is one of the major dealers. The gall bladder of Asiatic Black Bear has been erroneously credited with many beneficial qualities, especially for gastrointestinal problems, while bear fat is supposed to bring relief from wasp stings – at least, it was suggested as a curative in the 19th century to the Reverend Walter Weston, pioneer of mountaineering in the Japan Alps. Perhaps if people were provided with opportunities to watch bears in safety and to learn of their lifestyle, they would gain some respect for them and be willing to conserve them. It would be wonderful to think that the fortunes of the moon bear may rise once more in the future, rather than be eclipsed.

Ōguchi no Magami[1] – The Wolf

Japan's last wolf was exterminated in 1905, or so it is believed. There are some reports of wolves into the 1930s, and others have said that they survived until the massive deforestation and habitat destruction during the industrial recovery of Japan after the Pacific War. Constant reports of sightings of canids in wild areas raise suspicions in the public mind that wolves may still somehow survive in Japan. There have been concerted searches and attempts to lure any surviving wolves out of the forests by using playback of howls, but without success; one possible observation was reported in the press as recently as in 2019.

Once Japan's top predator, Grey Wolf, was fully integrated into the practical, cultural and spiritual landscapes of two of the three major ethnicities of Japan[2]: wolves were considered deities that were at home in the mountains and forests of Japan, and which were symbols of fertility; they had the power both to help and to harm people. Farmers even prayed to the wolf deity to guard their crops against hungry deer. To the Ainu of Hokkaidō, their wolf was the howling deity known as *Horkew Kamuy*, some even attributing their ancestry to a union between a deity and a white wolf. Prior to the arrival of the Japanese, the Ainu lived in a landscape occupied by wolves and deer, where wolves hunted deer, raised their young and, simultaneously, aided people and occupied sacred mountains and forests as deities.

As a result of human activity and persecution following dramatic deforestation and the introduction of livestock farming, both subspecies of wolf in Japan were exterminated, that found in Hokkaidō in 1896 and that found in Honshū in 1905. The wolf has long since been banished from Hokkaidō and Honshū but has left a lingering legacy of confusion. Once worshipped, wolves were also killed and hunted, especially after the spread of rabies in the 18th century. Following the Meiji Restoration of 1868 and the modernization of agriculture towards the end of the 19th century, the wolf was hunted mercilessly, often for bounties. Wolves were also poisoned with strychnine, because they were regarded as dangerous in Man's world, where they preyed on the new domestic stock being farmed. On occasion they even reportedly attacked people. Unfortunately, both in documentation and in the specimen records, further confusion arose in the naming of creatures in Japan, some of which were *Ōkami*[3] (or wolves) and some of which were *Yamainu* (or mountain dogs). For some authors, the confusion lay in the names used, for others it lay in the biological taxonomy used. A book on the subject entitled *The Lost Wolves of Japan*, while fascinating in its detail, exposes and further fans the flames of confusion. Were Japan's wolves true wolves, or were they wild dogs? Were Japan's wolves exterminated, or do they

1 The Large-mouthed Pure God.
2 Ainu and Japanese had for long lived with wolves and extensive folkloric connection; only Okinawans lived where wolves did not.
3 Meaning the 'great deity'.

still survive? Fossils of domestic dogs dating back some 9,500 years indicate that Jōmon people brought them with them as hunting animals when they migrated into Japan and may have allowed them to crossbreed with wolves after their arrival. Perhaps it was those hybrids that were the *Yamainu*.

The wolf deity is commemorated at certain shrines and in artwork [PUDQ].

Endemic Japanese Hares were prey for Japanese wolves [KREO].

One might be forgiven for daring to ask: 'Weren't we, in retrospect, rather more dangerous to the wolves than they could ever have been to us?' After all, we forced *them* to extinction, not *vice versa*. Elsewhere, such as in Europe, wolves were, until the Middle Ages, a genuine danger in winter, when they ventured out of the forest in search of food. The old Saxon name for January was 'wolf-month' (*Wulf-monath*), as this was the time when being killed by a wolf was most likely, which was undoubtedly why they appear so often as the evil element in folk tales. Human meddling with the ecosystem and its top predator has now resulted in a fast-increasing deer population. Their favourite browsing food grows best around forest edges, and as increasing numbers of fields have been hacked or bulldozed back into once continuous forest, this habitat, this food, and hence *the deer* have increased. No natural predator now remains to act as a brake on their growth in numbers. In the absence of the wolf, Red Fox, too, has increased in numbers; its larger cousin is no longer there to keep it respectfully in its place, numerically speaking. Unchecked fox populations, therefore, have now become the target of mistrust by the public because of the *Echinococcus* tapeworms which they sometimes carry, and inevitably they have become targets for the hunters' guns. Worse still, roaming feral dogs, not uncommon in Hokkaidō, are far more dangerous to man and his livestock. They are dangerous because of their inbred affinity for people, whereas wild creatures, such as foxes (or, in the past, wolves), by their nature, tend to avoid humankind. Because of our own emotional attachments, feral dogs (and feral cats) are often less ruthlessly dealt with than the various wild creatures still struggling to share their rapidly shrinking world with us. In Japan, however, feral dogs and feral cats are legally categorized as game animals and can be shot, although guns are not necessary to eradicate them as they can be effectively caught in traps.

Japan's largest canid, the wolf, was once found throughout the main Japanese islands. Larger than the indigenous Red Fox, but smaller than Grey Wolves of Europe and very much smaller than those of North America, it had a variable coat colour ranging from pale greyish-brown to dark blackish-brown. Its fur was of medium length, but thicker and heavier in winter. It had a long head, short, erect pointed ears, a long narrow muzzle and prominent canine teeth. The limbs were long, the feet quite large with broad pads and long claws, and the tail, which was of medium length, reached almost to the ground.

Those wolves that once occurred widely across Hokkaidō were larger (subspecies *C. l. hattai*; head and body 120–129 cm; tail 27–40 cm) than those once found throughout the mountains of Honshū, Shikoku and Kyūshū (subspecies *C. l. hodophilax;* head and body 95–114 cm; tail 30 cm). Unfortunately, no records of their weights exist. This size difference is considered a consequence of much later contact between the wolves of Hokkaidō and those of Siberia, whereas wolves in Honshū were more isolated and subject to a dwarfing process, not uncommon among mammals long isolated on islands.

This terrestrial, nocturnal and crepuscular carnivore lived socially, in extended family groups, from Japan's lowlands into forested montane areas. The Japanese populations of Grey Wolf took prey ranging in size from Japanese Deer, Japanese Hare and Mountain Hare down to fish, rats and voles, and may have scavenged on beached whales. It is thought that Grey Wolf was the main predator of Japanese Deer. The wolf may have helped to limit the population of the deer, which, since the extermination of the wolf, has become abundant and widespread in Hokkaidō, where it lacks natural predators – a problem that has arisen almost everywhere where man has extirpated the natural top predators.

Searching for Japan's Grey Ghost – The Japanese Serow

Japan is a ghostly country, as readers of Lafcadio Hearn's *In Ghostly Japan* know full well. Ghosts may await us near temples, shrines and graveyards, but one ghost, the Grey Ghost, haunts Japan's mountains. Searching for the solitary Grey Ghost requires a sharp eye, patience and persistence.

The endemic and elusive Japanese Serow, a stocky 'goat-antelope', is a grizzled, frosty grey, and very difficult to spot on the flanks of snow-covered mountains. It moves like a slowly shifting shadow, blending in against the backdrop of its habitat, bare and lichen-encrusted rocks, dark patches amid the snow, or dry vegetation, such that glimpsing one is like trying to catch a will-o-the-wisp. A forested hillside that appears devoid of life and movement through hours of scanning suddenly presents a serow standing stock-still in a clearing between the trees. How did it get there? Was it there from the beginning? Did it emerge from some hiding place, or perhaps from another, ghostly dimension? Like forest wraiths, serows are silent, secretive wanderers, browsing through their forest territories, moving stealthily so as not to attract attention.

Mammals, including bears, monkeys, deer, foxes and squirrels, are all readily recognizable, and familiar, to first-time visitors to Japan. This rather strange creature, however, is rarely heard of outside the country and is frequently confused with other species when sighted. Its coarse ashy, grey-black hair, with pale cheeks and chin forming a prominent beard, camouflages it well in the shady undergrowth of its preferred range. The compact form and long fur causes confusion among first-time viewers, some mistaking serows for wolves, others for bears and, occasionally, even for Wild Boar.

As if it were a creature in a medieval European bestiary, Japanese Serow has been variously, and rather unflatteringly, described as a cross between a cow, a donkey, a pig and a goat! It belongs, along with its five serow relatives, in the family Bovidae, even though its scientific genus means 'goat'. A more informative popular description is 'goat-antelope', for when glimpsed in the shady forest, or forging a way through deep snow, that is how it appears. It is surprising that it has not been dubbed the forest devil, for the serow's head (both male and female) is topped with a pair of short (about 13 cm long), inconspicuous, somewhat Satanic-looking, prong-like black horns. They are conical, curving backwards slightly, and tapering to sharp points bearing strong ridges, or rings, which are an indication of their age.

Weighing 30–40 kg, standing 70–75 cm tall at the shoulder and with a combined head-and-body length of about 70–85 cm, serows are solidly muscular, with a long muzzle, short rounded ears, a stout barrel body, powerful legs, dense, shaggy fur, and a short inconspicuous tail of just 6–7 cm. They are well adapted to cold winters, the damp of the early-summer rainy season and the coldness of the typhoon rains of their native home in the mountains. On cold winter days, when snow clouds clear and the sun shines, they are sometimes to be found standing completely exposed, one flank presented to the sun, their dark, sooty-grey winter coat soaking up the sun's warmth. Their thick coat seems to keep them so warm that on occasion they even overheat and lie in the snow or stand in what feels like bone-chilling montane streams, quietly cooling off while ruminating. Serows are physically well suited to their mountain habitat, where they must push through tangled vegetation, particularly wiry dwarf bamboo, and deep snow in search of food. They are found largely in deciduous broadleaf forests, particularly of

suitable nesting hole below the waterline in a mud bank, and laying her 400–500 eggs. There is no copulation, fertilization taking place externally. The female lays her pearl-sized eggs in paired strings, each resembling a pearly rosary, each globular yellow egg floating in a clear bead-shaped gelatinous envelope, which swells to about 2 cm in diameter. It is the males that exhibit parental care, remaining and guarding the egg strings at the laying site. Oscillatory movements of his tail serve to keep his vulnerable eggs well oxygenated, and the sheer presence of such a large parent must prove effective defence against the attacks of any predatory fish.

The late giant-salamander researcher Tochimoto Takeyoshi of the Hanzaki Institute explains the life cycle of his study species [MAB].

The eggs develop quickly over a period of 8–10 weeks before hatching. The aquatic larvae, measuring approximately 3 cm long when they hatch, have three pairs of fringed external gills, two fingers on each hand, and rear limb stumps. The larvae begin to disperse soon after hatching, in about November, but settle into a home range by the following May. Not until about four or five years of age and 20–22 cm long do they undergo a partial metamorphosis. At that time, the gills are absorbed, the body flattens, and the young change their behaviour. They must adopt a new lifestyle on the river bottom and somehow avoid being eaten by their larger relatives.

Although territorial aggression is not common among salamanders generally, Japanese Giant Salamander males are highly territorial, attacking and driving away all conspecifics, except gravid females. Apart from deaths caused by human activity, fighting between males seems to be the most significant cause of mortality, the overwhelming majority dying during the breeding season in September, most by having their heads severed!

According to Hans Gadow and subsequent narrators, the giant salamander was much esteemed for its very palatable flesh, supposedly tasting like chicken; hunting, however, was made illegal in 1952, when the species was made a Special Natural Monument. The main threat to it now, and the factor which continues to reduce both its range and its numbers, is the relentless impact of river-engineering projects, leaving rivers more akin to storm drains, which are exceedingly unsuitable for giant salamanders. Yet an unaltered salamander river could hold in excess of 350 individuals throughout their acceptable altitudinal (or perhaps water temperature-related) range.

Not everyone is inspired by amphibians. Even the great Swedish naturalist and codifier of the natural world, Carl Linnaeus, is well known for having been prejudiced against both them and those who studied them. As Hans Gadow wrote so succinctly, for the time, in his book *Amphibia and Reptiles*, '*One reason for the fact that this branch of Natural History is not very popular, is a prejudice against creatures some of which are clammy and cold to the touch, and some of which may be poisonous*'. If the enormous Japanese Giant Salamander was the first point of contact, then perhaps even more people might be prejudiced against amphibians and reptiles!

Domestic Mammals and their Ecological Impacts

Maneki-neko are found throughout Japan [MB].

Awareness of the impacts of invasive alien species on Japan's biodiversity has been gaining strength in the last few decades, although the problem itself dates back centuries. Such attention is typically focused on animals such as North American Raccoon, Bluegill and Largemouth Bass, and a host of other species introduced first as pets that subsequently escaped, or were released, causing environmental damage; far greater damage is caused, however, by introduced insects, plants, parasites of food crops and trees, and pathogens.

Historically, a number of other species, such as Horse, Cattle, Sheep, Dog and Cat, have also had impacts on the ecology of Japan ranging from minor to major. Domestic horses probably reached Japan at some point between the 4th and 6th centuries. The purpose of the initial introductions is unknown, but horses were certainly used subsequently for cavalry warfare by the samurai class, who became particularly skilled horseback archers. Horses also played a role in the Shintō religion. Sacred white horses[1] have been donated to, and stabled at, shrines for more than a thousand years. A reduced form of this tradition continues to this day in the purchase and hanging of votive tablets, known as *ema*[2], on special racks at shrines. Horses were not used in agriculture until the important period of rapid modernization – the Meiji Era (1868–1912). Instead, they were used as pack animals for cargo, and as transport for the upper classes.

Japanese breeds of horse were small – ponies really – with a large head and heavy mane. White horses were rare, while those with a dark dorsal stripe were common. Locally bred horses were tough and adaptable, and were rarely shod because they had very tough hooves, but, once European and North American horses were imported and crossbreeding occurred, larger draft animals were bred with different characteristics. Local breeds are known for their endurance, their ability to survive on poor food and in severe weather conditions, and they all share the characteristics of having extremely tough hooves. Currently, horses in Japan are being bred either for meat or for racing, hence some areas, especially in parts of southwestern Hokkaidō, have open ranchland. Horses seem never to have been common enough to have had a major impact on habitats in Japan – unlike cattle.

Domestic cattle (along with domestic pigs and chickens) are said to have been brought to Japan at some point between the 3rd and 6th centuries CE from mainland Asia, and this may have coincided with the introduction to Japan of rice cultivation and Buddhism. With the introduction of Buddhism in the 6th century[3] came a prohibition on eating meat, which lasted until after 1868. Cattle were therefore bred as draft animals and for agriculture, not for food. With the lifting of the ban on meat consumption much changed, cattle were now widely bred for meat and milk, and Japanese beef known as *Wagyu* became world-famous, despite the fact that Japanese people consume relatively little beef. Today, the majority of cattle are kept indoors but, because large areas of farmland are required to produce fodder for them, wide-open farmland for grass production can be seen in east Hokkaidō; even so, additional imported fodder is required.

While pigs and chickens were already being farmed in Japan during the Yayoi Period (1000 BCE to 100 CE), sheep appeared not to have been imported. A sheep seems to have been donated to the

1 *Shinme* = sacred transport of the god.
2 Donating a horse to a shrine continued until the 8th-century Nara Period, but, as it was extremely expensive, shrine-goers began instead to (and continue even to this day to) purchase wooden tablets, primarily at New Year, with a picture of a horse [ema] on one side; they wrote their wishes on the reverse, and hung the tablets as an offering at the shrine.
3 The year of 552 CE is commonly assumed.

Feral cats, of various breeds, occur throughout the country and are a major predator of birds and small animals [MAB].

Domestic horses are now scarce, but distinctive breeds may be found, such as in Hokkaidō and on Yonaguni Island [VD].

emperor as early as the 6th century, but sheep did not become a common domesticated animal, despite the government encouraging farmers to raise them during the 19th century.

The history of the arrival of human commensal species, such as Brown Rat and House Mouse, is harder to trace and their impacts have likely been through competition with native rodents and, perhaps, as seed predators of certain plants.

Today, domestic and feral cats are to be found throughout Japan, from urban areas to dense montane forests, and even on remote islands from northern Hokkaidō to the Ogasawara Islands, and they are extremely numerous. Being voracious predators, they consume birds, reptiles, amphibians and small mammals, and their considerable impact on such populations is observed wherever they have been studied. They have been kept as pets in Japan over many centuries, but a large number are abandoned when unwanted. Such cats living wild are a far cry from Japan's two genuine wild cat species, one occurring on the island of Iriomote and the other on Tsushima.

It seems likely that domestic cats were first imported, along with Buddhism, in the 6th century CE as a means of protecting sacred sutras from being damaged by rodents. Like Buddhism the cats came from India via China and Korea. Cats have been depicted in Japanese scrolls since the 11th century. With the closure of Japan for a prolonged period, it seems that further imports may have stopped, and inbreeding in the Japanese cat population led to mutations that now seem to define the Japanese cat breed, namely large eyes and ears and a short stub-tail. Today it is known as the Japanese Bobtail. Perhaps the iconic cat images of Japan *Maneki-neko* (beckoning cat) and Hello Kitty are even more famous.

The dog may have been the earliest of all domestic animals to reach Japan, as there seems evidence that skeletal remains dating back some 9,500 years to the Jōmon Period[4] have been found. Jōmon hunter-gatherers eventually settled widely in Japan, and clearly already had companion animals. These dogs were small and are generally believed to have been the ancestors of the current Japanese dog breeds. Later waves of immigrants also brought their dogs, this time somewhat larger breeds, some dating back to just two thousand years ago. Scattered skeletal remains suggest that during the Yayoi Period (2300–1700 BCE) dogs may have been eaten as food, something unthinkable earlier and later during Japan's history.

After the Meiji Restoration and Japan's opening to the West, many new breeds of dog were introduced and crossbred with Japanese dogs. Japanese indigenous breeds were used mainly for hunting, including the Akita, Hokkaidō and Shiba dogs, whereas later breeds were kept as pets or companion animals.

4 A period of Japanese prehistory between 14,000 and 1,000 BCE, during which the archipelago was populated by hunter-gatherers.

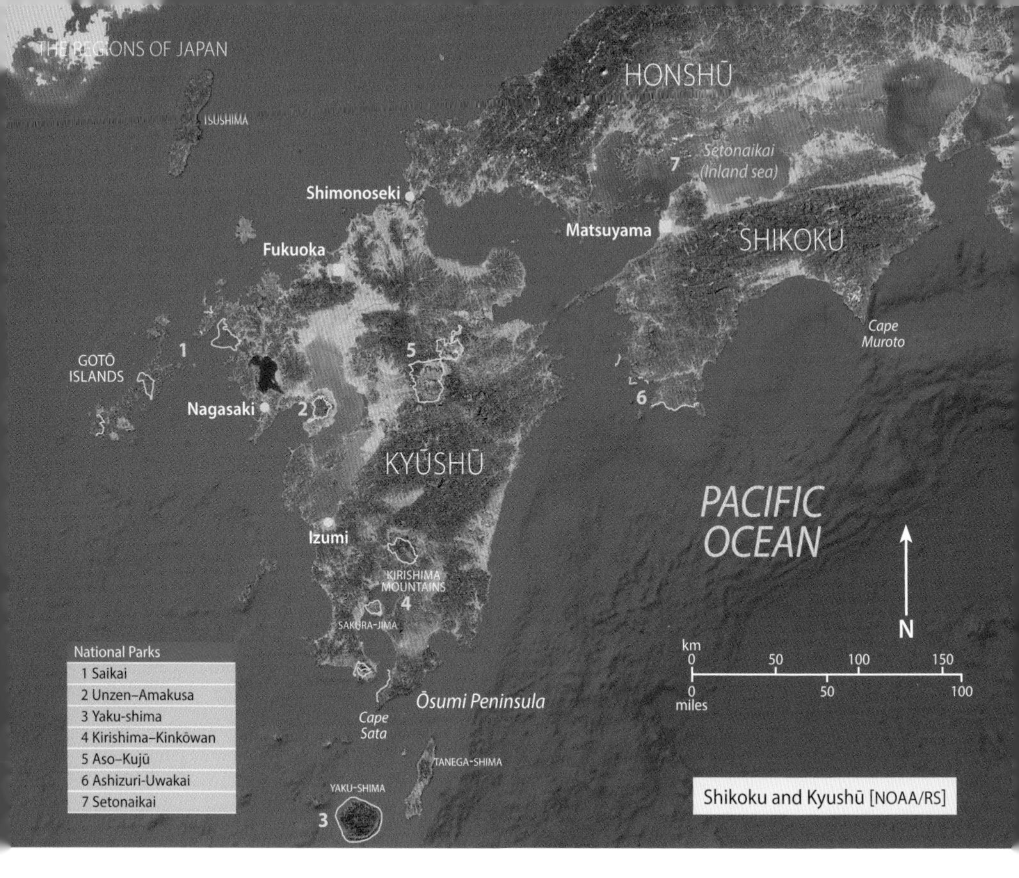

Shikoku and Kyushū [NOAA/RS]

Setonaikai, Japan's Inland Sea, has a distinctly Mediterranean climate [MAB].

Southern Lands – Shikoku and Kyūshū

Kyūshū and Shikoku, the two southern main islands of the Japanese archipelago, have milder climates than either Honshū or Hokkaidō. Japan's Inland Sea, or *Setonaikai*, separating Shikoku from Honshū, was formed as an ancient rift valley and contains more than 3,000 islands. Today it is distinct in Japan in having a decidedly Mediterranean climate. The area is rich in history, having served as the eastern end of the ancient silk route, bringing trade and culture to the region. *Setonaikai* was one of the first areas of Japan to be protected as a national park, in 1934.

Kyūshū's geological past has been as violent as that of Honshū's and it continues to be so today, with numerous volcanoes that are active. The heavily indented west coast, with its innumerable small islands and volcanic mountain peaks, is protected by the Saikai and Unzen–Amakusa National Parks. The beating heart of Kyūshū is an active volcano, Mt Aso, with one of the largest calderas in the world and Japan's second largest. Protected as the Aso–Kuju National Park and the Aso Global Geopark, it offers the most dramatic scenery on the island. The enormous and dramatic Aso Caldera spans 18 km from east to west, 25 km from north to south, and has a perimeter of 128 km. The central cone consists of five mountains: Naka-dake, Taka-dake, Neko-dake, Kijima-dake and Eboshi-dake; while nearby are the Asodani and Nangōdani crater basins, all the result of eruptions after caldera formation, creating spectacular volcanoes in a dramatic volcanic landscape. Mt Aso's four major eruptions span the period from 300,000 to 90,000 years ago, the most recent being the most powerful.

Today, the Mt Aso landscape supports a rare habitat in Japan, grassland, which has been maintained by regular burning for thousands of years, probably initially to increase seasonal food for prey animals such as deer and boar. At 22,000 ha, Mt Aso's grassland is among the largest in the country, forming a landscape of cultural significance, and one that supports rare plant species such as the perennial but critically endangered Kyūshū Jacob's Ladder, and rare and vulnerable butterflies such as Large Shijimi Blue and Scarce Large Blue, both of which are parasites of ant nests.

Whereas Aso dominates the centre of Kyūshū, another volcanic region dominates the south. There, the dramatic Kirishima range and the striking Sakura-jima volcano together are within the Kirishima–Kinkōwan National Park. Kinkō Bay is the enormous bight in southernmost Kyūshū created by the gigantic Aira volcano, which dwarfs that of Aso. The Kirishima area alone is centred on 23 volcanic mountains which are collectively known as Mt Kirishima, the highest three of which are Karakuni-dake (1,700 m), Shinmoe-dake and Takachihonomine. The area is rich in hot springs. From Kirishima looking south, the view is towards the vast Aira Caldera. This enormous caldera is thought to have been formed in a single mega-eruption around 30,000 years ago. Sakura-jima, one of the most active volcanoes in the country, is a cone of the Aira Caldera. It dates back some 16,000 years and stands near the northern end of Kagoshima Bay. Once an isolated island, lava flows from the 1914 eruption connected it to the Kyūshū mainland's Ōsumi Peninsula.

Grasslands, such as those on Mt Aso, are a rare habitat in Japan [MAB].

Today, both Kyūshū and Shikoku are more often battered by typhoons and heavy rainy-season

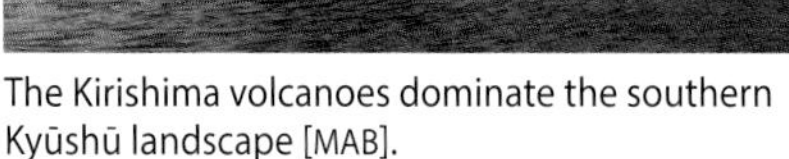

The Kirishima volcanoes dominate the southern Kyūshū landscape [MAB].

Active Sakura-jima seen across the Aira Caldera [MAB].

floods than are any other areas of Japan. Their natural history reflects their ancient connection to the Asian continent and their relative isolation from Honshū. More recently they have been affected by human impact. Kyūshū, in addition, was home to the earliest 'Japanese' clans and is steeped in history. Kyūshū's agriculture dates back further than that elsewhere in Japan, and the entire island has a long-established market garden feel to it, compact and ordered, yet it nevertheless supports some remarkable wildlife spectacles.

Wild Kyūshū

Japan's southern main island of Kyūshū has a pleasant temperate climate for much of the year and in summer the island is almost subtropical. Only in the high mountains is snow frequent. To the northwest lies the Korean Peninsula, by way of which Kyūshū was once connected to mainland Asia, and to the northeast lies Honshū and the remainder of Japan. Kyūshū's natural history shares links and similarities with both, and with the Nansei Shotō islands to the south.

Kyūshū's early settlement and its fertile soils have turned it into Japan's 'market garden' region. Nevertheless, nature survives here in the forested mountains, in the coastal wetlands and on the offshore islands. Intensely volcanic, Kyūshū has distinct highlands where there is a unique flora, including azaleas and other shrubs and flowers known nowhere else in the world.

Kyūshū is the indisputable focal point of azalea distribution in Japan, and the Kirishima region is home to two very special species, Kaempfer's Azalea and Kyushu Azalea. The latter is endemic to the Kirishima Mountains and, together, the two are major candidates for the ancestors of a long-cultivated and beautiful old Japanese hybrid azalea.

The attractive montane forests of southern Kyūshū are rich in species, many of them glossy or evergreen oaks such as Japanese Blue Oak and Japanese Willow-leaf Oak. This richness perhaps arises because of southern Kyūshū's relative warmth, the fertile volcanic soil and the incredible annual precipitation of around 224 cm in the region (up to 10 m in the mountains of Yaku-shima). At lower elevations there are Southern Japanese Hemlock, Evergreen Witch Hazel (also known as Isu Tree), Itaji Chinkapin, and various species of oak. Ascending towards the mountains there are Japanese Red Pine and towering Momi Fir, and above 1,200 m there are cool-adapted Mongolian Oak and Siebold's Beech, and even Japanese Monarch Birch. I say '*even*', because this cool-temperate species is currently at home in Hokkaidō, and at higher elevations in northern Honshū, but during the last glacial period its range extended as far south as Kyūshū, where today it survives as a relict only in these mountains. Where fast-flowing streams drain the mountains, the enormous Crested Kingfisher and the dapper Brown Dipper can be found. During summer in the humid montane forests of the south, Japan's most colourful bird, the elusive but gorgeous Fairy Pitta, can be found. Endemic species occur that are restricted just to Kyūshū and Yaku-shima, such as Yakushima Damselfly[1], just as Shikoku Damselfly is endemic to and occurs throughout that island.

1 This species has populations throughout Kyūshū and another on Yaku-shima.

Kaempfer's Azalea occurs widely in southern Japan [MOEN].

The Kirishima endemic Kyushu Azalea [TOOI]

Ruddy Kingfisher is a summer visitor to humid forested areas of Japan [MC].

The rare Black-faced Spoonbill is a scarce winter visitor to the Ariake Inland Sea [MAB].

Shorebirds migrating between Australasia and northeastern Russia (such as these Great Knot) stop to rest and feed on the shores of the Ariake Inland Sea [NS].

The forested island of Tsushima lies just north of Kyūshū [MAB].

During the spring and autumn migration seasons the coastal wetlands of Kyūshū, especially those around the Ariake Inland Sea, are crucial resting and foraging sites for large numbers of shorebirds travelling between Siberia and Southeast Asia. These include Asian specialities such as Great Knot and Grey-tailed Tattler.

Origami crane lei, symbols of hope, hang at the Hiroshima memorial [MAB].

Ten Thousand Cranes

An ancient Japanese legend promises that the gods will reward anyone who folds one thousand paper cranes by granting them a wish. A group of a thousand cranes, therefore, has come to symbolize a hope, a dream, and a wish fulfilled. And if 1,000 is good then 10,000[1] must be better. The only thing better, then, than 10,000 paper cranes is 10,000 real cranes!

That number of cranes in Japan was unimaginable 40 years ago, yet think of Japan, and think of cranes. Japan and crane images are virtually inseparable. They appear everywhere: as emblems on aircraft, on wedding kimonos, on brands of saké; in art, literature and mythology, not to mention origami – cranes hang in thousands like paper lei at many shrines. It is the 140-cm tall sacred, resident Red-crowned Crane that is most famous, renowned and revered as the Bird of Happiness. Undeniably beautiful, flocks of this species are so often pictured dancing in its snowy northern homeland of Hokkaidō that many visitors are unaware that at the opposite end of Japan, in southwestern Kyūshū, there is an equally astonishing ornithological sight, a

1 Ten thousand (*ichi-man*) is a significant number in Japan, as it is used as a standard unit when counting (rather than thousands as used in the west). ¥10,000 is also the largest denomination in Japan's currency.

Well over 15,000 Hooded and White-naped Cranes now gather each winter near Izumi, in Kyūshū [MAB].

gathering of migratory cranes. Thanks to concerted and prolonged conservation efforts, that small flock (a mere 300 in 1955; in the low thousands in the 1980s) has grown to become a phenomenal winter spectacle of more than 15,000 cranes.

Each winter, one of the greatest wildlife spectacles in East Asia occurs here in Kyūshū. Cranes steadily gather from their breeding grounds in northeast China and Russia, joining larger and larger flocks as they migrate by way of the Korean Peninsula, to spend the winter feeding in safety in the low-lying farmland around Arasaki, near the city of Izumi, in Kagoshima Prefecture in southwest Kyūshū.

The crane flock there is comprised mainly of just two species. The diminutive sooty-grey and white Hooded Crane, 96 cm tall and with a white neck and head, forms the bulk of the approximately 15,000-strong flock. Intermingled among them are good numbers of the taller, more elegant and delicately shaded White-naped Crane, which measures 127 cm tall. A few individual Sandhill Cranes and Common Cranes usually join the gathering each winter and, far less frequently, individuals of rarer species, such as Demoiselle Crane and Siberian Crane, appear from farther west. The stirring sight and sound at dawn and dusk of this enormous flock of birds gathering at their feeding and roosting grounds is a never-to-be-forgotten experience and for many naturalists is the sole purpose for visiting Kyūshū in winter.

Where Red-crowned Crane is a dramatic bird of bold contrast in black and white, with a red crown, White-naped Crane is subtly elegant, encompassing a range of shades from white to dark ashy grey. The drooping wing feathers of White-naped Crane that hang down and conceal its tail are white, while the back is pale dove-grey, becoming increasingly darker grey toward the foreparts. The long dark grey stripes running up the sides of the neck almost to the head can appear black at times, while the neck and head are white with some grey around the red face. It is arguably the most beautiful of all the cranes.

In the late 1800s, this attractive crane was thought to be the most abundant in Japan, when it occurred widely as a winter visitor in Kyūshū, Honshū and Hokkaidō, and even bred here. A century later, following decades of coastal and inland wetland reclamation and steadily intensifying agriculture, its wintering range had contracted and today it is considerably outnumbered by Hooded Crane, and Arasaki is now its *only* regular wintering site.

From October to March each year, both Hooded and White-naped Cranes gather at Arasaki. Whereas 60 years

Japan's iconic Red-crowned Crane is known as the Bird of Happiness [MAB].

The diminutive Hooded Crane is the commonest crane wintering in Japan [TK].

The subtly elegant White-naped Crane [MAB]

Several Sandhill Cranes join the crane gathering each winter in Kyūshū [MAB].

ago these cranes were shy, found only in small, scattered flocks totalling just a few hundred individuals, they now gather *en masse*. By winter 1992/93, the largest number of cranes in Japan in the 20th century was recorded when 8,258 Hooded, 2,102 White-naped and 12 individuals of other species gathered to total 10,372, finally breaking the magical 10,000 crane barrier. This staggering concentration of cranes has become an important tourist attraction, and by the same season approximately 500,000 people were visiting the site each winter. The number of cranes has continued to grow, and by 2018 had reached almost 15,000 birds, making Arasaki by far the most important crane site in Japan and one of the most significant in Asia. It provides a wildlife spectacle that has to be seen (and heard) to be believed.

Parts of the Izumi Plain, where Arasaki is situated, were reclaimed from the sea as early as the 1600s. Now wheat, beans, vegetables and rice are grown on land clawed back from the sea during centuries of labour. Perhaps cranes have always wintered there, originally at the once natural estuary and later on the new agricultural land. In 1921, the area's significance was recognized when it was designated a Place of Historical and Scenic Importance. In 1952, its status was raised to that of a Special Natural Monument, and in the winter of 1962/63 feeding of the cranes on a large scale was begun.

Certain rice fields are kept flooded throughout the winter especially for the cranes, so that they may roost in safety. They pack in closely overnight and then disperse during the day, some travelling tens of kilometres to feed. They all return to roost together each evening around dusk.

In the dim grey light just before a winter's dawn, a wash of sound emanates from rank upon rank of waking cranes. The morning hubbub at Arasaki ranks as one of Japan's great soundscapes as it builds to the bugling cacophony of thousands of cranes. At first light, the view out across the crane fields is of a solid dark line, like a battlefront of birds. They stir and call, reuniting sleeping families. As the flock wakes, the birds first preen; then they begin walking. Droves of cranes surge forwards like unruly mobs, ready to launch themselves into the air. The search for food is on. The flock suddenly swells enormously as birds set out in all directions over the fields. Soon the first skeins of cranes span the sky as it becomes light enough for them to fly, and their comings and goings are trumpeted across the main flock. Having gained height, they fan out, forming wavering lines and Vs bound for their favoured foraging grounds nearby. Some leave towards the lightening sky to the east and south, and may continue for several kilometres, but they will soon be back. The draw of their kind and of the food at Arasaki is irresistible. As the minutes of the new day tick past, the pink flush of dawn appears over the mountains, washing the recently monochrome wooded hills in the distance with colour. By the time the rim of the sun itself appears on the horizon, there are hundreds of graceful cranes in the air. Later they mass and rise in their thousands, swirling noisily aloft. The presence of food stirs them, until they resemble a swarm of angry bees and, as they jostle and fight, they surge forward across the fields. It is not just the sight of this huge mass of birds, but also the sound, the constant two-tone bugling of thousands of individuals, and the flurry of their wings overhead, that makes the experience so powerful.

On clear, spring-like days in late February, the massive gathering begins to fragment. Fine days find some cranes lifting, soaring, spiralling ever higher. Some departees are weak-willed and they return to a life of ease and plenty for a few days more at Arasaki, but then other urges overcome them. Instinct takes over when perhaps a day, or even a week, later the same birds, once airborne, plot a course northwards. Now, scientists track their courses, using satellite telemetry responsive to the tiny transmitters which a handful of individual cranes carry attached to their back like a rucksack. Those impersonal digital signals and data points plot a course that, from the cranes' perspective, reflects a pattern of land and sea, of urban deserts and, here and there, the isolated remnant of wetland, temporary haven and respite for weary wings. The route takes the cranes up through Kyūshū and over to the island of Tsushima, situated in the Korean Strait midway between Japan and the Korean Peninsula. Here is a true spectacle, with thousands of cranes coming and going, briefly resting and refuelling like squadrons of powered gliders, as they hasten northwards to breed.

Having left their winter haven in southwestern Japan, the cranes continue onwards, at first slightly west of north towards the Korean demilitarized zone. From there they turn northeastwards, tracing the

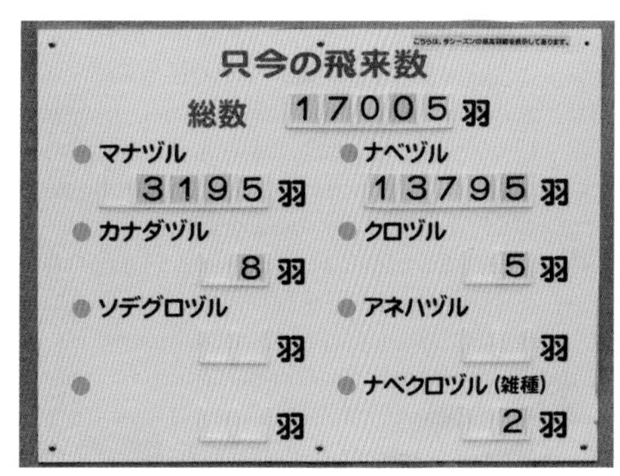

The signboard at the Arasaki Crane Centre indicates the species and numbers of cranes present during winter counts [MAB].

At dawn and dusk huge flocks of cranes move between their roosting and feeding sites in the vicinity of Izumi City [MAB].

In late winter White-naped Cranes commence their northbound migration [KT].

Siberian Cranes are a great rarity in Japan, but they sometimes join the crane flock in Kyūshū [ImM].

coast of North Korea, and thence across Manchuria, north almost as far as Khabarovsk, from where they fan out to find the remote Russian wetlands that are their summer homes. Once the adults have mated, nested and reared another family, they begin retracing their journey back to Arasaki during October, this time guiding the youngsters that they raised during the long summer days on the marshes and steppes of Russia or Mongolia. While instinct provides the direction for migration, cultural transmission from parent to offspring is vital for youngsters to learn the exact location of suitable staging sites.

The number of cranes visiting Arasaki each winter has increased enormously over the last half-century or so, but it is not clear how much of this relates to real growth in numbers (perhaps some) and how much is the result of a greater concentration of many smaller populations that previously wintered elsewhere (perhaps most). Is today's spectacular gathering, which is so huge when compared with that just a few decades ago, a conservation success story or is it a consequence of environmental disaster elsewhere?

There are conservationists who fear that there are just too many crane 'eggs' in this single habitat 'basket'. But how can such a concentration of birds be divided safely? How, in fact, can it be divided at all? So far there seem to be no satisfactory answers. For both species of crane, the Hooded and White-naped, there seem to be no other areas in Japan as appealing as Arasaki. But the proliferation of chicken-farming in the area over recent decades has dramatically increased the risk of transmitting avian flu to the wild cranes. So far disaster has been averted, but for how long? The magnetic appeal of the great gathering of cranes at Arasaki is enormous, overwhelmingly attracting cranes from across the Far East.

Riverside Life

The precise number of rivers and streams in Japan is unknown, but with more than 300 rivers in Hokkaidō alone, and streams not being counted, the number probably runs into several thousands. Because of Japan's narrow, mountainous topography its rivers are characterized by being relatively short and fast-flowing, with steep gradients. Japan's rivers are more often rapidly rushing cold cascades than meandering mild streams. They typically arise from springs in montane forests and in their upper reaches they erode deep V-shaped valleys. Then, in their lower reaches, they deposit the eroded rock to form the alluvial plains that have allowed Japanese agriculture to thrive. Today, many of these lowland plains have been overwhelmed by urban sprawl because of the cultural[1] and pragmatic[2] taboos against building in the mountains.

Of Japan's three longest rivers, the longest is the Shinano River, which flows from Nagano to Niigata, running over a distance of just 367 km, and has a catchment area of 11,900 km^2. The second longest is the Tone River of the Kantō Plain; while it is only 322 km long, it has the largest catchment area among Japan's rivers, at 16,840 km^2. It arises in Gunma Prefecture and, since 1654, it reaches the Pacific Ocean in Chiba Prefecture near Chōshi, although prior to that major realignment it flowed naturally out into Tōkyō Bay. No other Japanese river has been so heavily modified with embankments, dykes and dams, nor so heavily used for agriculture and industry, and by the more than 30 million inhabitants of the Tōkyō metropolitan area. The third longest river in the country is the Ishikari River of western Hokkaidō, measuring 268 km in length and with a catchment area of 14,330 km^2. Japan's rivers are highly modified, having been either canalized and diverted for agricultural irrigation, or dammed to store water and produce hydroelectricity.

Rivers are major attractions for wildlife, supporting a plethora of native fish species along with a great diversity of aquatic insect life, including caddisflies and stoneflies, damsels and dragonflies, along with a wide range of amphibians and reptiles ranging from tiny frogs to giant salamanders. The rivers used to support Eurasian Otter, too, but that is now extinct. River mouths and estuaries are important wintering grounds for hundreds of thousands of waterfowl, such as Eurasian Wigeon and Northern Pintail, Bewick's Swan and Common Shelduck, the species present depending on the region of the country, while the middle and upper reaches are home to wagtails, dippers and kingfishers.

---

1 Considered the domain of various deities, they were so sacred as to be unvisited except by ascetic monastic pilgrims until late in the 19th century.

2 The cultural taboo is backed by practical concerns about frequent landslides, mud slides, floods and other disasters wrought by the annual battering by monsoon- and typhoon-season rains combined with the monotonous frequency of earthquakes.

Japan is as much a land of rivers as it is of mountain. Here the Katashina River flows through the dramatic Fukiware Gorge [MAB].

The rush, chatter and babble of a stream on a summer's day are a great delight; the constantly shifting sounds make an entrancing soundscape and provide a wonderful source of entertainment for the wait-and-see naturalist. In autumn, drifting foliage brings flashes of colour as the water carries and spins fallen leaves far beyond their mother trees. In winter, ice rims stones at the water's edge and coats the branches of riverside trees as they dip and bob in the water; frost sends veils of ice creeping out slowly from the banks, groping steadily towards the middle as the mercury dips and periods of freezing become prolonged. Repeated cold snaps bring a thickening to the ice that forms in layers, and these may shatter, sounding like fine panes of glass, or produce crunching and solid thumping sounds when thick enough to bear weight. But this is a lethal 'pane of glass', leaving fish visible but unreachable from above, as the river's surface becomes like the tempting window of a closed shop. Cross-country skiing along the surface of a frozen river is a thrilling experience, and inevitably leaves me with thoughts of how riverine life survives beneath this ceiling of ice, until warming spring weather finally thaws the surface once more and brings the myriad colours of wildflowers to the riverside.

Bobbing at the waterside there are often Common Sandpipers and occasionally a Green Sandpiper, while in rocky-bedded sections of rivers with shingle bars there are Japanese Wagtail and Long-billed Plover. Along forested upper reaches there may be breeding Mandarin Ducks and, in wilder areas of northern Honshū and Hokkaidō, even Harlequin Ducks. Brown Dipper favours fast-flowing sections and, just about anywhere from source to sea, you may encounter either of Japan's two resident species of kingfisher.

The larger species of the two is Crested Kingfisher (37·5 cm long); its Japanese name, *Yamasemi*, means mountain kingfisher and that is where it is more frequently found. Its finely vermiculated black and white plumage appears grey at a distance and, as it takes flight, white feathers on the underside of its wings indicate a male and rufous-orange underwing-coverts a female. This is an exciting, although not particularly common, resident bird to look for when in mountainous areas up to about 1,300 m from Kyūshū north to Hokkaidō. Sometimes, its loud chattering "*chek chek*" calls draw attention to it before it is visible. This pallid pied bird, with finely barred plumage, has a monstrous beak and an enormous head crowned with a heavy crest that it agitates upwards, giving it a distinctly top-heavy look. This punk of Japan's riverine habitat is hard to miss as it perches in a riverside tree, or stops suddenly in mid-flight to hover over mid-stream before plunging beneath the water and returning with a fish. Easily distinguished from its diminutive turquoise-and-orange cousin (Common Kingfisher (17 cm long)), it whirrs by on grey wings at high speed. During winter, when temperatures are lower in mountain rivers, and fish less active or even locked away beneath ice and snow, Crested Kingfisher may be forced to move to lower elevations. During this season, therefore, they are more often found along the lower reaches of rivers, near springs and even at river mouths.

Japan's smaller kingfisher, despite its rather dull name of Common Kingfisher, is in fact a fiery jewel of a bird. This flying shard of blue-ice with turquoise, red and orange is more appropriately named *Isfugl* in Danish and *Eisvogel* in German, both meaning ice bird, so appropriate for a bird more commonly seen in Japan in winter.

This living jewel has a widespread presence in Japan along riversides, at lakes and reservoirs, and even in urban parks with ponds in winter. It is often overlooked because of its habit of sitting quietly and motionless on a waterside stump or bank where, despite its colours, it can blend into the background. It remains motionless until disturbed, or until some motion in the water leads it to dash out in pursuit of prey. When it flies it gives a hard, penetrating whistle, a short, shrill "*cheee*" as it hurtles off following a low-level track in a blurred blaze of brilliant colour across the water.

Common Kingfisher has a habit of escaping, leaving merely a haunting electric blue trace in the memory. But as a creature of habit, it has favoured perches, flashing back to perch atop a swaying reed, or on a weathered stump protruding from the water, bobbing its large head, dipping its bill, and flicking its stumpy tail, before sinking fractionally on its extremely short legs until seemingly frozen in the guise of the motionless hunter.

Narrow, steep-sided valleys are typical of Japan's montane landscape [MAB].

Long-billed Plovers favour shingle bars and islands in rivers [MC].

The near endemic Japanese Wagtail is at home along rocky-bedded and shingle-bedded rivers throughout the main islands [ImM].

River estuaries in Kyūshū support important numbers of wintering waterfowl, including Common Shelduck [NS].

This small kingfisher is a master angler, imbued with endless patience, but capable of plunging in an instant from its perch into the river below to catch a small fish idling beneath clear water. It shuns white water, however; the disturbance in that highly oxygenated habitat makes it hard for it to spot fish. This kingfisher's small size and its dependence on access to small fish make it particularly susceptible to periods of prolonged freezing weather and many succumb. Under these conditions, the diminutive turquoise, cobalt and blue kingfisher seeks out ice-free waters in the lowlands or in warmer regions, and so, as winter approaches, kingfishers descend from the higher reaches of rivers to milder areas. They are then more readily encountered as they visit the lower reaches of rivers, coastal estuaries, harbours and rocky shores, and even urban-park ponds in the heart of major cities. With winters becoming steadily milder, however, Common Kingfisher is one species that may benefit and become more numerous.

Where brightly coloured or boldly patterned kingfishers demonstratively flash and dive, the widespread Grey Wagtail and the near-endemic[1] Japanese Wagtail walk sedately at the streamside, slowly bobbing their long tails. Meanwhile, drab and dumpy, territorial Brown Dippers make their living along the roughest of waters, in fast tumbling mountain streams, and white water. They pelt up and down rivers at such high speed that they seem to blur into brown streaks as they aggressively chase away rivals from their patch, leaving behind their hard "*djit djit*" calls. They are not gentle.

At first sight, the dipper looks wrong for its riverine habitat – after all, it has neither webbed feet nor a streamlined appearance. In fact, this portly creature appears like a dark and overgrown wren

1 Almost its entire world range is in Japan, but it occurs also on the southern part of the Korean Peninsula.

more than a highly adapted aquatic bird. It is, in fact, the most specialized of river birds. Although its plain dark brown plumage is unrelieved by colour, it has a perkily cocked tail, and everything about this bird's behaviour is fascinating. It appears agitated, hurried and restless. A pugnacious character, proud of its waterfront property, the dipper will see off any intruder, for here property definitely means private.

During winter dippers engage in protective battles along their narrow territorial strips. These involve dashing flights and much bobbing and blinking. But most unusually of all, while the remainder of our resident birds are merely making the best they can of enduring winter, the dipper, truly in its element, sings. As if vying with the rushing stream, daring it to flow faster, splash higher, swirl more wildly, the dipper pours out a powerful song that is audible even over the roaring white-noise of the river. Long before winter loosens its grip, the dipper senses the turning of the seasons, and in song and flight marks out the borders of its riverside property, eager to draw in a mate.

One moment it is an airborne black arrow, the next a stocky bird stands splayfooted, bobbing and curtseying on a boulder in mid-stream. It flicks and turns, bobs and peers, its strong legs flexing and bending like an athlete doing push-ups. Then, with what appears to be suicidal abandon, it jumps into the water and is gone, leaving an expectation that its body, battered by rushing water, will be cast up to the surface bloodied and lifeless. But nothing appears, there is no trace of feather or wing, it is as if the dipper has quite literally been swallowed by the water.

The doughty dipper defies the strength of the stream, it seems even to revel in it. Dropping to the bottom, it turns to face upstream; it grasps the stream-bed litter of stones and gravel with its strong toes and flexed legs and works its way up against the current, literally walking underwater. The flow of water streaming over its head, short neck and broad back forces it downwards, while its stout legs keep it in place as it pokes and probes between and beneath the rocks in search of invertebrates, particularly the larvae of stoneflies and caddisflies.

As if miraculously there it is again, bobbing restlessly; a beaded droplet of water on its mantle is the only witness to its recent immersion. After each foraging bout, it releases its grip, points its beak and head towards the surface, bobs upwards and flies out of the water, often from exactly where it submerged, or farther upstream. It seems oblivious to the relentless battering of the white water over rocks, or the chuckling, gurgling flow around emergent boulders; this is a bird that seems even to like

Crested Kingfishers are dimorphic: males have an orange chest band and white underwings, whereas females lack the orange chest but have orange underwings [ST].

Common Kingfisher [ImM]

Brown Dipper resembles a large Eurasian Wren and has a song powerful enough to be heard over rushing water. It is largely restricted to fast-flowing upper river sections, but may visit the coast in winter [WaM].

waterfalls and is quite willing to fly through a falling cascade of water. Suddenly a fleck of white appears, like a minute curtain drawn across its bright and beady eyes, then it is gone; a few seconds later the white flicker passes across its eyes again as it 'blinks' with its third eyelid – its nictitating membrane. The dipper takes off, and this time it plunges from mid-flight into the water, not with the graceful swan-dive of a plunging kingfisher but with a solid plop, more like a thrown stone.

Females Choose so Males Compete

The natural world is full of examples of choice and competition, two of the driving forces of reproduction. Female vertebrates have no direct means of scrutinizing the genetic bank accounts of their prospective mates; no direct way of deciding whether this male is the best available or whether it would be better to wait for another to come along. Instead, they must rely on secondary characteristics, the appearance of health, the brightness of colours, the boldness of patterns, the diversity of songs, or the extent of adornments, which give an indirect indication of genetic quality.

Some males, for example those of Japanese Deer, compete by means of body size, antler size and rutting calls. Others use vocal methods, as demonstrated by the jovial Japanese Bush Warbler, a brown warbler of dense vegetation, which explodes into a song redolent of warmth and sunshine, and seems to admonish every lugubrious forest walker to 'cheer up' and 'think positively'. It competes by means of elaborate songs sung repeatedly, and loudly, during the breeding season to keep other males away from its territory and to attract females into it for mating. Yet other males bear cumbersome and costly adornments, male Indian Peafowl providing a classic example; it carries around an enormously long

Male Japanese Deer joust to establish their dominance and to attract females, and rely on antler size as a measure of their genetic status [PP].

Male Japanese Bush Warblers compete for territories and mates by means of extremely loud songs [KT].

Males of many species of birds rely on bright plumage and distinctive songs in the competition to win territories and mates; *left to right* Daurian Redstart male [MC]; Siberian Blue Robin male [JW]; Stejneger's Stonechat male [WAM].

train that it displays by fanning out the feathers before females. Male Indian Peafowl does everything it can in its own way to impress the choosy females, in order to further its own genetic lineage. In Japan Indian Peafowl is an introduced species, and found only in the far south, in the southern Nansei Shotō; a more common and widespread bird, however, the male Barn Swallow also carries extended tail adornments and bright colours to impress the females.

The true harbingers of summer in Japan are the birds whose songs seem to summon in the warmth and humidity that banish thoughts of winter. Warblers, buntings, chats, thrushes, all of these bring their particular voices to the chorus that can be heard best in cool early-morning air, when sounds seem clearer and to carry farther.

One summer songster with fantastic adornment is the male Japanese Paradise Flycatcher. It sports a blue bill, a bright blue eye-patch, a tail far longer than its body, a black crest, and an esoteric light show of a song. Scientists know it as *Terpsiphone atrocaudata* – a venerable name that takes its origins from the old Greek words *terpsi*, which means 'delighting in', and *phone*, which refers to voice or sound. Thus, this bird's generic name, *Terpsiphone*, refers to the pleasant songs of the paradise flycatchers. Its specific name, *atrocaudata*, is simply descriptive, for *atro* is derived from the Latin *ater*, meaning black, while *caudata* comes from *cauda* (tail), and thus *atrocaudata* means black-tailed. In Japanese, this 'black-tailed delighter in song' has a more meaningful moniker, a name that is evocative of hot southern summer days and warm summer nights. It is known as *Sankocho*. *Sankocho* translates as 'three lights bird', referring to the three lights of the moon, the sun, and the stars. If you once hear this bird sing, you will have no doubt about the accuracy of this name, for its distinctive song is rendered into Japanese as *Tsuki-hi-hoshi, hoi-hoi-hoi* (*tsuki* meaning moon, *hi* meaning sun and *hoshi* meaning stars): it is literally the three lights bird.

This beautiful summer visitor to southern Japan is startling in its adornment, but particularly striking are the male's extremely long tail streamers. The body size of the males and females is almost the same, but the male's extended tail feathers bring his total length up to 45 cm, compared with 17–18 cm for the female. Glimpsed in flight, he seems like a dark comet undulating between the trees. The males of several of Japan's flycatchers, such as the gorgeous Narcissus Flycatcher and Blue-and-white Flycatcher, rely on a combination of bright colours and beautiful songs to compete with their rivals and attract mates, but in this part of the world only Japanese Paradise Flycatcher has such extraordinary tail adornments.

Male Japanese Paradise Flycatchers have the longest tail of any native Japanese species [ImM].

Shifting Patterns of Distribution

We may view the natural world through the narrow window of time that is the human lifespan. It is easy to assume that what now is remains the same, or that change occurs only at a very slow pace unnoticeable during our brief lives. Yet changes can be seen all around us, occurring for a multitude of reasons in response to many factors, some of them human-induced, and some of the changes are very rapid.

Seawater temperatures and the strength of currents are shifting quickly. The distributions of marine organisms (many of which are specific to very particular temperature ranges) are shifting rapidly in response. Fish stocks and fisheries around Japan are moving northwards as ocean temperatures are warming. Species including the familiar Japanese Deer and Japanese Macaque are penetrating farther and higher into mountain zones as springs come earlier, autumns later, and snowfall decreases.

One group of birds already mentioned, the crows (see *p.* 230), is also affected. Just 40 years ago, Eastern Rook, a locally common flocking species sometimes accompanied by the rarer Daurian Jackdaw, was found only in winter and only in southwestern and western Kyūshū. Over the intervening decades its range has extended, and it is now a locally common winter visitor and migrant all along the Sea of Japan coast from western Kyūshū to southwestern Hokkaidō.

Flocks of Eastern Rooks are now a common sight in winter in western Japan, especially Kyūshū [BOTH KT].

So numerous are the changes in patterns of distribution that it becomes increasingly difficult to map species distributions meaningfully, although bird atlasing projects reveal fascinating changes in distribution over human timescales. Making it more difficult still are the introductions of non-native species. Oriental Magpie, for example, has an extraordinary distribution. In the past, it was restricted to, and resident in, a narrow range of western Kyūshū. Suddenly, it has become established locally in southwest Hokkaidō near the city of Tomakomai. It seems unlikely that this was natural colonization, but more likely that, instead, deliberate or accidental introduction occurred, although, as with many distributional changes, we may never know the real cause.

Daurian Jackdaws are sometimes found among flocks of Eastern Rooks [KT].

Some species of bird are crucial to the distribution pattern of plants, through their dispersal of plant seeds in their droppings. One such species is the ubiquitous Brown-eared Bulbul.

Bulbul Berry-snatchers

Superficially somewhat thrush-like, Brown-eared Bulbul grows to about 28 cm in length and has a longer bill and tail than the average thrush; it is much more conspicuous in the urban, suburban and rural environment than is any of the thrushes found in Japan. It is common in gardens during winter, and readily visits bird-feeders, where it will eat bread crusts and seeds, or halved *mikan* oranges given a chance. For most of the year it is a forest-dwelling omnivore.

The Japanese name for Brown-eared Bulbul is *Hiyodori*, one derivation of which is onomatopoeic, from the bird's "*hi-yo hi-yo*" calls. Another explanation dates from the Heian Period (794–1185), when barnyard millet (*hie*) was a common crop for which this type of bird (*tori*) appeared to have a particular liking. Court nobles of the period would keep these birds as pets, because they were (and are) sociable and easy to feed, and at that time, it is said, the name morphed from *hie-dori* to *hiyo-dori*.

Although it may be easy to dismiss bulbuls as rowdy avian neighbours, a closer look reveals an attractive bird that is mostly olive-brown, frosted with pale shades of grey, and with white-tipped frosted breast feathers. The feathers of the crown, neck, throat and breast are finely pointed and, when fluffed out in cold weather, they stand apart from one another, as if the bird is having a bad-hair day. This bird's most distinctive visual feature, and the one that gives it its English name, is, however, the bold chestnut-brown crescent that sweeps across the face like a comma from behind the eye down on to the side of the cheek. Broadest behind the eye, it tapers to a fine streak on the side of the throat, where it makes the lighter-coloured feathers of the chin look like a pale bib.

The bulbul's deep-brown eyes appear notably perceptive, while the slender, straight black bill is an effective tool; together they make the species highly effective at berry-plucking and bug-hunting. It is hardly a stretch of the imagination to say that, in ecological terms, Brown-eared Bulbuls are among the most important birds in Japan, and perhaps *the* most important. Although solitary and territorial during their early-summer breeding season, once that period is over the families begin to move. The small units of four to seven birds coalesce steadily into larger and larger groups. Then, in dozens or even hundreds, they migrate.

These seemingly choreographed flocks present an impressive sight, especially when passing over headlands, capes or mountains. Characteristically, these bulbuls fly closely together in an undulating manner, their wings appearing to flicker almost in unison in rapid flaps interspersed with the briefest of glides.

Whereas many birds found in Japan make megamigrations, disappearing from one season to the next, bulbuls are an anomaly. Unlike the swans that appear from the north only in winter, or the warblers and flycatchers that spend summer here before heading back to warmer climes, bulbuls are different: they are island-hoppers. In any one area of Japan, you will likely find the species present year-round, although those present in the different seasons are most likely to be different individuals. For example, bulbuls that breed in Hokkaidō head south in autumn to parts of northern Honshū, only to be replaced by birds breeding farther north that move in and remain throughout the winter. Meanwhile, populations farther south move on, and are replaced by their northern cousins. So each population is in turn replaced by the next.

In spring and summer bulbuls eat many different insects. During autumn and winter, however, they are marauders that scan forests, woodlands, parks and urban streets for shrubs and trees the berries of which they can harvest, often descending mob-handed to strip them bare.

Winter-flowering camellias are another source of sustenance for bulbuls, which plunge their beaks deep into the red blooms to feast from their deep reservoirs of sugary nectar. As they come up from

Brown-eared Bulbul is a common garden and woodland bird, foraging on flowers, fruits and seeds. Its wide gape allows it to swallow a wide range of tree and shrub fruits. Many gather to migrate conspicuously in tight flocks [*left* ImM; *right* YM].

between the petals for air, their faces are finely dusted with yellow pollen in a clear example of evolutionary symbiosis. That is because these nectarivores are not just daylight robbers; the rich 'aviation fuel' that they take from the plant is actually a reward for performing a fertilization service, carrying their face-dusting of pollen to their next flowery feast – like a lipstick kiss transferred between lovers.

Although bulbuls are not always beloved by farmers, as their feeding frenzies sometimes inflict damage on orchard and field crops, they are instead true friends of the forest. They may become the crucial salvation of forest plants unable to disperse and keep up with change in any other way during times of climatic change. That is because many birds help plants by dispersing the plants' seeds through their droppings, a 'service' fostered by the plants, which offer them attractive fruit, or nectar-bearing flowers, in return for their own reproductive reward. It is not surprising, therefore, that among Japan's fleshy-fruited flowering plants there is a seasonal correspondence between the phenology of fruiting and the abundance of the very frugivorous avian species on which they depend for dispersal, such as Brown-eared Bulbul. Plants are immobile after all, so who better to enlist as delivery workers than highly mobile birds?

Every rich package of fruit contains seeds. The factors that influence how effective a bird species is as a dispersal agent for those seeds include how far it moves, and in which seasons, and how wide is its gape.

Small birds, with a small bill, such as Japanese White-eye, can open their mouths only sufficiently to swallow small fruits; faced with larger fruits they will likely just peck at the flesh and so do nothing of use in return for the plant. Larger birds, with a larger bill and gape, such as Brown-eared Bulbul, can open the bill wide and swallow a greater range of large, medium-sized and small fruits. Once swallowed, the fleshy fruit is digested, providing energy, while the seeds that were contained in the fruit are carried in the bird's gut until evacuated in its droppings or coughed up if they are particularly large.

Research into bulbul behaviour shows that bulbuls not only consume fruits of an astonishing 53 different species of plant from 24 families but also that stripping the pulp off the seeds as they pass through the birds' digestive tracts improves rates of seed germination. Not only do the bulbuls provide seed transport, carrying smaller seeds farther than they do larger seeds, but they also improve those seeds' chances of germination. Short-distance migrants such as bulbuls, which hop from

Top to bottom Waxwings [SP]; Long-tailed Tits [WaM]; white-eyes [JW]; and thrushes [JW] will all eat berries in winter.

island to island, spread plant seeds through the forest, thereby aiding the reproductive dispersal of a wide range of shrubs.

In temperate regions of Europe, North America and Japan, researchers have found that one characteristic event in forests is the correspondence between plant fruiting periods and seasonal variation in the abundance of dispersal agents, in particular birds. In cool temperate forests, many plants produce their fleshy fruit just as large numbers of frugivorous birds are migrating south. Conversely, in southern warm temperate forests, fruiting periods occur in late autumn and winter, coinciding with the arrival of large populations of frugivorous birds for the winter. Research in central Honshū shows that the numbers of fleshy fruits fluctuate during the summer but increase rapidly during early September, before peaking in late October and mid-November. After that, the number of species producing fruit declines from late November onwards, and remains low during winter. The significant time correlation between numbers of fruiting plant species and numbers of frugivorous birds comes as no surprise – and a crucial frugivorous species is the somewhat demonstrative Brown-eared Bulbul.

Give a thought, then, when next you are awoken in the wee hours, that the noisy shreeping bulbul is actually a 'good thing'. Our subtly shaded but rather noisy local bulbul has clear bragging rights when it comes to being an important seed-dispersal agent. Its large gape allows it easily to swallow fruits of various sizes. Moreover, its island-hopping movements allow the bulbul to disperse the seeds of those fruits from one patch of habitat to the next and so, during times of rapid climate change, bulbuls may become the salvation of forest plants unable to disperse and keep up with change in any other way.

Avian Architects

Birds nest in an extreme variety of different ways. Those such as the Brown-eared Bulbul build a cup-shaped nest in the fork of a forest tree, but the range of avian nests takes in shallow scrapes in sand, and even complex architectural pieces made from mud.

The most abundant and widespread swallow of them all, Barn Swallow, also occurs in Japan. The Japanese consider having swallows nesting under their eaves a blessing, so it is a common sight to see little platforms set up to help to support the cup-shaped nests made of clay which are built by the swallows.

These summer visitors are insectivores that nest in both urban and rural areas. In the north they are associated particularly with cattle yards and cattle barns. Those swallows living in cities have occasionally devised novel ways of dealing with problems. One newsworthy pair of swallows in Kyūshū learned to hover in front of an electronic door sensor to gain access to a building's foyer, allowing the pair to nest inside. Some property-owners, faced with the regular arrival and departure of swallows and the rain of their droppings, have found the same ingenious solution to allowing the auspicious birds to remain while protecting their property from droppings, by hanging an open umbrella beneath the nest.

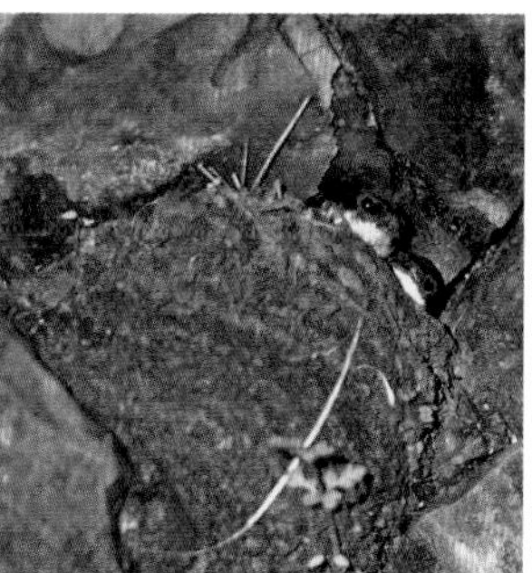

Left to right Barn Swallow is an accomplished architect, building an open, cup-shaped nest of clay [ImM]; the almost completely enclosed clay cup nest of Asian House Martin [KT]; Long-tailed Tits weave a complex nest of spider silk, moss, lichen and feathers [WaM].

Gaius Plinius Secundus, otherwise known as Pliny the Elder (23–79 CE), was among other things a naturalist and the author of the encyclopaedic work *Naturalis Historia*. He perceived that early human architects who built houses of clay took their inspiration from the natural world, in particular from swallows. It is not difficult to imagine some human proto-architect thinking: 'if a bird can build from wet mud, why can't I?' That humans can achieve architectural feats on dizzying scales should come as no surprise. After all, we have enormous brains, are manually dexterous and can call upon on a wealth of materials with a wide range of properties. We can use a vast array of tools and equipment, from minuscule needle-fine drills to mighty tower cranes. That no other mammals come close, architecturally, creates something of a conundrum, but that birds can be superior to social insects in their array of architectural marvels is amazing.

Just think of their consistent excellence in architecture, yet it is a field of endeavour for which birds have little time and *no* training. Birds must build their nests in a matter of a few days, relying entirely on their instincts. The final structure is then used for a matter of only a few weeks, after which it is abandoned, in most cases forever. Birds are, in a sense, stuck in one architectural groove: they have a theme, and that is to build nests and nothing else, year after year. But for those nests, they must gather all of the materials themselves and then build, and shape, the structure using only their bill, their breast and their feet. The range of structures and the materials from which nests are made are phenomenal; the typical avian nest, however, the classical avian architecture, consists of a cup made from vegetation, spider silk, mosses, fine threads and fibres to coarser materials, or even mud.

The building of mud nests is by no means confined to swallows – after all, flamingos and albatrosses make nest mounds from it – but Barn Swallow is a master architect in clay. Clay, as any potter knows, is a tremendous material, but it must be of just the right consistency, with just the right amount of water, to be malleable enough to make things from it. Yet, they get it just right! Barn Swallows breed widely throughout the warmer regions of Japan, and their mud-cup nests are therefore a common site in urban and rural areas alike, often in open garages or outbuildings.

Kawaguchi Magojiro (1873–1937), who collected *kikinashi* bird rhymes in the Hida district of central Honshū, was aware not only of the lingering effects of building a mud nest with a beak, but also of the ability of the Japanese language to approximate the rhythm and sounds of avian song.

'*Tsuchi kute; mushi kute; kuchi shibui!*'[1]

'*Eating earth; eating insects; mouth bitter!*'

Uguisu – Japan's Ubiquitous 'Nightingale'

Japan's fauna produces a tremendously rich array of sounds and songs, from the bizarre deep thrumming sound made by the spread tail of Latham's Snipe during its aerial display flight, and the crazy boiling-oil sounds of some of the cicadas, to the odd whistle of Japanese Deer and the pig-like nocturnal squealing of Okinawa Rail.

Some Japanese avian songsters produce simple, even monotonous *chansonette*, such as the predictably repetitious "*cuc-koo*" of Common Cuckoo, or the deep rhythmic and almost endlessly repeated "*boop boop*" of its lesser-known cousin Oriental Cuckoo, which can be heard throughout much of the country during summer. Other Japanese birds produce delightfully varied, rich, even musical songs. Although not so varied in its repertoire as Common Nightingale[2] of Europe and North Africa, Japanese Bush Warbler, often dubbed the nightingale of Japan, has a song that is both powerful and delightful. It certainly serves as a strong device in Japanese poetry, as does that of the European species in the

1 Japanese is perhaps unique in its capacity through *kikinashi* to reproduce bird songs and animal and nature sounds in meaningful words, such that the sounds of many species can be memorized through verse such as this one. The English translations do not always do them justice.

2 Frequently referenced in literature and poetry since Roman times at least, from Ovid's myth of Philomela's transformation from human form into a nightingale, and her representation of repressed grief, to the Romantic poets' use of the nightingale to represent an enigmatic symbol of both sorrow and joy.

literature of that region. The bush warbler provides a potent motif relating to spring and is included as one of the *kigo* (seasonal words) used in Japanese poetry, especially Haiku.

The bush warbler is a summer visitor to northern Japan and a resident farther south. The first songs of this species may be heard during February in Kyūshū, but not until April in Hokkaidō. A secretive inhabitant of dense foliage, this unadorned brown warbler is best seen before trees leaf out in spring, but it may be heard over several months of spring and summer. While its winter call is merely a hard dry "*tchak*" or "*chek*", its spring song, varied and rich-toned, consists of a long whistled "*pheeuw*", then an explosive burst of three notes, "*hou-ke-kyo*", or it gives a descending staccato series of notes following a rippling precursor: "*tirrrrrrr chepi chepi chepi che-pichew che-pichew che-pichew*". The effort and energy that it pours into this explosive song, and into manoeuvring repeatedly on to multiple suitable perches from which to deliver it, have to be seen to be believed.

They may be humdrum and drab when it comes to breeding plumage, but so great is the effort that bush warblers put into their courtship displays and songs that in summer breeding males develop much stronger musculature in their throat and legs. They have much thicker, more flabby skin on the neck and throat combined with an elastic, inflatable oesophagus, and a larger syrinx (avian voice box) than non-breeding males. In addition, they develop much heavier flexor muscles in their legs, a necessity to support their frequent antics and perch changes during courtship. Their efforts pay off. Because of its song and despite its lacklustre plumage, the bush warbler has nevertheless made its way into a wealth of literature.

Japanese Bush Warbler also figures in the world of cosmetics. It is strangely renowned for its droppings. Said to include a particular and potent enzyme, bush warbler droppings have for long been used in Japan as a skin-whitening and anti-wrinkle agent. They are noted also among wearers of kimono as useful for removing stains.

Chinese Hwamei sing richly varied songs [KT].

Asian Stubtail's song is high-pitched, ventriloquial and insect-like [KT].

Japanese Bush Warbler's song is loud, complex and energetic [KT].

Equally drab, and as notoriously skulking as Japanese Bush Warbler, is its diminutive cousin, Asian Stubtail. This mainly forest-dwelling bird produces a song of an entirely different quality, one so high-pitched and so continuously mechanical as to be beyond the hearing capacity of some listeners and to be mistaken for a cicada-like insect by others.

Competing now in the richness stakes is the song of the introduced Chinese Hwamei. An escaped (or released) cagebird species, the hwamei has proliferated and has now become an established part of the avian chorus throughout much of southern and central areas of Japan.

Lifestyle Choices – From Caching to Butchery

Bull-headed Shrike [YaM]

If you can imagine it, it almost certainly happens in the natural world. Some species live inside other species, on other species, in the extreme cold, within snow, within boiling sulphur springs, pretending to be other species, and even hunting other species and storing them for later.

This caching of food is practised, to varying extents, by at least 45 species of the crow family, including Spotted Nutcracker that occurs in Japan. Spotted Nutcracker has a special throat pouch that enables it to collect several food items and to carry them together to the area where it will hide them for later consumption. Not only do these food-hoarders bury seeds, they also remember where they hide most of them, and can return to find them later when food is less available – even when the area is covered with snow.

Caching seeds and nuts is one thing, but the caching of insect and mammalian prey is the behavioural domain of a very different group of birds, the shrikes. Although eight species of shrike are known from Japan, most are rarities and just one, Bull-headed Shrike, is widespread and common. It is resident throughout much of Japan, a winter visitor in the southwestern islands, and a summer visitor to northern Honshū and Hokkaidō. It occurs in open habitats around woodland edges, in farmland, in the scrub and open areas beside rivers, and even in suburban parks.

Shrikes in general, including Bull-headed Shrike, perch in the open on a high vantage point, such as a treetop, pole or wire, from which they search for prey. Bull-headed Shrikes eat mainly insects and other invertebrates and, upon spotting a large beetle, spider or other invertebrate, they swoop down to the ground to catch it. Occasionally, they will also catch prey while flying or will snatch it from vegetation. Bull-headed Shrike also occasionally captures small mammals, lizards, frogs and birds, including thrushes that are larger than themselves, and during winter, when their preferred prey is less common, they will even eat fruits.

Many things about this shrike are odd. For example in Honshū, they breed in the lowlands during February to May, then move up to the highlands, where they breed again during May to August. In Hokkaidō they appear only during that latter period, so perhaps some of those birds have already bred in Honshū before arriving in the north. Another odd and as yet unexplained behaviour is that shrikes neither bob nor wag their tail, but instead they twist it in a circle!

Their strangest habit of all is butchery. When food is plentiful, Bull-headed Shrikes continue to capture prey and store the excess for later consumption by impaling it conspicuously on a thorn bush or on barbed wire. With shrikes, such caching is believed to serve two functions. The obvious one is to save food for later when demand exceeds supply, and the second is as a sign of a territory boundary. It transpires that shrikes which prepare more food caches in winter sing at a higher tempo and are more successful in attracting mates during the breeding season.

Joining Bull-headed Shrike in winter, the Northern Shrike is a scarce visitor. It is larger, bolder, and shares the same habits and the same bandit mask – which is very appropriate for this northern butcher.

Wild camellias in Japan flower during winter, from December onwards, when they alone bring the beauty of colour to the forests. They may continue blooming through March and into April, by which time they are competing with other forest beauties, such as yellow forsythia, creamy-white magnolia and pale pink cherry blossoms.

The bright winter flowers of the camellia are irresistible to local resident birds such as Japanese White-eye and Brown-eared Bulbul, which visit to feast on nectar and so become pollen-carrying conscripts among the camellia trees.

Although rarely seen in flower, there is a more common camellia species in Japan, and that is *Camellia sinensis*. Planted in ranks of bushes, its leaves and leaf buds are trimmed mechanically to produce the aromatic green beverage beloved in Japan – known as tea!

Japanese Camellia is widespread in mild regions and an integral part of Japanese culture. Its blooms fall complete [TK].

Japanese Sasanka flowers closely resemble those of Japanese Camellia, but fall petal by petal [NZUE].

An important oil can be produced from the seeds of Japanese Camellia [MAB].

Japanese Camellia is common in gardens and in art [MAB].

Japanese White-eyes take nectar from Japanese Camellia, but in return serve the plant by cross-pollinating it with pollen from a previous feeding site [ImM].

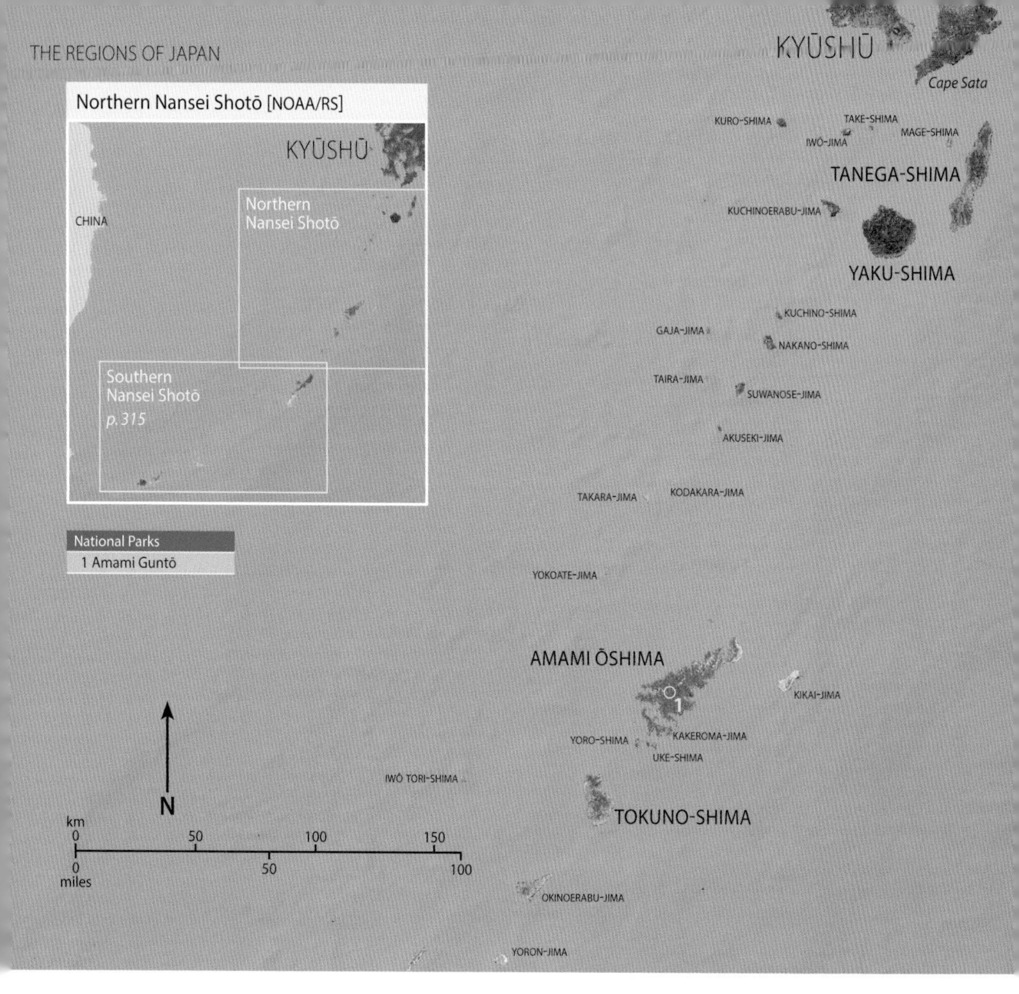

Coral reefs, limestone cliffs and subtropical vegetation mark the coastline of the Nansei Shotō [RC].

Northern Nansei Shotō

One of the defining characteristics of Japan is that it consists of several smaller archipelagos within a larger archipelago. The 1,200-km-long chain of islands lying between Kyūshū and Taiwan is one of those lesser archipelagos, known as the Nansei Shotō and sometimes called the Ryūkyū Archipelago, comprising nearly 200 subtropical islands.

Japan's straggling southern archipelago is located where the Philippine Sea Plate descends beneath the Eurasian Plate. The islands were created 15 million years or so ago, as a consequence of the formation of the Okinawa Trough. Over the last two million years, changing sea levels as a consequence of climatic variation, and the deposition of Ryūkyū limestone, have built up the island chain, along with the formation of the islands' immense coral reefs. Repeated separation from and reattachment to the Eurasian continent, and to one another, have resulted in two very different types of island: 'high islands' composed of pre-Tertiary rocks with hills rising several hundred metres above sea level, such as Mt Yuwan (694 m) on Amami Ōshima, and rivers; and 'low islands', formed of Quaternary Era[1] exposed coral-reef limestone, with plentiful groundwater.

Some 15 Mya, the island chain formed the eastern rim of the Eurasian continent, as a consequence of which the islands hold a similar composition of terrestrial species. As the islands became isolated from the continent by oceanic channels, the biota of those islands also became isolated and, over time, diverged from their source populations as a result of differential selection pressures and random genetic drift. While many of the species surviving today in the Nansei Shotō are relicts of those ancient times of connection to the Eurasian continent, the long isolation resulting from the barrier of the Kuroshio Current, separating the island chain from the continent and from other islands, has allowed a unique biota to evolve independently on these islands. As a consequence, Japan's southern archipelago is rich in flora and fauna, and offers a clear example of the effects of geographical isolation resulting in genetic change. The islands are often described as a living museum, displaying the same natural phenomena as shown in the Galápagos.

Periodic connection and isolation have set the stage here that has resulted in the evolution of endemic species of plant, insect, bird and even mammal, and so created a biodiversity hotspot in these islands. By way of illustration, the warm-temperate forests of southwestern Japan may support 50 to 70 species of tree per hectare, whereas a typical subtropical forest in the Nansei Shotō supports about 100 species. Although these subtropical forests may at first seem superficially similar to the evergreen broadleaf forests of the warm temperate zone, their tree-species diversity is actually much higher and their dominant canopy species are evergreen broadleaf species, such as Itaji Chinkapin, a member of the beech family, and Needlewood, a member of the tea family.

Rich coral reefs fringe many of the southwestern islands [KuM].

A Needlewood Tree in bloom [MOEN]

1 Beginning approximately 2 Mya and lasting until the present.

Over geological times these islands have been linked to one another, and at other times they have been disconnected from each other and from the main island of Kyūshū to the north and Taiwan to the south and have evolved their own unique, endemic species.

The Nansei Shotō archipelago is a string of biological 'pearls' so rich in species and so fascinating in their biogeographical differences that they warrant being far better known. Here, in long isolation, live a number of species endemic to these small islands. These coral-fringed islands support subtropical forests, with tree ferns and cycads, where one can search for the extraordinary and highly localized, such as Iriomote Cat[1].

In this archipelago, two regions in particular stand out as biodiversity hotspots. The first, the northern Nansei Shotō, which encompasses the islands of Tanega-shima, Yaku-shima, Amami Ōshima and Tokuno-shima and their associated islands, is the subject of this chapter. The second region, the southern Nansei Shotō, including the islands of Okinawa, Miyako, Ishigaki, and Iriomote and associated islands, is dealt with in the next chapter. The islands of the northern Nansei Shotō combine elements of Japan's main islands, including unique races of Japanese Macaque and Japanese Deer for example on Yaku-shima, with unique single-island endemics such as Owston's Woodpecker and Lidth's Jay on Amami Ōshima.

The forests of the Nansei Shotō south of Yaku-shima are subtropical in character. They are dominated by Japanese Chinquapin, Okinawan Oak and Japanese Bay Tree.

The fauna and flora of the whole region are distinctive, differing dramatically from the main Japanese archipelago to the north, and characterized by high levels of endemism as a consequence of prolonged

1 Although described originally as a unique species in a unique genus, the taxonomic position of Iriomote Cat (once *Mayailurus iriomotensis*) has caused much debate. Morphology suggests that it is unique, although more recent genetic research has led to its being considered a subspecies of the widespread Leopard Cat *Prionailurus* (or *Felis*) *bengalensis*.

Subtropical forest on Amami Ōshima [TsM], with endemic Owston's Woodpecker [*top inset* TK] and Lidth's Jay [*bottom inset* TsM].

insularization. Some species occur throughout the island chain and no farther north, although they are not endemic. For example, the charismatic dragonfly known as Variegated Flutterer ranges from South Asia (*e.g.* Sri Lanka) to the Nansei Shotō, while Orange-tailed Sprite ranges from Southeast Asia to Kyūshū. Other species occur throughout the island chain, sharing their range with the more widespread flutterer and sprite, but are endemic to the Nansei Shotō. Examples include some that are restricted to islands of the northern Nansei Shotō or to the islands of the southern Nansei Shotō, such as Amami Damselfly which is endemic just to the Amami Islands; Japanese Matrona and Ryukyu Damselfly, both of which are endemic to the Amami Islands and Okinawa Islands; and Yaeyama Gossamer-wing Damselfly and the Yaeyama Clubtail, both of which are endemic only to the Yaeyama Islands.

The situation of these islands, straddling the faunal boundaries between the Palaearctic and the Oriental Regions, could hardly be more provocative. Put an entomologist, a botanist, a herpetologist and an ornithologist together in these islands and they will hotly debate where the regional boundary should be drawn. Biogeographical maps of this region should be annotated with 'here lies confusion'. That such debate takes place is witness to this region's extraordinary natural history.

Whereas the Galápagos and Hawaiian Islands comprise collections of volcanoes which emerged from the ocean floor and have been isolated since birth, the Ryūkyū Islands are entirely different. All living things surviving on the former, namely Galápagos and Hawaii, isolated volcanic islands are descended from life forms that flew, rafted or were blown there. Additional factors, however, come into play in the Ryūkyū Islands, which did not arise uninhabited from the ocean floor. Until about 1.7 Mya, the Ryūkyū chain was part of an already inhabited great peninsula extending eastwards from the Asian continent.

This island arc has three unequal parts (northern, central and southern). As sea levels have fallen and risen, the three segments have sometimes been joined as one unit, sometimes isolated as three. At times of lowest sea levels, they have been partly reconnected to the north, to mainland Japan, and at times to Taiwan to the south, allowing the spread of plants and animals from those areas.

Top to bottom The widespread Variegated Flutterer occurs in Japan only in the Nansei Shotō; Orange-tailed Sprite can be found throughout the southwestern islands north to Kyushū; Amami Damselfly is endemic just to the Amami Islands; Japanese Matrona is endemic to the Amami and Okinawa Islands [ALL TsM].

At times of higher sea levels, what were once hills on larger islands became many separate islands. Confusing? Yes, but just imagine what that confused geological history means in terms of biogeography.

Living on these few small islands today are wide-ranging Southeast Asian species that just reach their northern limits here, such as Purple Heron and Cinnamon Bittern. Other species reach their southernmost limit here, such as Japanese Macaques on the island of Yaku-shima. Of particular interest are the endemic species. Some species are endemic to the whole island chain, such as Ryukyu Green Pigeon, while others are endemic to just one major sector of the island arc and occur nowhere else. Finally, some species are restricted to a single island, such as Pryer's Woodpecker and Okinawa Robin. That amounts not just to an enormous biodiversity in a small area, but to some fascinating foci of isolation and evolution. Isolation on an island from other populations of the same species over a long period of time results in genetic change, so that the isolated population becomes distinctive from the source population. The isolated population may be regarded as a race of the original species, a new subspecies, or even a new full species. Such new species on these islands are by definition endemic: they occur nowhere else.

Birds, bats and flying insects colonize islands rather easily; their flight enables them to disperse to the most isolated of islands. Reptiles may take longer to spread to new areas but, because of their ability to go for long periods without food or water, they can survive to reach even remote islands, such as the Galápagos. Amphibians on the other hand, having permeable skin, have little hope of surviving a sea passage. Large mammals, too, display a poor track record of long-distance dispersal across seawater. The ancient connections between the Ryūkyū Islands and continental East Asia, however, have enabled certain amphibians and mammals, such as the ancestors of our Amami Rabbit, to disperse through the region. During their subsequent long isolation, they have evolved into the distinct forms which we see today.

Because of its 20th-century history, Okinawa is the best-known of the Nansei Shotō. Far less well-known are Yaku-shima and Amami Ōshima. The latter is a gem of an island that lies to the north of Okinawa and which, like Okinawa, is a major centre of biological endemism. Here, in lush subtropical forests dotted with tree ferns and sago palms, occur many of the Ryūkyū

Top to bottom The endemic Amami Rabbit [TsM]; and Ryūkyū regional endemics Ryukyu Kajika Frog [TA]; Sword-tailed Newt [TsM]; and Narrow-mouthed Toad [TsM].

regional endemics such as Sword-tailed Newt, Anderson's Crocodile Newt [VU], Narrow-mouthed Toad, Ryukyu Kajika Frog and Hallowell's Tree Frog, as well as species more closely restricted to the Amami group of islands. For example, these islands are home to several ancient endemic amphibia, such as Amami Green Tree Frog, Amami Brown Frog, Amami Tip-nosed Frog [VU], Amami Ishikawa's Frog [EN], and the enormous Otton Frog, the last reaching 14 cm in length. Birds found nowhere else in the world except on this small island group include the extremely attractive Lidth's Jay, a relative of the crows, resplendent in dazzling cobalt and chestnut plumage, and the elusive, more sombre-clad Amami Thrush. Other near-endemic species, *i.e.* those confined to Amami Ōshima, Okinawa and the islands and islets in between, include Amami Woodcock [EN], Pryer's Keelback Snake (sometimes called the Ryukyu Water Snake), and Ryukyu Brown Frog.

Geology, geography and mammalogy move hand in hand here in the Ryūkyū Islands, with clear demarcations subdividing the archipelago into northern, central and southern parts. The mammalian fauna of the northern sector shares affinities with mainland Japan, while that of the southern sector shares affinities with Taiwan and East Asia. In contrast, the mammals of the central islands are highly distinct. Birdwise, we find the same pattern, with no (or very few) locally endemic species in the northern or southernmost island groups, and just the regional endemics occurring throughout; whereas in the central islands there are a number of endemic species, such as Okinawa Rail, Lidth's Jay, Pryer's Woodpecker and Owston's Woodpecker, and Amami Thrush. So, here on subtropical Amami Ōshima, surrounded by coral reefs and situated in the Ryūkyū Islands 400 km southwest of Kyūshū, lives an extraordinary collection of endemic species, the largest of which is a unique rabbit.

Island endemics: Amami Brown Frog (*top left*); Amami Green Tree Frog (*top right*); Otton Frog (*bottom left*); and Amami Ishikawa's Frog (*bottom right*) [ALL TsM].

The single-island endemic Amami Thrush [TsM]

The Ryūkyū regional endemic Anderson's Crocodile Newt [SM]

The Natural Marvels of the Nansei Shotō

Connection and isolation have played an important role in the Nansei Shotō, such that the Kerama and Tokara Straits effectively divide the archipelago into three parts. In times past, the northern section retained a connection with Kyūshū, whereas the southern section retained a connection via Taiwan to the continent. At other times, all islands have been connected, and at still other times of high sea levels all have been isolated.

Not all of the islands share the same species, because each has a unique history. Few of Japan's familiar mammals occur here. There are no squirrels, no foxes, and no Tanuki. Instead, the largest mammals are rather small subspecies of Japanese Deer and Wild Boar. There are, however, other mammals on the islands, and among them are several endemic rodents, such as Amami Spiny Rat, Okinawa Spiny Rat and Ryukyu Tree Rat or Ryukyu Long-haired Rat, and the extraordinary forest-dwelling Amami Rabbit. The largest predator of the region, Iriomote Cat, no larger than a Domestic Cat, was discovered only as recently as 1965 and lives on just one small island, Iriomote-jima.

Some animals of the northern and southern islands of the Nansei Shotō remain genetically similar to their source populations on the main islands of Japan or of southern Asia. Some others living on the islands of Okinawa and Amami Ōshima tell a different story. Genetic analysis has shown that Large Japanese Field Mouse belongs to a group of species that is widespread in the temperate zone. It occurs on all four of Japan's main islands and as far south as the northern Nansei Shotō, but it is entirely absent from the central and southern islands of that chain. Although those Japanese field mice on the various islands of the northern Nansei Shotō have been isolated for some time from those in Kyūshū, they still share many similarities with each other, and are not very distinct from those of other areas. At the other

The largest native land mammal of the Nansei Shotō is Japanese Deer [MAB].

Iriomote Cat is confined to Iriomote Island [MEIW].

end of the Nansei Shotō, in the extreme south, lives Iriomote Cat. Although distinctive, it is closely related to the widespread Leopard Cat of Asia, and research has shown that there has been a relatively recent exchange of genetic material between the continental and island populations. Although unique, Japan's Iriomote Cat is, therefore, a relatively recently derived species.

So, if the animals of the northern and southern islands of the Nansei Shotō remain closely related to those of the main islands of Japan or of southern Asia, from which they have only recently diverged, what of the animals living on the central islands of Okinawa and Amami Ōshima? Here, the situation is very different indeed. Amami Spiny Rat, Okinawa Spiny Rat, Ryukyu Long-haired Rat and Amami Rabbit are four very distinctive mammals of these islands. Research has shown their levels of endemism to be so high that they are considered to belong not only to distinct species but also to distinct genera (the next highest category into which species are grouped), and bear witness to the ancient isolation of these fascinating islands.

Ryukyu Spiny Rat of Amami Ōshima, Tokuno-shima and Okinawa is particularly special, as its lineage is as distinct as that of the various groups of mice, rats, voles and field mice. Ryukyu Long-haired Rat, however, has been found to be much more closely related to the better-known rat species that belong to the familiar genus *Rattus*. Despite occurring on the widely separated islands of Amami Ōshima and Okinawa, Ryukyu Long-haired Rat populations differ only slightly. This is a consequence of relatively recent gene flow between them, even as recently as during the latter half of the Pleistocene, and perhaps as recently as during the last glacial maxima 15,000–24,000 years ago. Conversely, the degree of divergence between the populations of Ryukyu Spiny Rat on the much more closely situated islands of Tokuno-shima and Amami Ōshima is very high.

It seems that, although populations of Ryukyu Long-haired Rats mixed across the land bridges formed between Okinawa, Tokuno-shima and Amami Ōshima during periods of low sea levels, spiny rat populations did not experience the same gene flow. So, what was it that kept these spiny rat populations separate, despite the islands on which they lived having been joined together? It seems that a major divergence between the spiny rat populations of Tokuno-shima and Amami Ōshima had already taken place, such that even when land bridges reconnected their ranges they were unable to breed together. Although to the untrained eye spiny rats from the Tokuno-shima and Amami Ōshima populations look identical, an examination of their chromosomes reveals an enormous difference. Whereas those from Tokuno-shima have just 25 chromosomes, those from Amami Ōshima have 45. Such a massive difference would have made it impossible for individuals from the two populations to breed together, and so they would have maintained their differences, even while the long-haired rats were blending theirs.

As genetic research has already revealed such a major difference between the geographically close populations of spiny rats on Amami Ōshima and Tokuno-shima,

Top to bottom Amami Spiny Rat [TsM]; Okinawa Spiny Rat [KuM]; Ryukyu Long-haired Rat occurs on both Amami Ōshima and Okinawa [KuM].

it is more than likely that future research will prove the population on Okinawa (which have 44 chromosomes) to be equally distinct. In fact, all three populations have been elevated to separate species on the basis of their genetic differences.

In addition to those two rats, the other highly distinctive mammal of the region is Amami Rabbit. This forest creature, although immediately recognizable as a rabbit of sorts, has, on closer inspection, so many distinctive characteristics that it is obviously not closely related to other rabbits and hares, either of Eurasia or of North America.

Amami Rabbit's lineage is an ancient one. The divergence of the genus *Pentalagus*, of which this rabbit is the only living species, occurred some 10 or 20 Mya, making it a truly ancient relict species. Its continued survival, along with both species of Ryukyu spiny rat and Ryukyu Long-haired Rat, in such a small cluster of islands is remarkable, and it forms a vital part of Japan's biodiversity legacy. Unfortunately, rapid forest destruction and the introduction of alien predators to its home islands may prove its undoing.

Japan is extraordinary for its size in that each region has endemic mammalian species, but whereas Hokkaidō, at the northern extremity of the Japanese archipelago, despite having the richest mammal fauna of any part of Japan, has only one small endemic rodent, Hokkaidō Red-backed Vole, Central Japan has many more endemics. So, what makes the Nansei Shotō exceptional? Although other regions of Japan have endemic mammals, in the Nansei Shotō a very high degree of endemism has survived for millions of years despite the very small area of the islands.

What remains unclear about the evolutionary history of the animals of the Nansei Shotō is whether their ancestors arrived in the Nansei Shotō during ancient times, at least 10–20 Mya, and diverged after they had arrived, or whether their lineages diverged first, somewhere outside the region, and they then reached what are now the Nansei Shotō Islands. Whether they arrived and then diverged, or diverged and then arrived, the animals of the Nansei Shotō are, nevertheless, very special indeed.

The mammals of the Nansei Shotō may be inconspicuous and rather difficult to observe, but their distributions, and the relationships between the different species and populations on these islands, provide a fascinating insight into the past history of the archipelago, with its past connections to the north into Honshū, to the south into Taiwan and the Asian continent.

It takes no stretch of the imagination to picture how rising sea levels flooded lowlands, separating larger islands into smaller ones, and marooning creatures on the separate parts of what was once a larger homeland. Similarly, it takes little mental effort to envisage small mammals, rodents, rabbits and the like, being unable to swim the deeper channels that were formed in this way between the new islands. In such isolation they might easily evolve slowly in time in tune with local conditions, and so end up distinctly different from their earlier close relatives. We might expect, therefore, that the closer islands are together the more likely they are to share the same fauna and flora and, conversely, the farther they are apart the more likely it would be that their fauna and flora would become more and more distinctive over time. These very patterns are observable among the mammals of the Nansei Shotō.

Amazingly, other groups of animals show the same pattern. Amazingly because, while it is easy to imagine terrestrial animals being divided by water channels, it is much harder to think of birds being affected by the same 'barriers', yet that is exactly what we find. Given that typical birds can fly, why then would they be so affected by the distribution of islands and the distance between them? Of course, by no means all birds are affected. Many species are migratory, and many migratory birds fly from one island to the next seemingly with relatively little effort. Many species of small bird pass through on migration, and they are quite obviously able to cross water, so how is it that some are unable?

Even within the same species, some populations migrate longer distances than others, while some populations may not migrate at all. Migratory avian populations and migratory species have longer wings than non-migratory ones have. Although non-migrants have wings for flight, they do not have wings suitable for migration. In particular, those birds living in a rich forest environment may have only small territories and so they may need to fly only short distances. In general, fewer predators live on islands, especially small islands, so the birds living there have less reason to fly; they no longer have predators to flee from, so their unnecessary ability to fly long distances may decline over time. In fact, among the birds that live on islands quite a high proportion are entirely flightless, having given up the ability. Typical among these are the rails. These sturdy, omnivorous, ground-dwelling birds do well on islands, or at least they did until humans introduced rats, cats, pigs and the like. Flightless island rails are especially vulnerable to alien predators. The endemic rail of Guam was completely wiped out in the wild by 1987, as a consequence of Brown Tree Snake being introduced to its home island. The snakes, which were alien to the island, were intended as controllers of rodents but, finding the native birds easy to catch, they consumed them quickly. Thankfully, captive breeding has saved the Guam Rail and made it possible to introduce these birds to a nearby snake-free island. Many other flightless island species, however, have disappeared before such conservation efforts could be put into action.

The single-island endemic Okinawa Rail is restricted to northern Okinawa [KuM].

The flightless Okinawa Rail is confined to the forests of the northern part of Okinawa, which it shares with the equally restricted, and even rarer, Pryer's Woodpecker. And on Amami Ōshima is the very attractive forest-dwelling Lidth's Jay, the secretive Amami Thrush and Amami Woodcock. The last-mentioned is seemingly more reluctant to fly than its widespread relative Eurasian Woodcock, but it can fly and it is found also on Tokuno-shima and in northern Okinawa, in addition to Amami Ōshima. But did Amami Woodcock evolve on one of these islands, and manage somehow to cross later to the others, or did it previously have a wider range on what was once a larger island that now has been divided into several islands by changes in sea level? Either way, here in these central Nansei Shotō islands there are at least five distinct bird species that occur nowhere else in the world, not even on the northern or southern islands of the same archipelago. That is an extraordinary coincidence, for they live in the very same central region, within the Nansei Shotō, where the most unique mammals are also found.

The northernmost and southernmost islands of the Nansei Shotō have few endemic mammals, and that pattern is repeated among the birds. Although the ranges of a number of birds, such as Watercock and White-breasted Waterhen, creep north just far enough to include the southernmost Nansei Shotō islands and nowhere else in Japan, none of the species there is locally endemic. Similarly, some species more typical of Kyūshū and farther north extend just down into the northern islands and no farther south, but again no birds are locally endemic to those northern islands. It seems that the basic pattern

Amami Woodcock has distinctive plumage features and a bare pink patch around its eye [KuM].

The widespread White-breasted Waterhen just reaches Japan in the Nansei Shotō [ImM].

Left to right Ryukyu Robin is a regional endemic [TsM]; Japanese Robin occurs widely and as far north as Hokkaidō [JW]; Ryukyu Green Pigeon is a regional endemic [TsM].

of distribution of endemic species on these islands is very much the same for both mammals and birds, even though the latter can fly.

Among the mammals, there are also those that occur on a number of islands in the Nansei Shotō archipelago but not beyond it. In other words, they are endemic to the archipelago as a whole, but they are not so restricted as some of the other species that occur on only one or two of the islands of the Nansei Shotō. Ryukyu Fruit Bat is one such example among the mammals. Once again, there are amazing parallels among the birds, species such as Ryukyu Robin and Ryukyu Green Pigeon having very similar, archipelago-wide ranges without being restricted to any particular islands. Further investigation may show that it is species which have more general, broader habitat requirements that have the wider ranges within the islands, while the more restricted species may be confined because of their narrower habitat requirements.

There is still so much to learn about these fascinating islands, and every piece of the puzzle that is found and fitted in to place makes them seem even more like 'the Galápagos of Asia'.

Now let us look at each island group in turn.

Yaku-shima – The Island In Between

The highest point between the peaks of the Japanese Alps of Honshū that exceed 3,000 m and the even higher peaks of Taiwan is a distinct, lonely massif: the mountainous island of Yaku-shima. Rising from the sea, approximately 70 km from Cape Sata, in south Kyūshū, and isolated since its formation 15 Mya, Yaku-shima has not merely a distinctive character, and a particular local climate, but also a unique blend of natural history. For this reason this small piece of coral-fringed paradise was made a national park and was further honoured in 1993 as a World Heritage Site, Japan's first World Cultural and Natural Heritage site, listed by UNESCO.

Formed by an upthrusting of subterranean granitic magma, it is now roughly circular, and conical. Yaku-shima is only 505 km^2 in area, with a 132-km circumference, yet it is pinnacled and turreted with 40 granitic peaks rising above 1,000 m. The highest mountain, Mt Miyanoura, reaches a challenging 1,936 m, although Kurio (1,860 m), Okina (1,850 m) and Kuromi (1,836 m) are not far behind, making Yaku-shima a miniature oceanic and forest-clad Alps, and a wonderful wandering ground for the avid hiker.

These 'mountains of the sea' generate their own climate – a very wet one. Warm Yaku-shima 'enjoys' an incredible rainfall and is renowned (jokingly) for having '35 wet days a month', creating superb waterfalls and rivers amid these misty, moss-covered peaks. Yet within a very short distance there are staggering climatic changes. Whereas the coastal climate is subtropical, mild, frost-free and relatively dry, with a mere 400 cm of rain each year and temperatures reaching 35°C, the subalpine mountain peaks, with their granitic boulder landscapes, experience the southernmost regular snows in Japan, and yearly precipitation amounting to a staggering 1,000 cm.

The diverse topography and climate of Yaku-shima leads in turn to a diverse and exceptional plantlife. Uniquely in Japan, plants at home in the subtropics and others equally at home in subarctic climates occur within a few kilometres of each other along a very short vertical distribution. Tropical plants, including hibiscus, bougainvillea, bananas, passion fruit, papayas and Japan's northernmost groves of banyan trees, grow near sea level around the coastline. Not far above grow typical temperate mixed forests, home to conifers and broadleaf trees, rich with blossom in spring and coloured foliage in autumn, but additionally carpeted with mosses. Higher still there are species at home in the cool-temperate zone. Above 700 m, the forests consist largely of the island's world-famous cedars and cypresses. These towering, lofty and long-lived trees create a calm cathedral-like atmosphere.

Of the 1,900 species and subspecies of flora recorded from the island 94 are endemic, while Yaku-shima's most memorable botanical feature is its ancient cedars. Even some of the 'younger' ones are more than 1,000 years old, while standing among (and before) them is what is reputedly the world's oldest tree, the Jōmon Sugi. This venerable tree, measuring 25·3 m high, 16·4 m around the trunk and an astonishing 43 m in circumference at the base, is reputed to be as much as 7,200 years old. It even outranks California's oldest Giant Sequoia, known as the General Sherman Tree, which is estimated to be 3,000–4,000 years old;

Mountainous Yaku-shima rises to 1,936 m. Its very wet and mild climate supports lush forest draped with mosses, lichens and liverworts, and ancient Cryptomeria trees, the oldest of which (reputedly the world's oldest tree) is known as Jōmon Sugi [*left* BOTH MAB; *right* BOTH AJ].

it outspans the 4,000-year-old Patagonian Cypress, and easily outstrips the 'ancient' 4,800- to 5,100-year-old Bristlecone Pine of the high mountains of California, Nevada and Utah.

While at mid-elevations Japanese Cedar, Jolcham Oak, Itaji Chinkapin and Isu Tree are typical of the laurel forests, at lower elevations and beside the coast there are Curtain Fig and Sea Fig and forests of Oval-leaf Mangrove.

In ancient times, the mountains of Yaku-shima were thought to be populated by gods, and the cryptomeria forests were considered sacred. By the late 16th century, however, their value as timber was already recognized, especially for the massive *Yakusugi* boles required for the building of temples in Kyōto, then Japan's capital.

Today, the serpentine roots of the ancient cedars embrace a forest floor of rocks and fallen timbers that are thickly moss-covered and slippery. In the ever-present dampness, the rich, loamy smell of the forest pervades the air through which the gnarled cryptomeria trunks struggle slowly skywards.

At higher levels, where dwarf bamboo is abundant, secretive diminutive Japanese Deer browse in small herds. At lower levels, the woods echo to the sounds of Japan's most vocal groups of Japanese Macaques. Here, in such dense forest, individuals call more often than in other habitats in order to maintain contact with other group members. Another unique feature of these macaques is their big cheek pouches. They nibble off large quantities of tree fruits, and these they store in their pouches for consumption later. Spitting seeds as they go, it is the monkeys that contribute to the spread of the next generation of trees. The southernmost population of Japanese Macaques on Yaku-shima share their range with the Japanese Deer. They live with them in a loosely mutualistic relationship, following each other, in a relationship reminiscent of that between Spotted Deer and Rhesus Macaque and Northern Plains Langurs of India. Both deer and monkeys live in social groups and quickly give alarm calls when disturbance or danger threatens; in the case of these species' pairings, deer and monkeys each understand and respond to the other's alarms. It is the deer that reap the further benefit, however, because monkeys are messy eaters. As they climb into trees they select only certain leaves or fruits to eat and send a steady rain of detritus, discarded leaves and the like, down on to the forest floor below. To the deer, the miraculous appearance of fresh food from above must seem like manna from heaven – delicious floral morsels falling from above. While the deer appreciate the rain of leaves, buds and flowers that the monkeys let fall, the monkeys appreciate advance warning of disturbance at ground level.

The heavy rainfall that regularly drenches Yaku-shima makes life miserable for the soaked macaques, which huddle together in family groups against the rain. The same dousing, however, is ideal for damp-loving plants such as mosses and ferns. There are 300 species of fern living on this island alone. Typical of Yaku-shima as a botanical stepping stone, 42 of them are at their northern limit here and 43 are at their southern limit. At the highest levels of all, the lush moss-draped conifer forests give way to Yaku-shima's astonishing mountain-top alpine zone, where subarctic plants capable of withstanding

Mangroves occur around the coast of Yaku-shima [AJ].

Loggerhead Turtles visit Yaku-shima to nest [OM].

buffeting winds and cold temperatures grow. When the rain clouds deign to disperse, magnificent views include all the peaks across the island, the forests below and, in turn, the turquoise ocean below them.

It was for this exceptional natural beauty, its outstanding natural phenomena, and the diversity of its flora and fauna that Yaku-shima was included in the World Heritage List. It is because of its height and geographical location that Yaku-shima has a unique transitional ecology, combining communities that reach their northernmost and southernmost limits. For example, a large number of tropical plants and butterflies, along with other tropical insects and animals, exist at their northernmost limits here, while certain temperate and alpine species occur here at their very southern limits. Neither Japanese Macaque nor Japanese Woodpecker, for example, occur farther south than Yaku-shima.

Loggerhead Turtle [VU] and Green Turtle [VU] visit to nest on the beaches, and the laying frequency of the former is so marked that the nesting beach was designated a Ramsar Site in 2005.

It is the coexistence, on one small island, of plants and animals from such widely different climatic zones that makes Yaku-shima so special. In profile, from the sea, it is Yaku-shima's mountains that stand proud but, up close, it is the great and venerable cryptomeria trees and the richness of life at this crossroads in the ocean that make Yaku-shima so very special. For Japanese Macaques, Japanese Deer and the many other animals and plants that live here, however, Yaku-shima is not merely special, it is home.

Amami Ōshima – Japan in Miniature

Detailed observation and analysis of a small part of the Japanese islands can reveal a representation of the patterns that are written large across the archipelago as a whole. There is an apparent nested pattern of distribution, in which species with larger ranges overlap species with smaller ranges, which in turn overlap species with smaller ranges within them, reminding me of nested Russian dolls, known as *Matryoshka*. Hence I refer to this situation as 'Matryoshka Distribution'.

When we look, for example, at Japan's southwest islands, we find a particularly high level of endemism in the central Ryūkyū region spanning the Amami and Okinawa Islands, separated from other regions by the Tokara Strait to the north and the Kerama Gap to the south. These deeper channels reflect the longer isolation of this central group of islands and help to explain the high level of endemism locally.

An examination of the avifauna of the Amami Islands, in Japan's northern Nansei Shotō, reveals what we might expect from an island with a much smaller area, namely that it supports fewer species in total than do Japan's main islands, although a similar range of species – some resident, some migrant, some widespread, some local and some endemic – can be found.

Analysis also reveals species with a wider distribution, for example species that are widespread throughout Japan from Hokkaidō to Okinawa, such as Japanese Pygmy Woodpecker and Brown-eared Bulbul. There are also species that range fully through Japan's southern archipelago, the Nansei Shotō, including the island of Amami Ōshima, but are absent from Japan's main islands, such as Ryukyu Scops Owl, Ryukyu Green Pigeon and Ryukyu Robin. Then, there are those species which occur only on Amami Ōshima and its very closely associated islands, and nowhere else in the world, such as Owston's Woodpecker, Lidth's Jay and Amami Thrush. In prolonged isolation these endemic species have evolved *in situ*.

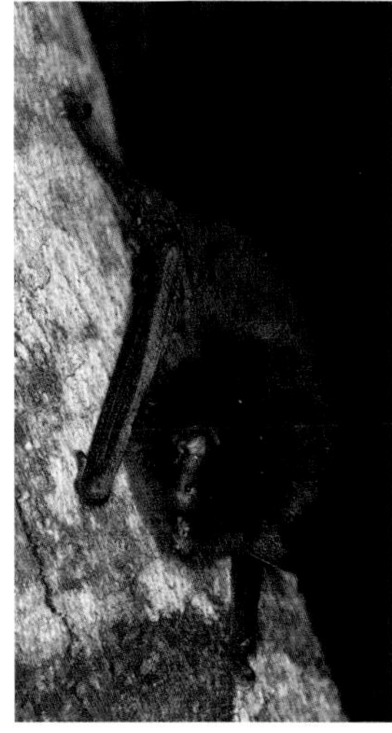

Regional endemics: Ryukyu Tube-nosed Bat (*top*); and Yanbaru Myotis (*bottom*) [BOTH IiM].

The Amami Islands are now registered as the Amami Guntō National Park, which includes the island's coastal coral reefs, its mangroves, coastal mudflats and rivers and its forest ecosystems. The island offers dramatic coastal scenery and sea cliffs of Ryukyuan Limestone. While some Islands have limestone caves, others support some of Japan's largest subtropical laurel forests. The park protects a host of unique taxa, including the endemic Amami Ishikawa's Frog, the endemic Amami Tip-nosed Frog, the regional endemic Ryukyu Tube-nosed Bat and regional endemic Yanbaru Myotis (which, despite its name occurs also on the Amami Islands), and the endemic Amami Damselfly.

Endemism in the Northern Nansei Shotō

Recent studies of the evolutionary histories of some of the endemic species living in the Nansei Shotō have added considerably to our understanding of the evolution that has occurred in this little-known archipelago. Mammals from the northern part of the island chain closest to Kyūshū (for example, Japanese field mice) and mammals from the southernmost part of the island chain closest to Taiwan (for example, Iriomote Cat) exhibit relatively low levels of endemism, whereas those from the more distant, central islands show high levels of endemism. Three mammals, namely Ryukyu Long-haired Rat, Ryukyu Spiny Rat and Amami Rabbit, are so genetically distinct from other populations of similar species that each is assigned to its own unique genus, indicating that these species have been living here for a very long time. All three are regarded as endangered, and each is protected as a Japanese Natural Monument. The lineages of these species have been traced back to their origins in the Tertiary Period (from approximately 66 million to 2.6 Mya); their continued survival in such a small island cluster is nothing short of miraculous.

Recent research on the mammals of the Nansei Shotō has shown that Ryukyu Spiny Rat, once considered a single species ranging across Amami Ōshima, Tokuno-shima and Okinawa, should be reassigned as three separate species: Amami Spiny Rat, Tokunoshima Spiny Rat and Okinawa Spiny Rat. It seems that the spiny rat populations on each of the three islands (Amami Ōshima, Tokuno-shima and Okinawa) have diverged significantly from one another, the population on Okinawa even having twice as many chromosomes as that on Amami Ōshima. The Okinawa population also has very large sex chromosomes, whereas sex chromosomes are absent from both the Amami Ōshima and the Tokuno-shima populations. This absence makes these two populations particularly odd, because sex chromosomes are normally essential in mammals for creating distinct males and females.

Research on sequence variation of mitochondrial DNA of Amami Rabbit has shed light on its long-disputed phylogenetic position, revealing that its lineage has been as independent as that of any other genus in the Leporidae.

The Ancient Rabbit of Amami Ōshima

With tyres spinning on a muddy gravel track through the forest, I tightly grasped the steering wheel with one hand; I was searching for – *rabbits*. In the other hand I held a searchlight, playing the beam around curves into the blackness unreached by the headlights – *not just any rabbits*. A third hand, to prevent my binoculars from crunching into my ribs as I bounced up and down in the vehicle while I drove the rutted trail, would have been useful; and even a fourth would have helped with changing gears, and occasionally wiping away the annoying trickle of sweat which ran down from my forehead into my eyes – *very special rabbits indeed*. Dense vegetation swept the sides of my van, and trees overhung the track, so my scanning pattern with the searchlight had to be a swift sweep around each corner, along each side of the track, under the overhanging brush, along the exposed branches and trunks overhanging the route, and up and down the banks at each side of the track. No, I was not in a jungle-warfare drivers' training camp; I was searching for the mysterious black Amami Rabbit of Amami Ōshima.

Unusual in many ways, the elusive rabbit on Amami Ōshima is, for starters, entirely *nocturnal*. Furthermore, it is a relaxed creature, alert but generally unhurried, ready to melt into the darkness at any moment. It is large, measuring up to 50 cm in length, and weighs up to 2 kg, appearing heavy and thickset. Just occasionally, these rabbits use forest tracks as routes, too, lolloping or hopping across them, lingering briefly under the dense overhanging vegetation at the edge, revealing red-reflecting eyes and rather short, pinkish rounded ears. Another feature always conspicuous for its absence is the tail. This rabbit has no flashing scut with which to signal, in fact nothing at all – it is tailless.

The extraordinary black rabbit of Amami Ōshima [TsM]

At first sight, it is the extremely dark blackish-brown fur that seems most striking, but coat colour in fact is extremely variable among mammals. More distinctive during a closer view is Amami Rabbit's proportions. It lacks the rangy, long-limbed appearance of any of the hares, and it does not share the delicate proportions of the European Rabbit. Instead, it has a rather large head with small ears and eyes, a rather short body and very stout hind legs. Its short, rounded ears are more like those of a pika than those of any other rabbit species, and tipping its stout legs are surprisingly prominent claws. All these points make for a most distinctive creature, and one that is instantly recognizable.

Few species come stranger than this mysterious creature, the largest wild animal on the Amami Islands. It was discovered in 1896, and described to science in 1900, and it occurs only on Amami Ōshima and adjacent Tokuno-shima. In the century since being described, the secret life of Amami Rabbit has eluded most biologists. So different is its morphology from that of other members of the family Leporidae that the ancestors of Amami Rabbit are considered to have diverged as an early side branch from the main leporid branch some 10–20 Mya, or about half as long ago as the pikas and the ancestral rabbits separated.

This rabbit occurs in dense hill forest. There, it excavates simple dens rather than the burrow systems of other rabbit species. Its strong hind legs and conspicuously heavy claws (at 1–2 cm they are unusually long for any rabbit) are ideally adapted to this difficult terrain and digging habit.

During daytime, Amami Rabbits retreat to holes among rocks or under tree stumps, or to burrows that they have excavated in the heavy soil of the mountain slopes. Each evening at dusk, they leave their dens and forage all night on the forest floor, then return at dawn. They move slowly about the forest, following traditional trails, browsing on ground-level vegetation, including growing and fallen leaves, ferns and grasses. All the while they are constantly alert, regularly sniffing and looking around, seemingly very sensitive to both sound and smells.

Amami Rabbit has evolved alongside venomous snakes, including Okinawa Habu [TsM].

Princess Habu commonly preys on amphibians [TK].

Just how did this extraordinary creature come to be so different from other rabbits, and how did it come to be so isolated in the Ryūkyū Islands?

While other rabbits may rear large litters of young several times in a year, Amami Rabbit is a very slow-reproducing animal, breeding just twice, from October to December and again during April and May, and then rearing only one young each time.

Amami Rabbit burrows are simple, short, and typically occupied by just one animal. The female rabbit digs a special birthing and rearing den about 1·5 m long, and ending in a chamber about 30 cm wide and high and 60 cm long and lined with vegetation. Amazingly, she seals her newborn youngster into it. Once every two nights she approaches, sniffing cautiously around the sealed den to make sure that all is well, and repeating a whistled call. Then, using her mouth and forelegs, she digs away the soil that plugs the den, creating an opening. Once her youngster comes to the entrance, the two nuzzle each other, and then, eventually, the female rises up on her haunches facing her youngster, enabling it to suckle. The female signals the end of suckling by lowering her front legs, and the baby returns to the safety of its den. She usually grooms near the entrance before closing the den again. Sealing is a prolonged process, requiring 20–30 minutes for her to scoop up loose soil, fill in the hole and tamp it into a hard plug with her feet. She then wanders off into the forest.

Each spring and autumn, the secretive forest rabbits of Amami Ōshima breed, rearing just one youngster each breeding season. Rearing takes 30–50 days, and, with feeding as infrequent as just once every two nights, Amami Rabbit's milk must be especially rich and creamy. As the youngster develops, it spends more time outside the den after suckling, eventually learning to dig its own way out. Finally, when the female decides that the youngster is large enough, she no longer seals the den when she heads off into the forest to forage. A couple of nights later, when the mother calls her youngster, it leaves the isolated birth den for the first and last time, and sets off with her. As mother and youngster wander together in the forest, they communicate by calling and by thumping the ground with their hind legs as a warning signal. If left well alone, even this species' distinctively slow rate of reproduction is just one more way that it remains in tune with the ancient forest environment of its island home.

The unique mammals of the Ryūkyū Islands have co-evolved with venomous snakes, in particular various pit-vipers (Okinawa Habu, Princess Habu and Tokara Habu), but not with predatory mammals. Rats accompanied human settlers to the islands and, since agriculture has spread so, too, have the rats, providing increased food for the Habu, which apparently increased in response. In order to control the snakes, local people introduced Small Asian Mongoose to the islands. This failed attempt at biological control has threatened the integrity of the natural forest ecosystem, because the mongoose, instead of eating the snakes, eats various indigenous and endemic mammals, although perhaps even that threat was not so drastic as clear-felling the forests, which was undertaken at an astonishing speed over the same time frame. Today, however, the forests are better protected, and efforts are underway to eradicate or at least limit introduced alien predators. Thus, the future for Amami Rabbit is looking brighter.

The slow pattern of Amami Rabbit's life ticks away with the seasons, marking the years of an ancient forest that has been isolated for millions of years, and which is home to a fascinating suite of creatures found nowhere else on earth. The ancient mammalian, avian, reptilian and amphibian lineages of the Nansei Shotō, and the natural forests that they inhabit, are not merely a unique Japanese legacy, they are symbolic of the significant biodiversity of the central Nansei Shotō, and of the need for effective conservation here; they are of global significance.

Introduced Small Asian Mongoose are a significant threat to endemic species in the Nansei Shotō [KuM].

Southern Nansei Shotō

Japan's southern extension – the Nansei Shotō – is a separate archipelago in its own right. As described in the preceding chapter, the numerous islands in the warm waters between Japan's southern main island of Kyūshū and the island of Taiwan differ greatly from the rest of Japan. They are biologically diverse, in fact representing a biodiversity hotspot. The present chapter deals with the southern extension of the Nansei Shotō, including the main island of Okinawa and its subsidiary islands such as the Kerama Islands, and the island of Miyako, which lies farther south and which is now dominated by sugar-cane agriculture; then, as Japan's last southern outpost, there is the group of islands known as Yaeyama, including in particular the islands of Ishigaki, Iriomote and Yonaguni.

Humpback Whales migrate each year to spend the winter around the Kerama Islands [MAB].

This is Japan's finger in the subtropical pie. While Hokkaidō freezes for part of each year, Okinawa swelters. A more extreme contrast to subarctic Hokkaidō is hard to imagine. Here are flowers, butterflies and trees more commonly associated with the tropics. White, coral-sand beaches and intense sun add to the image. Offshore, there are coral reefs with myriad colourful fish. One may be lucky and find turtles, too, or even migratory Humpback Whales in the warm waters there. The local vegetation includes ancient-looking living fossils known as Japanese Sago Palms or Sago Cycads, and forests that are lush and evergreen with broad leaves. The wildlife here is very different from that in any other part of Japan, including even the islands just to the north.

These islands have had a fascinatingly chequered history. They have experienced long periods of isolation as islands but, when sea levels have fallen, they have been connected to the continent via Taiwan. At other times, higher sea levels have separated them, not only into the islands that exist now but into even smaller fragments. When the islands were connected, species reached them from Asia and moved between the islands. When the islands have been separated, during the long periods of isolation species have evolved and diverged, and unique endemic forms are the result. New species are still being discovered in this hothouse of biodiversity. A new bird species, Okinawa Rail, was described as recently as the early 1980s.

Zamami, in the Kerama Islands, is part of the Kerama Shotō National Park [OCVB].

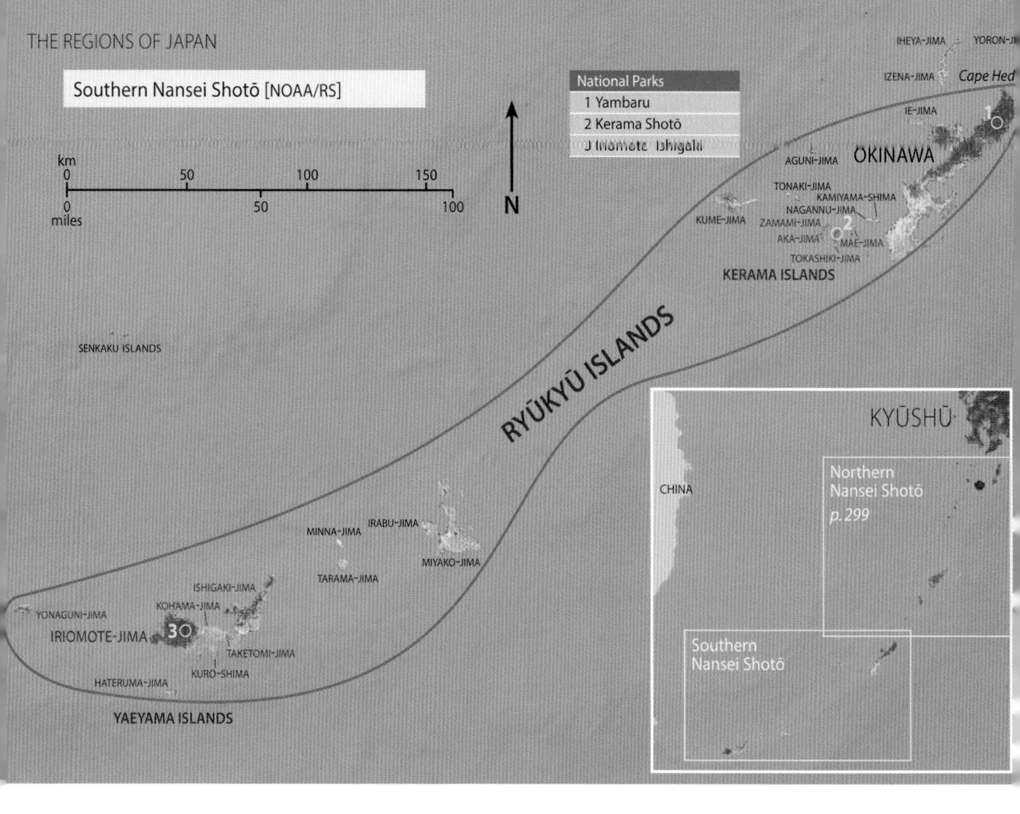

The northern hills of Okinawa sustain important forests supporting rich biodiversity [SM].

Endemic frogs: Okinawa Tip-nosed Frog (*top left*); Holst's Frog (*top right*) [BOTH KuM]; Ryukyu Brown Frog (*bottom left*); and Okinawa Tree Frog (*bottom right*) [BOTH SM].

The hilly northern third of the main island of Okinawa is known locally as Yambaru, and internationally as having an elevated level of biodiversity even within the biodiversity hotspot of the Japanese archipelago. Yambaru National Park spans the rugged coastline at Cape Hedo, at the north end of the island, Gesashi Bay's mangrove forest (the largest on the island), and subtropical hill forests. There the typical climax species is the evergreen oak known as Itaji Chinkapin, with occasional stands of the tree ferns known as Brush Pot Trees, on the island's highest peak, Mt Yonaha (503 m). Despite representing less than 0·1 percent of Japan's land area, the forests there are home to such rarities as Okinawa Rail, Pryer's Woodpecker and Yambaru Long-armed Scarab Beetle (Japan's rarest beetle), and support a quarter of all of Japan's frog species, including its most beautiful endemic, Okinawa Ishikawa's Frog [EN]. This is all in an area that makes up less than one tenth of one percent of Japan's land area.

In 2014, the subtropical archipelago lying just 40 km to the west of Okinawa's mainland was registered as Japan's 31st national park, known as Kerama Shotō National Park. The 30 islands of the archipelago belong to the villages of Tokashiki and Zamami. The warm waters around these low hilly islands, surrounded with delightful coastlines of rocky crags, white-sand beaches and coral reefs, are winter breeding grounds for Humpback Whales and marine turtles. For people, they are ideal for snorkelling and diving.

In the far southwest of the long chain of Japan's islands, about 430 km from Okinawa, are the islands of Iriomote and Ishigaki, part of the Yaeyama Islands, and registered as the Iriomote–Ishigaki National Park. These islands have more extensive mangrove forests than anywhere else in the country, these being particularly evident along Iriomote's two major waterways, the Urauchi and Nakama rivers. These forests support huge Looking-glass Mangrove trees with their bizarrely snaking buttresses. They are home to the endemic Iriomote Cat, various endemic reptiles including Sakishima Habu[1], and endemic amphibians (such as Yaeyama Harpist Frog, Greater Tip-nosed Frog [NT], Utsunomiya's Tip-nosed

1 A pit viper endemic to most of the islands of the Yaeyama Archipelago.

Frog [EN], Eiffinger's Tree Frog, and Owston's Green Tree Frog), and endemic insects including the endangered Yaeyama Bayadera dragonfly, while the beaches around them are the nesting grounds of both Green Turtle and Loggerhead Turtle.

The Sekisei Lagoon, between Ishigaki and Iriomote islands, supports one of the largest coral reefs in Japan, measuring 20 km from east to west and 15 km from north to south, and is home to a wide array of tropical fish. The seas there hold Reef Manta Ray, Whale Shark, Smooth Hammerhead Shark and Great Hammerhead Shark among others, and delicate Black-naped Terns nest on tiny raised coral islets. West of Iriomote-jima is a tiny island that supports breeding Bridled Tern, Brown Noddy and Brown Booby. Iriomote-jima's forests contain many species of broadleaf evergreens such as various species of chinkapin and bay, and endemic species such as Yaeyama Palm, this last species in an endemic genus known only from the islands of Ishigaki and Iriomote.

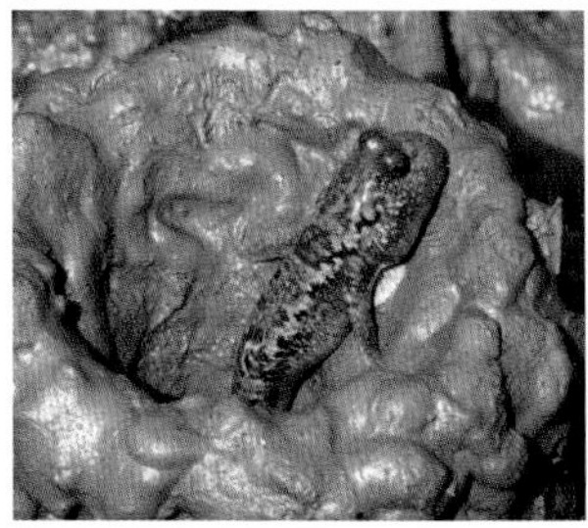

Top to bottom Hermit Crabs now utilize beach detritus or natural shells for their mobile homes [MAB]; Mudskippers are common in areas with mangrove [MAB]; elegant Black-naped Terns nest on rocky islets offshore [KuM].

Rails, Woodpeckers and Robins – Endemism in the Southern Nansei Shotō

By the late 19th century, Japan's birds and mammals were better known than those of any other country outside Europe and North America. It was easy to believe during the 20th century that all was already known of what Japan offered. Then an amazing story unfolded: a species new to science was discovered on Okinawa, one of the most heavily populated islands in the land, and not a small cryptic warbler but a large, boldly marked rail – Okinawa Rail.

A number of ornithologists had missed the opportunity and accolade of its discovery. Japan's great sound recordist, Kabaya Tsuruhiko, had recorded the voice of this species while sound-recording in the area in the mid-1960s, but the then unknown vocalizations were interpreted as novel calls of the rare Pryer's

Manta Rays can be found in warm waters around the Nansei Shotō and Ogasawara Islands [OVTB].

The endemic Ryukyu Box Turtle [*left* KuM] and Black-breasted Leaf Turtle [*right* SM].

The endangered endemic Yambaru Long-armed Scarab Beetle [*left* KuM] and endemic Okinawa Ishikawa's Frog [*right* SM].

Mangrove forest in the Yaeyama Islands [MC]

Green Turtles nest in the Southern Nansei Shotō [OVTB].

Confined to northern Okinawa Island, Pryer's Woodpecker is one of the rarest species in Japan [KuM].

Okinawa Rail occurs in much the same range as that of Pryer's Woodpecker, but is more flexible in its habitat requirements [SM].

Woodpecker, which also inhabits those forests. In the early 1970s, the renowned American researcher of woodpeckers, Dr Lester Short, spent time in Yambaru on researching the woodpecker; he noted a rail sighting, but attributed it to a different species known in Japan only from the Yaeyama Islands much farther south and with different plumage features. Either of them, and several others, too, who had made sightings, or taken photographs in the mid-1970s, might have discovered this remarkable species – but did not. That discovery fell to a local resident who found a specimen beside the road and despatched it to the Yamashina Institute for Ornithology, then situated in a leafy suburb of Shibuya, Tōkyō. Imagine the surprise of the ornithologists there as they unpacked the parcel and were confronted not just with a rare specimen, but with the very first specimen of a new species. A team set off for the area intent on trapping one, and their success in catching eight more rails led to the species being officially recognized and described in 1981. Back then, nothing was known about the numbers, habitat or behaviour of this new species. All that was known was that the specimen and the few known records were from the northern third of Okinawa, from the forested hills known as Yambaru. It was the first entirely new bird found in the country for more than a century.

It is surprising that this large colourful bird, smartly plumed in slate and silver, with red legs and feet and a large yellow-tipped red bill, could have remained unknown on such a populated island, albeit in the shady forests in the least populated northern third of the island. Today

Ryukyu Scops Owl ranges throughout the Nansei Shotō and south to the northern Philippines [KuM].

The endemic Pryer's Scops Owl [KuM]

The endemic Okinawa Robin [RC]

The endemic Kuroiwa's Ground Gecko [KuM]

The endemic Namie's Frog [SM]

Okinawa Rail is far better known. Its distribution has been mapped, its morphometrics studied, and we know not only that it is flightless but also that it inhabits the highest latitude among all extant flightless rail species. The rail's flightless nature, seemingly putting it at risk from ground predators in its evolutionary history, had led it to clamber, using its strongly clawed toes, high into trees in which to roost at night, and to descend the next morning by jumping down while wildly fluttering its wings.

Then came news of the mongoose. Small Asian Mongoose had been introduced on to the island by people who thought that it would help to reduce the native reptile fauna, particularly a venomous snake. Instead, this predatory alien species ate just about anything, including both adult and young rails. The introduction of the mongoose, combined with the presence of plentiful feral cats, meant that the days of the rail, despite its roosting in trees at night, seemed numbered as mongooses can climb trees; and, in any case, whenever the rails descend to forage or to nest they are at increased risk. As a consequence, rail numbers fell and their range contracted northwards through the 1990s.

Since 2002, great efforts have been made to control the numbers of the rail's new predator by trapping the mongoose, and trapping appears to have borne fruit and led to an increase in the rail population. The elusive, and largely nocturnal, rail had changed its habits and became more diurnal, but then ran afoul of feral dogs. It is a classic example of how avian species on islands tend towards flightlessness in the absence of mammalian predators, but quickly succumb once such predators are introduced. Today the rails seem as much, if not more, at risk from traffic accidents, as traffic has increased substantially on the roads that now crisscross Yambaru.

Yambaru, the northern part of Okinawa Island, is a biodiversity hotspot. It is home not merely to many species, but also to many *endemic* species. The area hosts an array of endemic plants and insects, amphibians, reptiles, birds, and even mammals. At night, endemic frogs, such as a trio of larger species, Okinawa Ishikawa's Frog, Holst's Frog [VU] and Namie's Frog [VU], along with endemic lizards such as Kuroiwa's Ground Gecko [VU], spice up any search for a roosting rail or a Ryukyu Scops Owl. At dawn, the chorus includes the delightful song of the endemic Okinawa Robin and, sometimes, the sharp tapping of a foraging endemic Pryer's Woodpecker. During the day, one may find the endemic Okinawan Tree Lizard [VU] and endemic dragonflies such as Yambaru Damselfly, Ryukyu Damselfly and Okinawa Sarasayama Dragonfly. Yambaru is a region that repays visiting at any time of year, during day and night; who knows what new faunal delights will be discovered there!

Japanese Deer

The indigenous Japanese Deer comes in many forms and sizes, small, medium and large, depending on the subspecies. The subspecies in Hokkaidō is the largest, those on Yaku-shima and the Kerama Islands off Okinawa, in the southwest, being the smallest. Individuals may range in colour from pale greyish-tan to warm chestnut-brown and even dark, blackish-brown, depending on the subspecies, gender, age and the time of year.

Resembling small European Red Deer or North American Wapiti, all Japanese Deer, regardless of subspecies, share a black-rimmed rump, a white bottom and a short white tail. The legs are darkest, almost black, with a white spot on the rear of the lower hind legs. The young, females and some stags (males) show rows of pale spots on the flanks, haunches and back. The head is long, with a black muzzle; the medium-long ears are slightly tapered and pointed. They exhibit marked sexual dimorphism. Stags can weigh 50–130 kg, reach 90–190 cm in length and stand 70–130 cm at the shoulder, whereas hinds (females) are smaller, weigh 25–80 kg, measure 90–150 cm in length and stand 60–110 cm tall, although there is a tremendous range in size depending on the subspecies. Hinds have a long slender neck, whereas that of the stags is stockier. Mature stags carry large antlers. The limbs are long, while the feet are small, each foot ending in four toes, the middle two of which are hoofed, and leave strong prints; the hind two toes are shorter, more pointed, but often leave no impression at all.

The natural range of Japanese Deer extends from Vietnam to eastern China, the Korean Peninsula and the maritime provinces of the Russian Far East, while in Japan the species can be found throughout all four main islands, and on various islands off Kyūshū and Okinawa.

Seven subspecies are currently recognized in Japan. Some genetic evidence, however, suggests a major division between the northern forms (*C. n. yesoensis* of Hokkaidō and *C. n. centralis* of Honshū) and the southern forms (*C. n. nippon* of Shikoku, Kyūshū, islands of the Inland Sea and on the Goto Islands, *C. n. yakushimae* only on Yaku-shima, *C. n. megashimae* only on Mageshima, *C. n. pulchellus* only on Tsushima and *C. n. keramae* of the Kerama Islands.), which may mean that there are in reality two different species.

Body size and antler size vary regionally between subspecies and with level of nutrition. Only males carry antlers, and these increase in their number of points as the animal ages: yearlings have one point, young males of two or three years of age have two or three points, and older, adult males have four or more points on each antler.

The deer live socially, especially in winter, when they are commonly found in single-sex herds. They become more solitary during the summer, although females and their young remain together. Japanese Deer range through various wooded and grassland habitats, from broadleaf evergreen forests in the south to forests of deciduous broadleaf mixed with evergreen in the north. They also favour forest edge and, particularly in Hokkaidō, they are frequently encountered in grasslands, meadows and wetland reed beds. In Hokkaidō, they migrate seasonally so as to avoid areas with deeper snow.

Japanese Deer is a ruminant that grazes and browses on a wide range of plants, which include grasses, such as dwarf bamboos, the leaves, twigs and bark of trees, and fallen seeds and nuts. During winter in snowy areas, the deer are reduced to browsing on tree bark, and their feeding signs are readily distinguished by the fraying of the bark at the upper end of the section where it was first nibbled and then torn from the trunk.

During May and June, the deer moult out of their paler, greyer winter coat into their browner summer coat, and then moult back again during September and October. In late September the annual rut begins, and this continues through October and November. During this season, males become more aggressive, scent-mark, roll in mud, and vocalize more frequently. Their calls consist of a powerful, far-carrying and rather haunting two-note whistle. It is typically heard on autumnal evenings, but can be heard at other times of the day and of the year, too. Their rutting and calling behaviour are part of their competition with other males and an attempt to attract females. The male's fine set of antlers is used mainly for bluff and bravado, although sometimes also for outright aggressive clashes. The clattering and rattling sound of the antlers of jousting stags is dramatic indeed. The males strut and pose with their fine racks of antlers, hoping to triumph over their competitors in the race to pass on their genes. In their polygynous mating system, larger-bodied and larger-antlered, dominant males establish territories, gather and defend harems, and so obtain most matings. After the seasonal rut, usually by March and April, when the extra burden of their spreading antlers no longer serves a purpose, the stags cast them, like a deciduous tree shedding its leaves. These fallen antlers then become an important source of minerals, and are consumed by rodents, foxes, badgers and deer, or eventually decompose and leach back into the soil. The growing of a new set of antlers each year and then dropping them distinguishes deer from other animals with horns, such as sheep, goats, cattle and antelopes. Those animals, unlike the deer, have simple, non-branching structures of bone, covered with keratin, on their forehead and they keep them throughout their lives.

In spring each year, usually during late May after gestating for around 230 days, females give birth to a single fawn (twins and triplets occur, but very rarely). Males may live for up to 15 years and females up to 20 years, but many do not survive even long enough to begin breeding from two years of age.

When disturbed or alarmed, deer stand alert with ears erect, watching carefully, and they may stamp a fore foot. If they continue to sense danger, they raise the tail and fluff out the white fur of the bottom to a considerable size, making themselves very conspicuous from behind. On fleeing, they give a short, sharp alarm whistle while dashing away with tail raised and rump fluffed out. The young, in herds or with their mothers, also give strange wheezing calls. The combination of raised tail and fluffed white bottom makes for a strong visual signal, especially in shady forest or poor light, allowing disturbed herd members to follow one another easily. The sharply whistled alarm call always gives the watcher a jolt of surprise, even those who have heard it innumerable times.

Various subspecies of Japanese Deer occur in the archipelago, the largest of these in Hokkaidō, the smallest on the Kerama Islands, off Okinawa, and on Yaku-shima. The deer (male *left*; female *right*) on Yaku-shima share their range with the southernmost Japanese Macaques [BOTH IiM].

Left Japanese Deer will dig down through snow to reach food in regions where snow covers the ground; *right* when food is limited, Japanese Deer resort to stripping tree bark [BOTH MAB].

Japanese Deer commonly segregate into single sex herds (here a group of females and young) [WaM].

Wild Boar

Wild Boar [YaM]

Wild Boar is not a party-pooper, nor a domestic pig gone feral. This truly wild species has the widest distribution of any pig. In addition to ranging across Europe and North Africa, and across Eurasia to India and Indonesia, it is found widely throughout central and southern Japan from central Honshū to the Ryūkyū Islands.

A medium-sized stocky, even rotund pig, Wild Boar has an almost neckless appearance, with a broad head, short rounded ears, and a long snout, the tip of which is partly prehensile. The males have four tusks and are around 20–30 percent heavier than females, but size varies from region to region, those in the southwest islands being the smallest and lightest. The larger subspecies (*S. s. leucomystax*) occurs widely in central and western Honshū, Shikoku and Kyūshū, but not at all in northern Honshū or Hokkaidō. The smaller subspecies (*S. s. riukiuanus*) is restricted to the Nansei Shotō, from Amami Ōshima to Iriomote Island. Measuring 110–160 cm long and standing 60–80 cm tall at the shoulder (males), these animals range between 50 kg and 150 kg in weight, depending on subspecies, region, gender and age.

The coat of Japan's Wild Boar contains coarse bristles and varies in colour from pale greyish-brown to dark blackish-brown, the legs being typically darkest. The short, straight tail is covered with short hairs. Piglets are a paler, warmer orange-brown with stripes. Their limbs are powerful, each ending in four toes, leaving strong prints very similar to those of the larger Japanese Deer. In snow, Wild Boar plough like manic bulldozers, leaving deep furrowed impressions.

Terrestrial, nocturnal and crepuscular, Wild Boar live solitarily, in small all-male groups, or as extended family groups consisting mainly of females and their young. This species lives mainly in natural evergreen and deciduous broadleaf forests, but can be found also in secondary forests, at forest and woodland margins, and in agricultural land (rice and other crop fields) where these are adjacent to woodland.

The omnivorous Wild Boar eats plant and animal foods taken from the surface or from just below the surface of the ground. Its diet includes roots, tubers, leaves, stems, fruits, nuts, invertebrates (especially earthworms and snails), amphibians, reptiles, and ground-nesting birds. In autumn, boar are attracted particularly to concentrations of food, such as fallen acorns and fallen fruit. Extensive areas of disturbed surface soil are typical signs indicating where Wild Boar have been rooting for food.

Males mature at two years of age, but females are ready to breed at one year and typically give birth once a year in spring, although rarely they may produce a second litter in the autumn. Following a gestation period of 120 days, females may give birth to a litter of up to ten young, although in the northern subspecies litters average fewer than five young and in the southern subspecies fewer than four. Most young Wild Boar do not reach their first birthday. Those which do, however, may go on to reach ten years of age.

Japan's smallest Wild Boar subspecies lives in the Nansei Shotō [TsM].

Ploughed ground is evidence of the presence of Wild Boar [MAB].

Southern Mixing Grounds

The many islands and channels of the Nansei Shoto and their geological histories have created a complex mixing zone. Many species at home in mainland Japan or Northeast Asia reach their southern limits here and mingle with those more frequently associated with regions farther south, which have their northern limits in this zone.

Among the herons, egrets and bitterns, for example, Malayan Night Heron [NT] and Purple Heron reach the northern limits of their ranges in the Yaeyama Islands, while Cinnamon Bittern reaches north through the Nansei Shotō Islands as far as its northern limit of Amami Ōshima. More northerly breeding species, such as Grey Heron, Black-crowned Night Heron and Yellow Bittern, move south to these islands, but only for the winter.

Each year, Chinese Pond Heron, something of a rarity in Japan, wanders northwards and individuals reach these islands. Conversely, Black-faced Spoonbill, a globally rare species breeding on the Korean Peninsula, winters here in small numbers. The islands are truly a biogeographical crossroads.

Left to right Cinnamon Bittern reaches its northern limit in Japan in the Nansei Shotō; Chinese Pond Heron wanders regularly northwards to the southern Nansei Shotō; Eastern Yellow Wagtails are scarce winter visitors to the southern Nansei Shotō, but breed in northern Hokkaidō [ALL TsM].

Left to right Purple Heron reaches its northern limit in Japan in the Nansei Shotō [KT]; Malayan Night Heron occurs in the Yaeyama Islands each winter [MC]; small numbers of the rare Black-faced Spoonbill winter in the southern Nansei Shotō [TsM].

In the northern half of their Japanese range, the deer must contend with snow in winter. Deep powdery snow is to a Japanese Deer what a stage covered with fluffy feather pillows would be to a top-ranking ballerina. Both lead to loss of grace and floundering for slim-footed deer and ballerina alike. The narrow-hoofed deer sink and flounder, so they prefer areas with shallower snow. By way of contrast, on the southern island of Yaku-shima they must contend with heavy rain.

On Yaku-shima, where the deer share their range and habitat with Japanese Macaques, and live with them in a loosely symbiotic way, the deer are used to the miraculous appearance of fresh food from above, thanks to the messy eating habits of the monkeys. 'Miraculous provisioning', however, is not restricted to the deer living on Yaku-shima. With the passion that human visitors in Japan have for feeding living creatures at tourist spots, those deer that overcome their shyness sufficiently to approach tourists are guaranteed handouts. This happens frequently at Itsukushima shrine on Miyajima, and at Nara, although unfortunately many of the deer end up consuming litter and plastic, too.

Japan's Reptile Fauna

The gradient in climate from northern temperate in Hokkaidō to subtropical in Okinawa is reflected by patterns of biological distribution. Yet, while the gradient in richness in the Northern Hemisphere is typically towards the equator, as is found among the reptiles and amphibians of Japan, the pattern is strangely reversed for the mammals: the farther north one goes in Japan, the greater the number of mammal species, Japan's northern prefecture Hokkaidō supporting the most. Perhaps other factors are involved here, such as the area of the prefecture and its diversity of habitats, but clearly much remains to be understood even in a well-studied country such as Japan. As mentioned, when looking at reptiles we find that the opposite pattern is the case: there are fewer species in the colder northern regions and many more in the subtropical south. Despite those southwest islands having a much smaller land area, they have a much greater diversity of reptiles.

Japan's reptile fauna includes various marine turtles, freshwater turtles, several geckos and

Top to bottom The non-venomous Japanese Ratsnake is widespread from Hokkaidō to Kyūshū but absent from the Nansei Shotō [MAB]; Ryukyu Odd-toothed Snake [SM]; Japanese Striped Snake [JoW]; Tiger Keelback [MAB].

Top left Although venomous, Japanese Mamushi is rarely aggressive [HT]; *top right* the endemic Japanese Coral Snake [TsM]; *bottom left* the endemic Ryukyu Green Snake [SM]; *bottom right* Okinawa Tree Lizard [KuM].

ground geckos, lizards, numerous skinks, and a wide range of snakes including pit vipers and sea snakes. Forty-three of Japan's reptile species are endemic, and most of these occur in the south.

The most frequently encountered of Japan's native reptiles is the non-venomous Japanese Ratsnake. While its English name seems rather unfortunate, its Japanese name of *Aodaisho* (literally 'green commander') is far more imposing. It ranges from Hokkaidō to Kyūshū but is absent from the southwest islands, and is as likely to be found in urban parks as in rural areas. This is the largest of Japan's main island snakes, reaching lengths of 1–2 m and with a diameter of about 5 cm. Although frequently a mottled dark green, this species varies, some individuals being pale yellow-green. As its name suggests, it is primarily a rodent-hunter.

Venomous snakes, too, occur here. Japanese Mamushi[1], which grows to less than a metre in length, is a widespread species occurring from Hokkaidō to Kyūshū and Yaku-shima in a range of habitats, from marshes to rocky hillsides. Although venomous, it is rarely aggressive. Japanese Mamushi, Japanese Coral Snake and Okinawa Habu are the three most venomous snakes in Japan. The last-mentioned is endemic to the southwest islands, and grows to 1·2–1·5 m on average, some individuals reaching as much as 2·4 m, making it Japan's largest snake. This terrestrial and almost entirely nocturnal species is most likely to be found in forest and farmland edge habitats, and has a reputation for being irritable.

From the alarming to the charming, few reptiles are as attractive as another of the Nansei Shotō's endemic species, this one known as Kuroiwa's Ground Gecko. This endangered species is in fact very likely to consist of two separate species, those on the Amami island group being genetically very distinct from those on the islands of Okinawa to the south. Their genetic divergence dates back to the middle Miocene, about 14 Mya, which was even before the formation of the straits that have constantly separated these island groups since the early Pleistocene, about 2.5 Mya.

1 Known also as the Japanese Pit Viper. The specific name, *blomhoffii*, is in honour of Jan Cock Blomhoff, director of the Dutch trading colony at Dejima, Nagasaki, Japan, from 1817 to 1824. (See also *p. 12* for more on the three scholars of Dejima.)

Forests Where the Tides Flow

Saltwater brings death to most trees, yet there are those – the mangroves – that are specially adapted to it. Where warm-water rivers merge with the sea, there is an ebb and flow, and not just of the tide. Rivers carry silt and detritus downstream towards their estuaries and the sea beyond, depositing these materials steadily seaward in an endless invasion of the marine realm by the terrestrial world, in a process known as accretion. Unwittingly fighting back against the process of accretion, the next winter storm or summer typhoon may undo the work of months in a matter of hours, eroding away the accumulated silt and washing it out into the deep ocean, but only for the process of coastal accretion to recommence, as if with infinite patience.

Along the flanks of the river's estuary an army awaits. This is a sluggish army moving only slowly seawards. No other force moves in this way – not in a rapidly advancing phalanx intent on attack, but moving inexorably, generation by generation, to take over newly created territory. This is an army that marches not on its stomach, but on its stilts along the margins, holding deep into the mud, grasping at it, and stabilizing it until the land beats back the sea. This army consists of mangrove trees. At high tide, the trees of the mangrove militia stand in ranks with their roots and trunks amid swirling deep saltwater; their dark leathery green leaves shine glossily in the harsh subtropical noonday sun. At low tide these botanical amphibians stand surrounded by sand-mixed mud; their stems twist and rise drunkenly, supported by a network of snaking stilt and prop roots.

The surface of the ultra-fine mud sediment becomes slick as the tide drains away; skating here is easier than walking. Dig beneath the seemingly benign grey surface of the drying mud and what rises from it can only be called a stench. Within a few centimetres of the surface, the sucking mud is dense, black, and emits the foul odour of decay. This perpetually waterlogged soil holds little free oxygen, and is laden with anaerobic bacteria that release among other things methane, nitrogen and sulphides that contribute to that gagging, rotting stench. That anything grows here is incredible. That an army of forest trees can march steadily through this muck is little short of miraculous. These are the conditions of a harsh environment in which moisture, salt concentrations and temperatures swing back and forth between extremes that would kill most species within hours.

While most plants are excluded by just these conditions, mangroves somehow thrive. Here, in this unstable, tide-washed salt-laden 'land', they have the competitive edge over other plants. They cope with the alternate wetting and drying of their roots caused by tidal inundation, and as halophytes their resistance to salt concentrations (which may rise as high as 90 parts per thousand) makes them tougher survivors than other plants. They may not grow massive trunks, or reach high into the sky, as do many other tropical or subtropical trees, but instead their many stilted roots lend them physical stability in a fine, shifting soil that would soon undermine other species – that is if it did not kill them through its lack of oxygen!

The cloying silt that excludes most species is something that the mangroves can turn to their advantage. Spongy tissues in their roots aid them with gaseous exchange, but more unusual and critical to them is their array of exposed breathing tubes. All plants require oxygen for respiration, including in the tissues of their underground roots. Like a swimmer relying on a snorkel, stretching up from the inhospitable water into the atmosphere, mangroves have breathing tubes that stretch up, like grasping fingers, from the inhospitable glutinous mud, giving them access to life-giving air. These aerial roots, known as pneumatophores, and the stilt-like prop roots, are modified in such a way as to be capable of absorbing gases directly from the atmosphere and transporting them to the roots buried below mud. They are so vital that a single mangrove tree may have as many as ten thousand of them bristling up above the mud for 10–30 cm.

Only 54 species of tree worldwide are considered true mangrove species, each adapted slightly differently to this saline environment with anaerobic soil, but they come from very varied taxonomic backgrounds, being defined more by their ecological adaptations than by their evolutionary relationships. This is an example of convergent evolution, whereby unrelated organisms evolve similar

Mangrove plant species are adapted to daily tidal inundation by saltwater and occur in Japan's warm regions. *Top left* Black Mangrove [BOTH TsM]; *top right* Kandelia Mangrove [BOTH TsM]; *bottom left* spear-shaped mangrove propagules self-plant in soft mud, but may also be transported by the tide [TsM]; *bottom right* branched mangrove roots support the trees and serve to stabilize coastal soils and block tidal surges [KuM].

Among the mangroves Barred Mudskipper [*inset* KuM] is capable of surviving out of water for part of the tidal cycle [TsM].

characteristics in order to thrive in a particular set of environmental conditions and do not represent a single natural taxonomic group. Some avoid taking in salt in saltwater by filtering most of it out at the surface of their roots; that, combined with a degree of salt tolerance not found in most other plants, allows these salt-excluding species to survive with ten times more salt in their sap than most plants can tolerate. Other mangrove species take in saltwater, but excrete concentrated salt from glands on their leaves in a process that still leaves them having to tolerate salt concentrations in their sap ten times that of the salt-excluders.

These extraordinary trees form forests where they can, in saline coastal sediments and in estuaries in tropical and subtropical regions around the world, but with an uneven distribution such that Asia alone has more than 40 percent of the world's mangroves. Despite their toughness in the harsh conditions of the intertidal zone, mangroves are sensitive to cool temperatures that prevent them from colonizing the temperate zone. At their extremes they extend north as far as southern Japan and as far south as New Zealand's North Island and Australia's Victoria coast. Although mangroves have for long been cleared to make way for coastal development and for shrimp ponds, or been burned as fuel, the immeasurable value of this biome, with its many mangrove trees, as a protective 'bioshield' for the coast and what lies immediately inland of it was brought into sharp relief by the devastating force of the 2004 Southeast Asian tsunami in areas unprotected by mangroves.

Where mangroves grow, they slow the flow of water and form a buffer between land and sea; they drain power from onshore waves, and accumulate ever more silt around their roots. As they grow and spread towards the ocean, they leave behind drier, richer areas that other plant species can colonize in succession.

Their extensive above-ground roots and pneumatophores and their complex mesh of fine rootlets within the soil all serve to trap and hold sediments as they spread tidewards. Their root system helps them to stabilize and build up soils and to block tidal surges, resist erosion during high-energy tropical storms and typhoon floods, and even help to abate tsunami. An additional benefit is that their trapping of sediments allows coral reefs to develop in clear water offshore. Where mangrove forests are retained, onshore damage from tsunami is greatly reduced. Even when damaged by storms, mangroves are remarkably quick to recover as they produce abundant propagules which, having already germinated before implantation, grow rapidly. The growing trees mature quickly, further speeding the process of regrowth and recolonization.

Even in the typically richly biodiverse tropics, where terrestrial forests may contain thousands of tree species, it is not unusual for mangrove-forest diversity to be as low as three or four species. While there may be few trees in the swamps that we call mangroves, this habitat supports a tremendous array of other species, from barnacles, bats and birds to sponges, crabs and mudskippers. Mangroves are true keystone species.

Mangrove trees have an array of fascinating adaptations to their environment, even utilizing the flow of the tide to help them to disperse their seeds. Most mangroves are hermaphrodites, with their flowers pollinated almost exclusively by small insects, birds and bats. Whereas most plants produce seeds that must disperse before germination in soil, many mangrove tree species produce seeds that germinate while still attached to the tree, in a process known as vivipary. What look like long green cigars hanging from the branches are in fact the already germinating fruits. These ready-to-go propagules are actually seedlings already capable of photosynthesizing to produce their own food by using the sun's energy. Once mature, they fall from the mother tree, where the tide can carry them long distances. Able to remain dormant for over a year, to survive in saltwater and to resist desiccation, some propagules are eventually washed on to a suitable substrate, where they lodge in the mud and take root.

Japan's mangrove forests are at the northern limit of their Indo-Pacific range and include a number of species, particularly: Black Mangrove, Narrow-leaved Kandelia, Spider Mangrove, Grey Mangrove, two species both known coincidentally as Looking-glass Mangrove, Mangrove Apple, and Mangrove Palm. They form an exotic landscape of various species growing modestly to 5–8 m in height and mostly confined to the country's subtropical southwest islands from Amami Ōshima to Iriomote-jima, although some grow as far northeast as Shizuoka Prefecture, Honshū. Along with coral reefs, they attract many tourists to the warmer southern parts of the country. Japan even has a protected mangrove forest: Nagura Anparu, on Ishigaki Island, Okinawa Prefecture. A wildlife-protection area since 2003, it was designated a Ramsar Site in November 2005. Nagura Anparu may spread only some 157 hectares at the mouth of the Nagura River, but it is an important home to about 60 species of crustacean and many species of bird, and is popular for recreational sightseeing and birdwatching.

Various crabs forage across the tidal mudflats in mangrove forests: Milky Fiddler Crabs (*top left*) seem dwarfed by their own claws; Asian Soldier Crabs (*top right*); Orange Fiddler Crabs (*middle left*); Orange Mud Crab (*middle right*); Tetragonal Fiddler Crabs (*bottom left*); and Dussumier's Fiddler Crabs (*bottom right*) [ALL TsM].

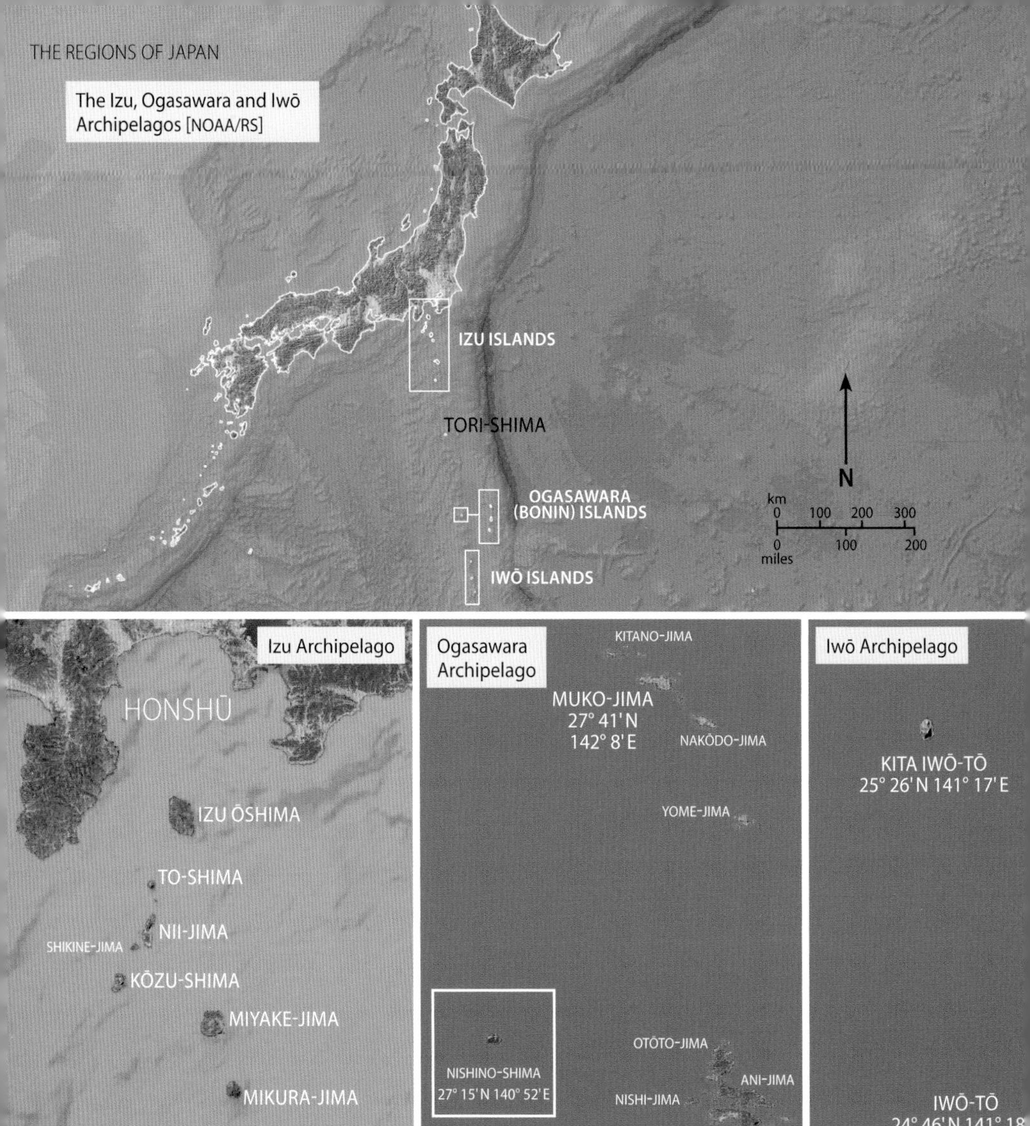

The Izu, Ogasawara and Iwō Archipelagos [NOAA/RS]
IZU ISLANDS
TORI-SHIMA
OGASAWARA (BONIN) ISLANDS
IWŌ ISLANDS
N
km
0
100
200
300
0
100
200
miles

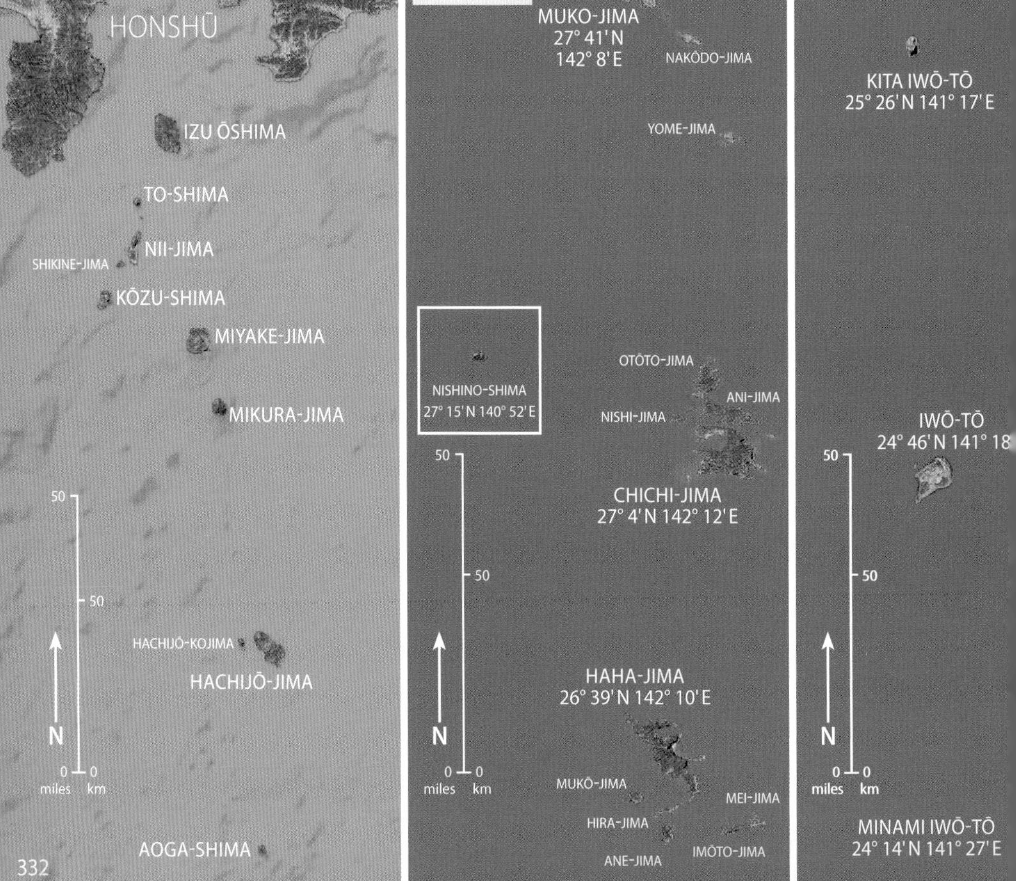

Izu Archipelago
HONSHŪ
IZU ŌSHIMA
TO-SHIMA
NII-JIMA
SHIKINE-JIMA
KŌZU-SHIMA
MIYAKE-JIMA
MIKURA-JIMA
HACHIJŌ-KOJIMA
HACHIJŌ-JIMA
AOGA-SHIMA
50
50
0
0
miles
km
N
Ogasawara Archipelago
KITANO-JIMA
MUKO-JIMA
27° 41' N
142° 8' E
NAKŌDO-JIMA
YOME-JIMA
OTŌTO-JIMA
ANI-JIMA
NISHI-JIMA
NISHINO-SHIMA
27° 15' N 140° 52' E
CHICHI-JIMA
27° 4' N 142° 12' E
HAHA-JIMA
26° 39' N 142° 10' E
MUKŌ-JIMA
MEI-JIMA
HIRA-JIMA
IMŌTO-JIMA
ANE-JIMA
Iwō Archipelago
KITA IWŌ-TŌ
25° 26' N 141° 17' E
IWŌ-TŌ
24° 46' N 141° 18
MINAMI IWŌ-TŌ
24° 14' N 141° 27' E

The Natural Marvels of the Izu, Ogasawara and Iwō Islands

Tori-shima lies between the Izu Islands and the Ogasawara Islands and is home for Short-tailed Albatrosses [TR].

Japan's geography is perhaps best described as consisting of archipelagos within archipelagos, not just islands but clusters and strings of them. Even metropolitan Tōkyō consists of an island chain. Situated on the mainland of Honshū is the huge urban island of Tōkyō itself, but head south out to sea and you can travel for 1,000 km and more and yet still be within Tōkyō because of the nature of Japan's administrative boundaries.

Stretching south from central Honshū, there are three different island groups: first the Izu Islands, then the Ogasawara (or Bonin) Islands, and finally the Iwō Islands. These are all more remote by far than any other of Japan's offshore islands, because they are the tips of oceanic volcanoes that have never been connected to the mainland. As a consequence, their flora and fauna are unique, consisting of successful long-range colonists and their descendants. While the Izu Islands (especially Miyake and Hachijō) offer great birdwatching, the Ogasawara Islands provide some of the finest whale-watching in the country. The islands are a wonderful example of natural heritage combining geodiversity and biodiversity in a small-island context.

The climate of the Izu Islands is generally milder and more oceanic than that of the mainland, and the islands have a delightful world-in-miniature feel. From a natural-history perspective they are home to a number of endemic species. If one approaches them early in the morning by ship, islands such as Miyake-jima are encircled by swarms of seabirds, for these islands are the breeding haunts of hundreds of thousands of Streaked Shearwaters. The warm waters offshore are popular with Indo-Pacific Bottle-nosed Dolphins, making dolphin-watching and even swimming with dolphins a distinct possibility. Onshore, in the evergreen laurel forests that clad the flanks of the local volcanoes, there are special birds, including Izu Thrush and Ijima's Leaf Warbler [VU], both of which are most easily seen on Miyake-jima.

The rugged and isolated Ogasawara Islands are truly oceanic islands [MOEN].

Left to right Hundreds of thousands of Streaked Shearwaters breed in the Izu Islands [JoH]; Brown Booby is frequently encountered in warm-water areas [MAB]; Matsudaira's Storm Petrel is a breeding endemic [YM].

Left Forest on the Ogasawara Islands [OVTB]; *top centre* the endemic Bonin Skimmer [OVTB]; *top right* Pillow Lava in the Ogasawara Islands [MAB]; *bottom centre* the bioluminescent Green Pepe Fungus [OVTB]; *bottom right* the rich marine life around the Ogasawara Islands attracts visitors for snorkeling, diving and whale-watching [OVTB].

Farther south still, between the Izu and Ogasawara Islands, there is a remote volcanic rock jutting from the sea known as Tori-shima – literally bird island. This spot on the planet is world-famous because each year the beautiful Short-tailed Albatross returns here to breed. From near extermination early in the twentieth century, this species' numbers have climbed back into the low thousands and it has reached relative security again thanks to concerted efforts by scientists, local government and conservationists.

The island chain does not end here, though. Far beyond Tori-shima, and 1,000 km south of central Tōkyō, lie the Ogasawara Islands. These tropical Pacific islands are a far cry from downtown Tōkyō, although still linked by local administration. In their extreme isolation they have few species of bird, but Bonin Honeyeater is endemic here, as is the recently described Ogasawara Greenfinch. For botanists, however, Chichi-jima and Haha-jima are remote island paradises, with many more tropical species and endemic species to interest them.

With extreme isolation comes extreme vulnerability, and life on isolated islands is especially weak in the face of outside threats in the form of introduced species. The Japanese main islands are vulnerable and deeply impacted by introduced species. One extremely influential species, accidentally introduced among imported timber from North America during the early history of biological invasion into Japan, is a pathogenic nematode known as Pine Wood Nematode, which causes Pine Wilt. Since the first was recorded in Nagasaki, in 1905, it has spread throughout Honshū, Shikoku and Kyūshū, causing tremendous damage in pine plantations. If even the large main islands are vulnerable to the likes of feral cats and dogs, Largemouth Bass and Bluegill, and a whole host of introduced plants, insects, amphibians, reptiles, birds and mammals, just imagine, then, how much more vulnerable are Japan's offshore and ultra-remote islands with their smaller populations, reduced genetic variability, and susceptibility to catastrophic events.

Long-isolated island regions such as New Zealand, Japan, Indonesia and Madagascar support tremendous biodiversity with extraordinary degrees of endemism. Endemic species are the truly unique ones found only on those isolated islands and nowhere else on earth, and they include species in every group of organisms imaginable. On Mt Chibusa on Haha jima, in the Ogasawara Island group, there is a tiny, jewel-like creature, an almost entirely transparent terrestrial snail[1]. That species of snail is just one of approximately 100 snail species endemic to the subtropical forest on that mountain alone, which rises to a peak of only 462·6 m! The islands have witnessed the exceptionally rapid evolution of a suite of endemic land snails in the genus *Mandarina*, such that separate species are found on separate islands, and also in terrestrial, semi-arboreal and arboreal habitats on each island.

That same isolation over the evolutionary timescale has rendered all island-dwelling species, whether endemic or not – such as Oriental Stork, Crested Ibis, Red-crowned Crane and Short-tailed Albatross, and that extraordinary see-through snail – vulnerable to environmental perturbations. They are especially susceptible to those perturbations thrust upon them by us: climate change, habitat degradation and fragmentation, habitat destruction, and the introduction of alien competitors, predators and disruptors, such as rats, cats, Goats, American Bullfrogs, mosquitoes, longhorn beetles, fungi and fish, to mention a mere handful of the species that we humans have transported to out-of-the-way places, including the Japanese archipelago.

Each of those four large and long-lived avian species mentioned above was considered extinct in Japan at some point in the 20th century. Each of the four is now on the long slow road to recovery from a perilously close brush with extinction. This is a 'bad news–good news' story. That each species nearly disappeared

Top to bottom Izu Thrush can be found on Miyake-jima and Hachijō-jima [MC]; Owston's Tit, a relative of Varied Tit, is confined to the southern Izu Islands [ET]; Ijima's Leaf Warbler is endemic in the Izu Islands, but winters outside Japan [ET]; the endemic Izu Robin, a close relative of Japanese Robin [ET].

1 Transparency is very rare in the terrestrial realm, but has evolved independently in a number of marine organisms.

during the 20th century is due entirely to human activities such as hunting, intensified farming, excessive use of agricultural chemicals, habitat degradation, or outright habitat loss. That each species is now on the up-and-up is also the result of human activities – intensive conservation effort – combined with the natural resilience of individual wild creatures driven to survive and reproduce. Each of those species occurs in East Asia and nowhere else on earth, and all of them live on islands in Japan.

Ani Island Mandarina Snail is just one of several endemic land snails on Haha-jima [MOEN].

The Ogasawara Islands consist of more than 30 islands stretching almost 150 km from Muko-jima in the north to Ane-jima in the south, although only two (Chichi-jima and Haha-jima) are currently populated by people. The islands have been produced by a combination of oceanic plate subduction and submarine vulcanism. Isolated oceanic islands, such as the Ogasawara Islands, are so far removed from the nearest mainland that few species are capable of reaching them naturally. Those few species that do make it to such places and survive may spread and diversify, but typically such islands have a depauperate species diversity because only a few species reach them. For example, there is only one native terrestrial mammal, and no woodpeckers, on such isolated snippets of land, and there are no snakes and no amphibians either.

All living organisms in the Ogasawara Islands ecosystem are the descendants of pioneers that both survived the journey by air or by sea and were able to survive on the islands once they arrived. The flora of the Ogasawara Islands consists of 441 species of vascular plant, of which 161 are endemic. There are no native amphibians and only two native reptiles, the endemic Ogasawara Snake-eyed Skink and Micronesian Saw-tailed Gecko, and the only terrestrial mammal is the rare Bonin Flying Fox. In contrast to the paucity of other vertebrates, birds have colonized the Ogasawara Islands more successfully, although only two of several endemic avian species, Bonin Honeyeater and Ogasawara Greenfinch, survive, all of the others having gone extinct. Vagrant birds, well out of their normal ranges, reach these islands, perhaps the strangest being Corncrake, but few survive. Of the invertebrates found here, 25 percent of the insects are endemic and 95 percent of the land snails are endemic.

When we humans provide assisted passage, deliberately or accidentally, to species that have not previously reached such islands, we frequently unleash an apocalypse on the unsuspecting indigenous inhabitants. Witness Small Asian Mongoose, introduced in Okinawa in 1910 and which now threatens all of the native birds and mammals already there. Likewise, Brown Tree Snake has similarly devastated the native fauna of Guam since it was introduced there in the 1940s; and Green Anole Lizard, introduced in the Ogasawara Islands in the 1970s, is now wreaking ecological havoc there (see *p. 349*).

Japan, although very late to embrace the issue of biosecurity, is now attempting to reduce, control and prevent further damage by introduced alien species. This is demonstrated by the hard-working mongoose-busters on both Okinawa and Amami Islands attempting to reduce that species' impact on native frogs, reptiles and birds, and the hunters active in removing the vegetation-destroying feral goats and the all-consuming feral cats that had become established on the Ogasawara Islands.

Environmental management today is a complex field, requiring public understanding of the fact that certain indigenous and endemic species are so endangered that they require conservation management to help their populations recover. Meanwhile, certain native species, such as Wild Boar and Japanese Deer, have become so common that there is a public clamouring for their population control. In between these two extremes are the out-of-control introduced exotic species in marine, freshwater and terrestrial environments causing considerable environmental harm. It is difficult to explain to the general public why hunting one species may be good and hunting another is bad. As natural historians, we may value biodiversity as the universal goal, but those values may not be shared by the public at large. It is not about the emotional attachment to the rights and wrongs of hunting, or about the emotive significance of the species being discussed, nor whether it is a 'cute' introduced raccoon or a lizard. It is all about understanding the crucial significance of functioning ecosystems and the negative impact that the species in question may be having on the ecosystem, whether that particular species occurred there naturally or not, and whether its numbers are in natural balance with its environment or not. Conservation management is certainly an uphill battle, but it is one that is vitally important in the fight to maintain the functioning ecosystems on which we all depend.

The beautiful subtropical Seven Isles of Izu attract walkers and wanderers, divers, snorkelers, dolphin swimmers, whale-watchers and birdwatchers. Peruse a map of the ocean south of central Honshū, and draw a line stretching from the tip of the Izu Peninsula on the southern Honshū coast northeast to the tip of the Bōsō Peninsula to close off the Sagami Bay. Mid-way across, this line will bisect Izu Ōshima, the largest of these so-called seven islands, with its prominent 764 m peak known as Mt Mihara. Scan down the map several hundred kilometres or, better still, sail south past the large resort island of Ōshima to the string of islands that lies ahead to the south (islands named in italics are currently populated). First we pass *To-shima*, then Udone-shima, followed by *Nii-jima* (with Han-shima and Jinai-to), *Shikine-jima*, and *Kozu-shima*, next comes *Miyake-jima* (famous for its camellias, hydrangeas and unique birdlife) along with Onohara-jima (actually a cluster of islets), then steep-flanked *Mikura-jima* (with its virgin forest of box trees and chinquapins home to thousands of nesting Streaked Shearwaters), Inamba-jima, and *Hachijō-jima*. Although *Hachijō-jima* is the turn-around point for the regular *Tachibana-maru* ferry from Tōkyō, yet more islands lie beyond it. Next comes Hachijō-kojima, followed by *Aoga-shima*, the Bayonnaise Rocks (known also as Myojin-sho), then Sumisu-to and Tori-shima (world-famous as the breeding grounds of the rare Short-tailed Albatross). Finally, some 650 km south of Tōkyō, and 76 km beyond Tori-shima, we reach the basaltic pillar of Sōfu-iwa (known also as Lot's Wife), which rises sheer-sided from the deep ocean. Now, you may see that there is a conundrum here: *nine* of the Seven Isles of Izu are populated, and in fact, if we add in the list of unpopulated islands, we find that

The Ogasawara Islands are remote and volcanic, true oceanic islands that have never been connected to a mainland – most of the Ogasawara Islands are uninhabited [MOEN].

the mysterious total of these Seven Isles is, in reality, at least seventeen, so why are they called the Seven Isles? I am still searching for an answer.

From the low perspective of a small ship, these islands are more prominent, and appear more isolated than ever, their shapes etched against sea and sky, as if cut like *Kiri-e*[1] from the seascape and the air. These islands, glimpsed from a ship, impress with their isolation, and often with their tropical tranquility (unless a typhoon is brewing); but these many islands of Izu create a far stronger and lasting impression, one of turmoil and drama. Their dramatic shapes are those formed by the eructation of lava and ash from the bed of the sea, the belching of noxious fumes, and the erosion of wind, waves and rain.

Welcome, then, to one of the most volcanically active regions of Japan. Here, volcanoes not only erupt on land, but also from the sea. Islands are formed here, by numerous submarine eruptions. For all the apparent drama and violence of their histories, the islands of Izu are, in fact, but a small part of the Izu–Bonin–Mariana arc. Here, geological forces play to transform the landscape, as tectonic plates converge along a boundary that extends southwards for more than 2,800 km from Tōkyō to beyond Guam, and along which we find the Izu, Bonin and Mariana Islands. If we could strip away the waters of the Pacific Ocean and see the ocean bed beneath the waves, we would find it pocked by many more volcanic islands submerged below the current sea level. These were formed by intense earth-melding forces as the enormous Pacific Plate subducted beneath the much smaller Philippine Sea Plate.

Geologists do not mess about when naming things: they call this narrow strip extending south of Tōkyō the 'Izu Collision Zone'. This is no car smash or train wreck, or even an imaginary clash of the Titans; we are talking full-on collision between two tectonic plates. Here subduction takes place at a rate of 2–6 cm a year; it is hardly surprising, then, that the hardy residents of these islands have an unenviable familiarity with earthquakes.

Today, the formidable forces that act both to form and to tear apart these islands are held in check by a very tenuous cultural leash. They are now bound together, but only lightly, in an administrative entanglement known as the Fuji–Hakone–Izu National Park. First designated in 1936, the national park included only Mt Fuji and Hakone, but it was extended in 1955 to add the Izu Peninsula. Then, more recently still, the Izu Islands were added to the national park in 1964. The proximity of this 1,227-km^2 park to the nation's capital of Tōkyō ensures that it is the most visited national park in the country, although admittedly most of these visits are to the Mt Fuji area. Nevertheless, the park is essentially a monument to the earth's natural forces and the turbulent geological history of the islands that we now call Japan. Stretching from sublimely iconic Mt Fuji in the north to lush green Hachijō-jima in the south, and incorporating lakes, calderas, lava flows, gaseous vents and submarine volcanoes, this extensive and disjunct national park is a paradise for those with an interest in geology, especially vulcanology.

The Izu Islands themselves represent a group of emergent submarine volcanoes the oldest of which erupted about two Mya, but their activity continues to this day. Miyake-jima, in the midst of the group, with its unique natural history and beautiful scenery, supports a range of avian species that has evolved here in the splendid isolation of these oceanic islands. As recently as 1983, a great fissure eruption with lava flows burst from the flanks of the island's core – Mt Oyama. That event devastated large areas of the west and southwest of the island, changing the scenery dramatically, destroying forests, overwhelming villages and schools, and wiping out a large and significant body of freshwater at Shinmyo-ike. A more recent eruption of Mt Oyama occurred in 2000. This was followed by more than 17,500 earthquakes in just a matter of months, and led to the formation of a deep caldera in the mountain's peak and considerable outpourings of toxic sulphur dioxide gas. That massive gas leak led to a state of emergency being declared and the forced evacuation of the island for five years.

Almost unnoticed, given the scale of the geological events unfolding in the archipelago, another event was set in action in the 1980s that would lead to the devastation of the island's ecology, but this was caused deliberately by the introduction of an alien predator – Japanese Weasel. This weasel went on to exert a

---

1 The Japanese art of paper-cutting.

Left The rare endemic Bonin Flying Fox [KuM]; *top centre* the endemic Bonin Honeyeater [KuM]; *top right* Styan's Grasshopper Warbler is more easily seen on the Izu Islands than elsewhere [ET]; *bottom centre* introduced Japanese Weasels have dramatically impacted on the native fauna of the Izu Islands [MC]; *bottom right* the warm waters of the Ogasawara Islands are rich in marine life [MAB].

dramatic impact on the fragile island ecosystem and to decimate populations of the island's endemic birds. Thus, the changes to the landscape and to the avifauna since 1980 have been dramatic, to say the least.

In the extraordinarily geologically active Japanese archipelago, life continues with a subliminal sense of imminent potential disruption, perhaps leading to the shared stoical approach to the events threatening beneath our feet and into the future. Miraculously, from the perspective of biodiversity, the forests around Tairo-ike have survived the devastation of both of the most recent eruptions, and today they harbour the majority of the Izu Islands' endemic bird species that draw naturalists from around the world.

Early summer mornings at Tairo-ike are best, before the heat builds and before the cicadas become noisy. Early on, the lush evergreen broadleaf forest surrounding the lake is alive with bird song. Here, one can hear a dawn chorus of a unique suite of birds. Long isolated here, they have evolved into locally distinct subspecies and species. A constant series of high-pitched notes is the somewhat monotonous sylvan song of a migratory insect-eating warbler known as Ijima's Leaf Warbler. These flit very actively, like forest nymphs, through the trees just beneath the canopy, rarely holding still to offer a definitive look, let alone a chance to photograph them.

The endemic, chestnut-cheeked and chestnut-bellied Owston's Tit is resident here and is far more sluggish. It feeds frequently on the acorns of the island's evergreen oaks, and takes its time to hammer these seeds open, often perching on a branch and holding the seed between its feet as it does so. Its harsh wheezing "*dzu-dzu-pee*" calls draw attention to it as it moves slowly through the dark forest. The Izu Thrush, rich in black, brown and chestnut, is a gorgeous creature that is unfortunately not so common as it once was (before the introduction of the weasels). The thrush commonly forages on the ground, where it sometimes meets another resident, Izu Robin.

A bird that occurs elsewhere at several places in Japan, but which is not so easily seen as on Miyake-jima, is Japanese Wood Pigeon. Its strange lowing or mooing calls reverberate through the forest, and frequently it can be heard clattering off from a hidden perch or be seen sweeping darkly across the forested hillsides flanking the lake. With patience, you may find the pigeon perching atop dead snags, surveying its forest habitat. Several widespread Japanese species also make these islands their home, and these include Japanese Pygmy Woodpecker, which abounds on the island; its brief buzzy calls are commonly heard as it forages along branches and twigs, while the brightly attired Japanese White-eye and Oriental Greenfinch are even more common.

While most of the highly sought-after birds may be found at Tairo-ike, no birder's visit to Miyake-jima is complete without visiting the capes of the west coast, especially Cape Izu with its prominent lighthouse. Here, in the tall grasses that abound just inland from the shore, one can hear the stuttering song of Styan's Grasshopper Warbler [VU] await them patiently, and they will eventually pop up into view or even make a brief aerial display flight.

Offshore from the same capes, watch seawards in the early mornings and the island will be ringed by hordes of Streaked Shearwaters gliding past just at wave-top height. Larger streamlined shapes flying more directly and higher are Brown Booby, but keep a sharp eye open – you never know what might fly past next on this island famous for its birds.

For those fascinated by the effects of the earth's forces, it is well worth visiting the various geological sites around Miyake-jima for dramatic and sobering views of the effects of the fissure eruption of 1983 and the gas eruption of 2000. A boardwalk near the coast at Ako allows visitors to walk through the 'field' of sharp-edged lava, beneath which lie destroyed buildings, and where one can see the remnants of a school that was overwhelmed by the disaster. Elsewhere, a narrow road winds high on to the flank of the island, allowing visitors to reach an overlook that affords views seawards to the other islands in the Izu group, and landwards to the currently sleeping mountain. Long may it remain dormant!

The Storm Rider's Return

Tori-shima. literally 'bird island', at the southern tip of the Izu Island chain, is the tiny above-water portion of an active submarine volcano in the Pacific Ocean, standing 394 m high and lying some 600 km south of Tōkyō. With a circumference of 6·6 km and an area of just 4·79 km^2, the island is now an unpopulated gem and virtually impossible to access. A more precarious breeding site could hardly be imagined. The remoteness of this location, situated in the annual Pacific typhoon track, makes it the perfect home for Short-tailed Albatross[1], yet in 1949 that species was declared extinct.

Once these gorgeous golden-headed gooneys were numbered in millions. Sailors told tales of a great white pillar of birds rising from the ocean south of the Izu Islands; and explorer Hattori Tōru, who visited the island in 1888, described their numbers as truly incredible[2]. Grainy footage shot by Prince Yamashina Yoshimaro (1900–1989)[3] in 1930, long after the albatrosses' heyday, still shows carpets of breeding birds, hundreds in each scene. Hunters, for the feather trade, strode among the placid birds known as *Ahodori* (meaning 'stupid bird' in Japanese for its completely trusting nature), wielding clubs and devastating the bird colony, a sickening example of human contempt for other species and their evolution in extraordinary environments. Today, the island's few permitted visitors, researchers mostly, keep their distance except when wielding ringing pliers and measuring the birds, their eggs or their chicks, for the *Ahodori* has become a shy island recluse.

Short-tailed Albatross, the largest North Pacific seabird, has an impressive wingspan exceeding two metres[4], weighs 4–5 kg and measures 84–94 cm from bill tip to tail tip. The adult has a golden head, white body, back and tail, and mostly dark wings and terminal tail band. The bill is disproportionately large and pink. The youngest birds are all dark brown, but whiten progressively with age. All have the characteristic pink beak, which distinguishes them from the two other North Pacific albatrosses, namely Laysan [CR] and Black-footed.

1 Formerly known as Steller's Albatross (formerly *Diomedea albatrus*).
2 In his *Tori-shima Bakadori no Hanashi* (The Story of the Fool-bird on Tori-shima).
3 Dr Yamashina was the founder of Japan's famous Yamashina Institute for Ornithology.
4 213–229 cm.

Left to right Short-tailed Albatross is steadily recovering from near extinction [TR]; Laysan Albatross is one of three breeding species of albatross in Japan [MC]; Black-footed Albatrosses breed on Tori-shima and in the Ogasawara Islands [JoH].

These long-lived birds (estimated 40–60 years) return to their islands to breed when 5–10 years old, laying just one egg a year, which they will incubate for 64–65 days; the chick will take a further 120 days to fledge. Typically, they return to their colony in mid-October, lay and incubate during late October and early November, their egg hatching in late December and early January. The adults leave during April, by which time the dependent chick weighs more than either parent. For about a month, the chick is forced to fast as it completes its feather growth and muscular development without further support from its parents. Then, during May or June, it leaves the island and heads out into the Pacific, relying entirely on its own instincts, to make a loop around the North Pacific and Bering Sea, allowing it to find its natural prey of squid, fish, krill and other crustaceans.

Volcanic Tori-shima is a precarious breeding site for the rare Short-tailed Albatross [TR].

Like an avian cross between a windsurfer and a powered hang-glider, the Short-tailed Albatross (and its relatives: the other albatrosses, shearwaters and larger petrels) prefers not to expend its own energy when it can use that of the wind. This long and narrow-winged bird, with high aspect-ratio wings[1], exploits the fact that the wind is not uniform at all heights above the ocean. It takes advantage of the differential in velocity of different air masses – those low over the ocean surface in wave troughs and those higher in the air. In this style of flight, known as dynamic soaring, the albatross swoops down towards a wave trough and into the lee of a wave, relying on its own current speed and momentum, sometimes skimming the water with a wing tip; and then it makes a turn up into the wind, and wheels back up high into the air as it is suddenly exposed to a head wind, giving it tremendous lift and energy as the speed of the wind increases over its wings. Turning again, it can glide downwind, repeating a distinctive wheeling pattern of flight. In this way it can travel thousands of kilometres from its breeding colony while expending relatively little of its own precious energy.

Designated now as a Special Natural Monument and fully protected as such, *Ahodori* was once the focus of an extraordinarily destructive industry, that of Tamaoki Han-emon. Between 1886 and 1922, the Tamaoki Company housed 300 people on Tori-shima, built schools for the children and even a light railway, all for one purpose, that of obtaining and transporting albatross feathers! These large seabirds with their dense plumage produced the ideal pure white material for stuffing pillows and quilts, suitable even for export. Tamaoki Company workers each killed between 100 and 200 birds per day, and Dr Yamashina estimated that by 1902 at least five million birds had been killed, providing an example of humanity's inability to manage itself or other species sustainably. This slaughter of one species serves as a metaphor for humankind's treatment of the living planet on which we depend for survival.

In terminal decline because of overharvesting, the albatrosses of Tori-shima were further vulnerable to nest-trampling by introduced cattle, but more significantly to the island's volcanic nature. In 1902, all 125 of the people then living on the island were killed in a massive volcanic eruption, although that did not stop Tamaoki from re-establishing his company in the following year. Nevertheless, the harvest was unsustainable and the decline of the birds was inevitable. Dr Yamashina estimated that only about 2,000 albatrosses existed on the island in 1930, there were fewer in 1932, and fewer than 100 in 1933. A last-minute scramble for feathers before a hunting ban could be enforced seems to have been the penultimate nail in their coffin, the final one coming with the catastrophic volcanic eruptions of 1939, when the island was evacuated, and 1941, when much of the island was covered with new lava.

By spring 1949, when Oliver Austin Jr, ornithologist of the General Headquarters Supreme Commander for the Allied Powers, sailed south to the Ogasawara Islands via Tori-shima, he found no albatrosses there during what should have been their breeding season. The golden gooney was gone, presumed extinct. And there the story of another species driven to extinction by human greed, expansion or resource extraction might so easily have ended had this particular island species not been a long-lived oceanic storm-rider.

Spin the clock forward six decades, and there is a world of difference. *If* Tamaoki Han-emon's goal was to destroy the Short-tailed Albatross for personal gain, then he almost succeeded. If this example of human greed and lack of understanding of species in their natural ecosystems was unique, then it would remain as a salutary lesson in how not to manage our planet, but unfortunately such avarice continues unabated to this day. Luckily, in the case of the albatross, another Japanese individual, Dr Hasegawa Hiroshi, was inspired in a different direction, and made it his life's work to help to resurrect this bird for itself, and for future generations. His efforts were aided and abetted by particular aspects of Short-tailed Albatrosses' biology: they live for several decades, are supremely loyal to their nesting sites, and do not even attempt to breed for at least five years, and they can survive indefinitely at sea.

---

1 The aspect ratio of a wing describes the ratio of the wingspan to the mean chord, and is equal to the square of the wingspan divided by the wing area. A long, narrow wing (such as that of the albatross) has a high aspect ratio, whereas a short, wide wing (such as that of an Eastern Buzzard) has a low aspect ratio.

Tori-shima's position, in the midst of the annual typhoon track, made it an ideal location for the establishment of a manned frontline meteorological observatory. This was set up in the late 1940s, and so humans returned to the island. It was in January 1951 that the observatory's Yamamoto Shoji stumbled upon a tiny group of albatross colonists on the steep scoria slopes of Tsubamezaki (Swallow Cape). Following the rediscovery of this tiny colony, consisting perhaps of birds that had wandered at sea for years, meteorological observatory staff made voluntary efforts to eliminate feral cats and to stabilize nests at the breeding site. In the early 1960s, Dr Yamashina sent researchers annually from the Yamashina Institute for Ornithology to monitor, document and ring the birds. In 1965, however, the weather observatory was closed and Tori-shima became unpopulated once more, and access became almost impossible.

Hasegawa Hiroshi is one of several albatross researchers who have dedicated themselves to the recovery of the beautiful Short-tailed Albatross [TR].

Nearly a decade later, in April 1973, British ornithologist Dr W. L. N. Tickell and Yoshii Masashi from the Yamashina Institute, who visited Tori-shima by British naval vessel, were able to count 24 Short-tailed Albatross chicks, and confirmed that the population was growing, albeit slowly. More importantly, Lance Tickell inspired the young Hasegawa Hiroshi to take up the baton for the bird, pointing out to him that, as the Japanese people had been responsible for the near extermination of the albatross, it was their responsibility to oversee the bird's recovery. Inspired and motivated, Hasegawa Hiroshi took that responsibility deeply to heart. He is a species champion for what is arguably the North Pacific's most beautiful seabird. Since his first expedition to Tori-shima, in 1976, he has devoted decades, and more than a hundred subsequent expeditions, to defending and promoting *Ahodori*. He has made saving *Ahodori* his personal calling and his life's work. Greg Balogh of the US Fish and Wildlife Service[2] said of Hasegawa that "...he has done more to raise the awareness of the plight of the Short-tailed Albatross than anyone on the planet" and "he is like a modern-day samurai fighting for conservation...".

Dr Hasegawa began with ringing and breeding studies, but recognized that the steep slopes of volcanic debris that made up the Tsubamezaki nesting grounds were inherently unstable, and he pushed hard for erosion control and stabilization. Japan's then Environment Agency (now Ministry of the Environment) at last stepped in and began erosion control, initially using native grasses in 1981–2, and subsequently earthworks to help to direct run-off water safely away from the site. Another tiny colony of albatrosses had been discovered on a pair of precarious islands, Minami-kojima and Kita-kojima (in the Senkaku Islands), on territory that is now, unfortunately, the subject of dispute between Japan, China and Taiwan. However, Dr Hasegawa was able to confirm breeding there in 1988.

The need for protection on Tori-shima was continual, as each season wrought more erosion. So, next came Operation Decoy, using playback and decoys placed there since 1993, to tempt new albatross breeders to a different, safer nesting location on the island; and, with pairs beginning to nest there in 1995, a second colony was established, although still on the same island.

Historically, there may have been as many as 14 breeding colonies of Short-tailed Albatross south of Japan and in the East China Sea, but, like that on Tori-shima, these colonies were easily raided, depleted, and then exterminated by feather- and egg-collectors. Prehistorically, fossil evidence from Bermuda suggests that the species even ranged into the Atlantic Ocean before the Central American Land Bridge arose. The current anomalous North Atlantic gap in albatross distribution may once have been filled by this species.

2 Biologist and former lead for the US Short-tailed Albatross Recovery Team in USA.

Bullets in Brown

Brown Booby [YaM]

Japan's location on avian migration routes between southern and northern seas, its position in the track of numerous typhoons, and its many isolated offshore and oceanic islands make it a paradise for seabird-lovers. One hundred and fifteen of the 750 or so bird species recorded from Japan are seabirds – albatrosses, petrels, shearwaters, storm petrels, boobies, cormorants, gulls, terns, skuas and alcids.

'Sleek elegance' is a befitting way of describing a catwalk model in the fashion world, suitable even for an ultramodern city tower or a Japanese bullet train, and appropriate on the race circuit for describing a pleasingly aerodynamic two-seater convertible. It is a surprisingly relevant expression, too, for describing certain birds, Brown Booby for example, the common large seabird effortlessly keeping pace with ships sailing the ocean route south to the Bonin Islands.

Most birds convey a clear impression of being feathered; there is something distinctly, well – feathery about them. Feathers are, after all, the single defining characteristic of all surviving 'dinosaurs', the living members that belong to the Class Aves. Most birds make no attempt to hide the fact that they are clearly wearing a feather coat. Some bird species, though, seem, in their extraordinary elegance, to go beyond the norm. With these the plumage appears more like plush velvet; their outer covering is so finely smoothed as to appear more like one continuous surface, as of a model's fine fabric jumpsuit or a base-jumper's wingsuit, than the isolated tips of countless thousands of tiny, overlapping contour feathers growing from follicles in the skin. The supremely aquatic divers or loons, the seemingly silken waxwings and, among the seabirds, the sleekly elegant boobies are clearly in this elevated category of avian species.

Brown Booby banks and rises slowly, lifting on the sea breeze like an elongated leaf blade carried aloft, showing off its long, smoothly tapering bill, which is pale yellow like old ivory. There are patches of delicate, pale periwinkle-blue bare skin around the base of the bill and enclosing the large dark eyes of the male, while the same skin is a delicate shade of pale sulphur-yellow on the female. Its deep brown eyes are unusual among birds (other than among predators) in being placed so far forward as to provide binocular vision. Most birds have eyes located on the sides of their head, giving them excellent all-round sight, but little binocular vision, but not the extraordinary boobies. The bird's plush, skin-like, brown plumage – the upper surfaces of its wings, its head, breast and tail – all appear somewhere on a

Brown Booby breeds on various islands in the Ogasawara group [MOEN].

scale between strong coffee and chocolate-brown. Subtropical sunlight reflecting back off the sea's surface renders the plumage beneath its wings and on the underside of its body as of the purest white. A mere brief stretch of its wings and it banks away; a twist of its long wedge-shaped tail and it sails back again.

Brown Booby is frequently encountered between the Izu Islands and the Ogasawara Islands [MAB].

The effortless way in which boobies appear to keep pace alongside ships masks the fact that this is a bird capable of considerable speed. Although typically cruising well within any Japanese 40 km/hr urban speed limit (males average 27 km/hr and females 34 km/hr), Brown Booby is easily capable of flouting 60 km/hr rural speed laws, as it accelerates to over 90 km/hr. The booby's sleek and velvety profile serves a double purpose, for not only is it aerodynamically adapted for speed in the air, but it is also aquadynamically adapted for swiftly penetrating the surface waters of the pelagic ocean.

The tropical boobies and their temperate relatives the gannets are spectacular divers, not because of the depths they reach – they are easily outpaced in that regard by the auks and the penguins – but because of their methods. At home in the air, they bank and glide, soar and sweep above the ocean, usually at a height of up to about 12 metres, until their acute binocular vision allows them to spot and pinpoint prey beneath the sea's surface. Their leisurely horizontal flight then changes dramatically. Suddenly, they are all focus, their attention refined and honed to the single point that is their prey beneath the waves. Their elongate and pointed wings, until now beating steadily in a horizontal plane, promptly swivel. The birds turn their bill tip downwards, and then, with a few high-powered beats of their wings, spanning a metre and a half, they surge vertically downwards towards the sea. A few metres above the ocean's surface they change posture once more as they slide their outstretched wings backwards so that they touch and are in line with the body: now they are the avian equivalent of a swing-wing fighter jet aircraft – long, slender but deadly darts. At speeds that would send a Japanese road-speed patrol during its regular 'safety drive' campaigns into paroxysms, the boobies hit the water like powerful missiles, sending up spouts of water vertically as they disappear beneath the waves.

The pantropical Brown Booby occurs around oceanic islands in the Atlantic, Indian and Pacific Oceans, favouring coral atolls and volcanic islands such as those in the warmer regions of southern Japan, where the boobies nest on open slopes or ridge tops. There, males and females share the work involved in incubating their clutch of two eggs a year, these taking more than 40 days to hatch, and a further three to four months are needed for their young to fledge. During this long breeding period the adults range out from their nesting grounds in search of prey, which includes all manner of small fish up to 40 cm in length, and particularly various species of flying fish.

When the boobies spot prey that is close to the surface, they dive at a gentle angle, entering the water smoothly and as easily as a razor-sharp knife slicing meat. They leave barely a ripple and bob back to the surface quickly, having submerged only a metre or so. If their prey are schooling more deeply, their plunge-dive, reaching speeds of over 90 kph, will end vertically and spectacularly, sending bursts of spray flying high into the air. They dive several metres down and, propelled by their beating wings and kicking feet, pursue their prey beneath the waves. Eventually they emerge and, if fortunate, will have prey to carry back to their nest mate or chicks.

Once their December–March breeding season is over, Brown Boobies do not so much migrate as disperse away from their breeding grounds, and then may appear well beyond their normal breeding range. They are often to be seen around the Izu Islands and off the coasts of Kyūshū, and they are easily seen when paying a visit to the Ogasawara Islands early in the year.

Now retired[1] from Tōhō University, Dr Hasegawa's personal goal has been to pull *Ahodori* back from extinction's brink, to see it re-established at several colonies, and for it to reach 5,000 breeding pairs by his 70th birthday. Greg Balogh considers that "there is huge potential for the species at its current breeding sites, possibly allowing it to approach historical levels in the millions, with linear or exponential population growth likely to continue", but with the caveat that this will be possible only if the carrying capacity of the ocean is still there. From just 25 albatrosses on Tori-shima in 1954, by 2008 the total population of the species was estimated to have reached 2,364 individuals, with 1,922 of these located on Tori-shima (with over 420 breeding pairs there) and 442 birds on Minami-kojima, providing clear evidence of an enormous increase since the albatross's rediscovery and the onset of conservation efforts. As Greg pointed out as we talked of the international efforts to save this species: "What has been achieved is fantastic, it's one of the great species conservation stories going on right now."

The apparent calm of the albatross colony is broken by the deafening sounds of the sea and the wind, and during the breeding season by the impressive vocalizations of the birds. During displays they trumpet and moan, warble and gargle, bill-tap and bill-clatter, stand face-to-face and rock high on to their tiptoes. When I first visited Tori-shima aboard *La Madonna* with Dr Hasegawa in 1990, and saw the volcanic rubble that made up the flanks of Tori-shima, the ragged red-brown cliffs and the partly stabilized scree slope on which the birds breed, I was overwhelmed by how precarious it all seemed. The island appears raw, recently forged from the earth, annually eroded by wind, rain and the sea, yet the birds visit briefly each year in their unceasing attempts to survive.

The open ocean, of course, is where these birds spend most of their lives and, although it was for long presumed that *Ahodori* dispersed from Tori-shima and rode the winds northwards as far as the Bering Sea, we now know, thanks to satellite tracking, just where they go. The breeding females prefer areas offshore from Japan and Russia, whereas the non-breeding birds and adult males wander farther, heading off to the Aleutian Islands, the Bering Sea and coastal North America. In 2010 therefore, a Short-tailed Albatross which I pointed out off Adak Island, in the Aleutian Islands, when bound for Kiska by ship, was likely from Tori-shima, and perhaps was one of the very birds that Dr Hasegawa had ringed. Thankfully, Japanese breeding birds that migrate to Alaskan waters almost entirely miss the plastic garbage trap of the North Pacific gyre, so they are not prone to ingesting plastic waste, but such floating debris is a threat to the few Short-tailed Albatrosses that have recently begun to colonize Midway Island, in the North Pacific.

Fully protected in Japan as a National Natural Monument since 1958 and as a Special Natural Monument since 1962, Short-tailed Albatross breeds on an island that is similarly enshrined in current conservation law. Although legal protection is in place, however, natural calamities may intervene. In North America, where it is also recognized as an endangered species, the Short-tailed Albatross (or STAL) recovery plan considered the establishment of a breeding colony on a 'safe' island to be a vital criterion required for its removal from the USA Endangered Species List. Muko-jima, in the Ogasawara Island group, was the chosen target. The STAL team, with support from Friends of the Albatross within Alaska's commercial fisheries, convinced Ted Stevens, the late senior senator for Alaska, to provide federal funding for albatross recovery. That, combined with funds from the North Pacific Research Board, was matched dollar for dollar by Japan's Ministry of Environment, which in 2007 approved the translocation of chicks by the Yamashina Institute some 300 km by helicopter in the hope of establishing a new safe colony. Greg Balogh, who has been twice to Tori-shima and once to Muko-jima, explained how at first the Japanese side seemed reluctant to "engage in translocation, but, having been invited to the Hawaiian Islands to participate in translocation experiments with Laysan and Black-footed Albatrosses there, they then got on board the translocation train and barrelled out of the station, we could barely keep up – it was heart-warming to see and very impressive." Greg feels honoured and privileged to have visited an island that few Japanese citizens can hope to see and, having observed, first-hand, translocated birds that had been hand-fed and

1 March 2014.

hydrated by Yamashina Institute staff, commented that "to see one of the greatest species recovery actions in play and to see how well coordinated it was, was truly impressive".

With the successful establishment of a second colony on Tori-shima by means of decoys, a newly self-seeded third colony on the island, and the foundation of an entirely new colony by the translocating of chicks to Muko-jima, conservation efforts for the golden-headed Storm Rider are paying off. Moreover, as Greg Balogh told me: "… it is only going to get better as the birds translocated to Muko-jima lead to a viable new colony in the years ahead."

Numbers of albatrosses at the three colonies on Tori-shima surpassed 3,000 in 2011, and Hasegawa Hiroshi, busy between visits to the island, told me that there were 609 nesting pairs on Tori-shima in the 2013–14 season. He commented that, if the weather was not severe that winter, he expected to find about 425 chicks on his next visit in the spring, leading to a post-breeding population size of about 3,550 individuals on Tori-shima alone; by 2018 that number had reached 4,200. Despite retiring from Toho University, Dr Hasegawa's wish was to "continue monitoring the population until 2019, when the total population size will be about 5,000 individuals to fulfil my dream".

Flying Foxes and other Bats

Of the 170 or more species of mammal known from Japan, 37 are bats. Both of the two major orders of bats, the Megachiroptera and Microchiroptera, are well represented here, although the Megachiroptera occur only in warm tropical areas. Japan's southern tropical island groups, the Nansei Shotō and the Ogasawara Islands, are home to many special species, including a brace of fruit bats in the family Pteropodidae. While one of them, Ryukyu Flying Fox, is restricted in Japan to the Nansei Shotō, essentially from Okinawa south to the Yaeyama Islands, there is an equivalent species, Bonin Flying Fox, found only in the Ogasawara Islands.

As recently as the 1970s Bonin Flying Fox was considered extinct, and even now the fate of this mostly black fruit bat hangs in the balance, and seeing this species was, until recently, well-nigh impossible. On Chichi-jima, in the Ogasawara Islands, 1,000 km south of Tōkyō, night safaris in search of it are now possible.

Bats are not everyone's cup of tea. Bizarre mental images abound, from those of DC Comics' Gotham City's Caped Crusader to those of blood-sucking vampires of South America, yet the majority of the world's 1,000 or so bat species are small, even tiny, and they subsist on a diet of insects. These Microchiroptera have the amazing ability to echolocate as they fly in near or total darkness. They emit calls, building up a sonic map of their surroundings by listening to the echoes of their own voices. Their outpouring of ultrasounds at different frequencies allows them to distinguish objects from prey, and to judge how far away they are by listening to how long it takes the sounds to bounce back. It is difficult even to imagine being able to 'see' one's surroundings with one's ears.

The endemic Ryukyu Flying Fox [SM]

Large-footed Myotis is a widespread species [IiM].

When we humans are young, our acute hearing allows us to 'feel' the chiropteran ultrasounds of echo-locating bats more as a sensation within our brains than one heard through our ears; we can literally tune in to the world of bats. Now, more than 30 years beyond childhood, I count as an unexpected blessing every year that I can still feel those sensations. Because, as we humans age, we tend to lose access to those higher frequencies, and bats use so many that are well above the response of the human ear that to 'feel' any of them is a bonus.

Bats are a bonus in many ways, not least because they are major predators of insect populations that might otherwise be overwhelming but also because certain bats are important pollinators of a wide range of plants, especially in the subtropical habitats of Japan's far south. In addition to those insect-consuming Microchiropterans, there is another group, the Megachiropterans, or Megabats. Larger in size, the latter are also fewer in number; Japan supports just two species[1], those in the subtropical Ryūkyū and Ogasawara Islands. Most of these Megachiroptera have lost the power to echolocate, and have large eyes and long snouts, earning them the sobriquet 'flying fox'. See one up close and that name becomes obvious. Their long fur and their facial features well deserve the 'foxy' epithet. Like that of other bats, their flight is a marvel. Long bony digits spread a soft wing skin so broad that they are capable of sustained flight. In their airborne capacity they are like birds, yet little about their anatomy resembles that group. These are mammals, armed with teeth, fur and the capacity to lactate, but, unlike the microbats, they have charisma on their side. In the case of Bonin Fruit Bat or Bonin Flying Fox, they also have rarity value.

Known here as *Ogasawara Ōkōmori*, Bonin Fruit Bat was first described to science in 1829 when the islands were first being explored. Long isolated on this remote subtropical archipelago, the species not only has a restricted range, limited to Chichi-jima, Haha-jima, Kita Iwō-jima, Iwō-jima and Minami Iwō-jima, but also is classified as Critically Endangered. This forest-dwelling bat has a range of less than 100 km^2, and it is threatened by continuing habitat degradation and loss. For long considered extinct, it was rediscovered only in 1986. The current population is estimated at fewer than 300 individuals, and if one could gather them all together the entire world population of this unique species could probably hang from the ceiling of a single eight tatami-mat room with space left over for individuals to move around. Setting off into the darkness to look for one seems worse than looking for the proverbial needle in a haystack.

Aptly named, fruit bats depend on plants, eating pollen, nectar and fruit; the Bonin Fruit Bat is no exception. One of its favourites

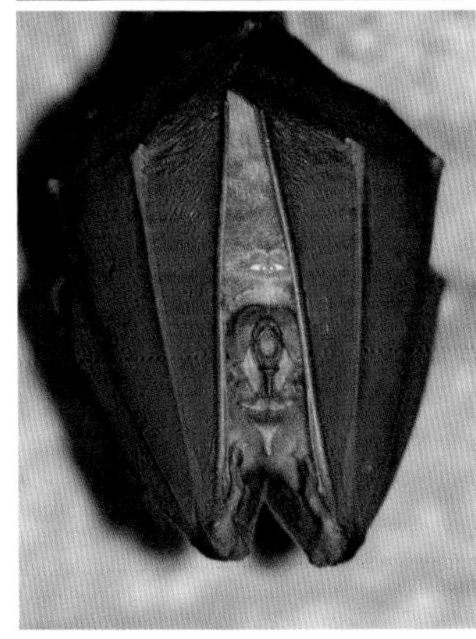

Top to bottom Okinawa Little Horseshoe Bat is an endemic of the southern Nansei Shotō [IiM]; the East Asian Little Bent-winged Bat [IiM]; the endemic Japanese Long-eared Bat [VD]; Greater Horseshoe Bat [VD].

1 A third species, Okinawa Flying Fox once occurred on Okinawa and then became extinct (it has not been reported since specimens were taken there in the 19th century.

Octopus Tree is endemic to the Ogasawara Islands [*left* OVTB; *right* MAB].

is an endemic palm-like tree colloquially known as *Tako-no-ki*, or Octopus Tree (named for its growth habit, with aerial prop roots giving it a superficial resemblance to an octopus). The fruit bat is an important seed and pollen dispersal agent for this and other native plants. While the endemic fruit bats feed on the fruits and nectar of this endemic tree, another endemic species, this time the leaf-mining Pandanus Bark Beetle (first described only in 1998), consumes its living leaves, and a further endemic species, the fungus *Kodonospora tetracolumnaris* (first described only in 1993), grows on its dead leaves. One tree – four endemics!

During the winter months, the fruit bats gather at communal roosting sites and form dense, ball-shaped clusters, making them much easier to locate than in summer, when they are widely dispersed. Radio-tracking research has shown that, whereas summer roosts are used by solitary bats and nursing females, in winter, as temperatures fall, the bats gather to thermoregulate at night and as many as 100 will gather at a single roost. These roosting clusters are of three types: mixed groups of subadult males and females, all-male groups, and groups mainly of females with just a few males. It is in this last type of winter roost that all the mating action takes place, and the males there will fight to exclude other males in what is known as female-defence polygyny.

Introduced Aliens – The Devil Wears Green

Bright sunlight casting a penumbra beneath fan-palm leaves exposes the unexpected, a silhouette on a palm leaf above as if a sinister *Kiri-e* creature had been excised from green cloth. This motionless, elongate creature, slender with a long narrow head and a fine whiptail extending farther than the length of its body, is a Green Anole, an invasive alien species (IAS). The delicateness of this creature's toes is notable. They are not merely elongated and slim, but they bulge laterally, becoming much wider than at either the base or the tip. Strange toes indeed, but they are crucial to this animal's arboreal lifestyle.

Green Anole is an invasive alien species that preys on the endemic insect fauna of the Ogasawara Islands, including Ogasawara Cicada [MAB].

Pause long enough to scan the branches and leaves of the trees around, and you will likely find one or more anoles. The endemic Ogasawara Snake-eyed Skink, however, is much harder to find. Green Anole lizards go by various names, such as American Anole, Carolina Anole and

The endemic Ogasawara Snake-eyed Skink has become a rarity [OVTB].

Japanese Skink is widespread in the main islands of Japan [HT].

Red-throated Anole, and even American Chameleon (because they are able to change colour from brilliant green to brown). Native to North America, Green Anole ranges through the southeastern coastal states from Virginia south to Florida and from Alabama west to Texas. Unfortunately, the species has been introduced to the Hawaiian Islands and to the Ogasawara Islands, where it is wreaking havoc among the native faunas, a pattern that has been repeated over and over on islands around the world as alien species have been introduced.

When introduced into naive ecosystems, such alien species as the anole on Ogasawara, or the lantanas *Lantana* (in the verbena family Verbenaceae) introduced into India, may find themselves face-to-face with local species that are completely unprepared with suitable behaviours to protect themselves. In such situations, the introduced species may outcompete the native species for food, space or other resources, or may, in the case of the anole, directly prey upon them. The introduced species become invasive, spreading rapidly, and overwhelm the local ecosystem – and the results are devastating.

Although not a true chameleon, Green Anole, 12–20 cm long, changes colour in relation to its level of activity, the light conditions and its background. It ranges in colour from bright emerald-green to dull brown; the larger, brighter males sport red dewlaps that they can extend and use for signalling. Green Anole is arboreal in nature, and it is those oddly bulging toe pads that make this lifestyle possible. Like those of a gecko, the toe pads have numerous fine-tipped hair-like scales allowing them to climb easily, not only up trees but even up metal.

Although sometimes described as a delightful small lizard and as a popular choice for an easy-care reptilian pet, on the isolated Ogasawara Islands they are not a delight, but a menace. Until the 1960s and 1970s, these remote and biologically sensitive islands lacked tree-climbing lizards. It is thought that the anoles either arrived accidentally among American military equipment or were imported deliberately as pets and later escaped or were released. It was only in the 1980s that they crossed from Chichi-jima to Haha-jima, further extending their range and their devastation of islands that are so biologically important that scientists frequently compare them with Ecuador's Galápagos Islands, because of the way the local fauna and flora have evolved into many unique local forms. Since their arrival, anole numbers have exploded from zero in the early 1960s to 2–6 million today, and those millions of lizards have had a long party at the expense of the native insect life that once thrived there. Female Green Anoles mature rapidly. Within about eight months after hatching, they are able to breed and can produce up to about ten eggs per year throughout their 4–8-year lifespan, giving them a tremendous capacity for population increase. They and their offspring are voracious, living off the indigenous life of the islands.

Green Anoles now scour almost all habitats in the Ogasawara Islands for insects on which to prey. They are directly implicated in the demise of no fewer than five endemic dragonfly species, the endemic Ogasawara Blue Butterfly[1] (a blue lycaenid), an endemic cicada, a long-horned beetle, and various small beetles and wood-borers. That is a hefty rap sheet. Not surprisingly, the anoles are now

1 It was relatively common until the 1970s, but its range is now reduced to just one small island in the Ogasawara archipelago.

high on the wanted list of alien species on the islands (along with goats, cats, rats, and a host of invasive plants). Because of the anole, dragonflies declined dramatically during the 1980s, and are now either extinct or nearly so on the two main islands of Chichi-jima and Haha-jima. The endemic butterfly has been wiped out everywhere on Chichi-jima, and survives now only in some areas of Haha-jima; and the once noisy chorus of the endemic cicada has been silenced for ever. Anoles have been found to be predators also on the islands' endemic bees, and as such they are affecting the pollination and fruiting of local plant species, many of which are endemic.

Introduced Red-eared Sliders are now abundant in ponds and waterways in warmer regions of Japan [MAB].

With the Ogasawara Islands inscribed on UNESCO's World Heritage list in 2011, Japan's Ministry of the Environment and local islanders have received a boost, and signs are encouraging that control of Green Anole and other invasive introduced species, such as East African Land Snail and New Guinea Flatworm, may be possible, although returning the lost endemic insects to the islands is impossible. Intensive trapping with specially designed sticky traps is helping, especially around the ports, and there is hope of both reducing the anole population and preventing its further spread to unpopulated islands. The fate of the remaining indigenous insects and snails of the Ogasawara Islands lies now in our hands, and, thankfully, public awareness is now on their side. Nevertheless, constant vigilance is necessary, as indicated by the arrival of the invasive alien Big-headed Ant, which reached Haha-jima in 2014, and which has been implicated in the reduction of land snails there.

On the Ogasawara Islands, Green Anole and Feral Goat have wrought havoc. On Japan's southwest islands of Okinawa and Amami Ōshima, it was the introduction of Small Asian Mongoose and Japanese Weasel, both IAS in this context (although the weasel is indigenous to Central Japan), that placed the native fauna under stress. On the Izu Islands, the introduction of Japanese Weasel dramatically reduced the populations of native and endemic birds. Some introduced species seem less harmful, possibly benign, although their long-term impacts are rarely clear. While the introduced Red-billed Leiothrix and Rose-ringed Parakeet seem not to threaten native species directly, introduced turtles do. Common Snapping Turtle, which arrived here via the booming pet trade in the 1960s, is most definitely an IAS. It can grow to be a metre long from nose to tail and weigh up to 35 kg. With its powerful beak-like jaws and highly mobile neck and head, it has become a serious predator of native fish and birds; furthermore, it can cause serious injuries to humans, too. These turtles are highly fecund and spread easily. In one lake alone, Lake Inba in Chiba Prefecture, the population has exploded to an estimated 16,000! Japan's list of IAS includes a wide range of mammals, reptiles, birds, amphibians, fish, crustaceans, insects and plants. Each IAS has different impacts; some compete for space, some for prey, some are predators of native (often endemic) species, and some transfer diseases to native populations, thereby threatening biodiversity. The introduction of Louisiana Crawfish and Signal Crayfish from North America into Japan in 1927 has negatively affected populations of the endemic Japanese Crayfish throughout the country's streams and rivers through competition and predation, and in 2018 the first transference of crayfish plague to Japanese Crayfish populations in Japan was diagnosed.

Much closer to home, domestic and feral cats cause untold damage to the environment, both urban and rural. With a pet cat population estimated to be ten million, the numbers of stray and feral cats out hunting are unknown, but no doubt enormous. As predators, they have a dramatic impact on all forms of life in Japan and they roam throughout the country, from isolated islands to montane national parks.

Japan came late to an understanding of biosecurity, but it is making rapid strides in dealing with some of its invasive alien species. Is it time for Domestic Cat to be listed as an IAS?

The Mammals and Birds of Japan

An Overview of Japan's Mammals

Orca are readily seen off the Shiretoko Peninsula in the Nemuro Strait [WaM].

Diversity in adversity and isolation defines the fauna and flora of Japan. Japan's geological history has been chequered with periodic connections to and disconnections from the Eurasian continent over geological time. Times of sea-level rise have isolated parts of the archipelago as long ranges of mountains, or even as isolated mountain-top islands (the Nansei Shotō today), where species have evolved *in situ*. When sea levels have fallen, long-isolated lands have been reconnected to each other and, sometimes, to the nearby continent. Populations of mammals from the continent have been able to migrate across these 'land bridges' and expand their ranges, allowing populations of once isolated species to meet and mix once more.

Japan's geological history, and its great length (3,000 km), great altitudinal range and multiple climatic zones combined with its great range of habitat types, mean that today it supports a surprisingly large range of mammal species for a largely temperate country of its size.

Thanks to its complex geological history, the Japanese archipelago supports a diverse range of ecological environments, as already described. Hence, it is inhabited by a wide range of mammals, from Northern Fur Seals off Hokkaidō to Dugong off Okinawa, from Brown Bears on Hokkaidō to Asiatic Black Bears on Honshū, and from Japanese Hare on the main islands to Amami Rabbit on the Amami Islands.

Brown Bears occur only in Hokkaidō and are most easily seen along the Shiretoko Peninsula [WaM].

More than 131 non marine and 43 marine mammal species have been recorded so far in Japan, giving Japan a surprising total of 174 mammal species, a significant proportion of which are either rodents or bats. These represent extremely diverse types, including whales, dugongs, bats, primates, carnivores and rodents, and 50 of them are endemic to the Japanese archipelago. Rates of endemism are especially high among the smaller mammals. Japan's diverse marine mammal fauna ranges from the subtropical waters off the southernmost islands, and the warm seas around southern Japan, to the cold Oyashio current flowing south down the Pacific coast and the frigid waters of the southern Sea of Okhotsk. Some of the non-marine mammals include wide-ranging Holarctic species, such as Red Fox and Brown Bear, found across northern North America and the great span of northern Eurasia. Other non-marine mammals are Palaearctic species, such as Eurasian Red Squirrel and Siberian Flying Squirrel, which range from Scandinavia to Hokkaidō. Finally, some are East Asian species, known both from Japan and from adjacent areas, such as Tanuki and Japanese Deer. Most especially there are the endemic Japanese species. These are of two types, those ranging widely throughout several of the main Japanese islands, such as Japanese Macaque and Japanese Squirrel, and those occurring only in isolated localities, particularly on offshore islands, such as Amami Rabbit.

Tanuki, an east Asian species, is widespread in Japan [MAB].

Regional variation in Japan's mammal fauna

Japan is one of the most diverse zoogeographical regions in the world, including boreal to subtropical zones. Most of Japan (Hokkaidō and Central Japan) falls within the Palaearctic Region, while southernmost Japan, particularly the southwestern islands lying south of the Tokara Strait, belongs to the

Japan's Leopard Cat population is confined to Tsushima, where it is known as Tsushima Cat [IiM].

Oriental Region. The distinctions between these two provide an indication as to the main origins of their faunas, and the zoogeographical boundary between them is known as Watase's Line. Thus, the northern region of Japan supports mammals mainly of temperate and subarctic origin, while the southern region of Japan supports mammals originating in the subtropical and tropical areas.

Looking more closely at Japan in the Palaearctic Region, and considering only terrestrial mammals, we recognize three readily identifiable groupings. First, there is Hokkaidō, in the north, then the three main islands of Honshū, Shikoku and Kyūshū, and thirdly there are the Tsushima Islands (lying between Kyūshū and the Korean Peninsula). Hokkaidō shares nearly half of its mammal fauna with regions farther to the north, particularly Sakhalin and the Russian Far East. This indicates Hokkaidō's past connection to the East Asian part of the Eurasian Continent via a land bridge to the north. This land bridge, connecting Hokkaidō via Sakhalin to northeast Russia, is thought to have endured for some 60,000 years or so, during the last glacial period, allowing Eurasian species, such as Northern Pika and Mountain Hare, to occupy lands as far south as Hokkaidō. Hokkaidō also shares just over half of its mammal fauna with Central Japan, but the much deeper Tsugaru Strait, south of Hokkaidō, remained open during most of the last glacial period, denying many northern species access towards the south.

Looking to Central Japan now, approximately 50 percent of the non-marine mammals found there are endemic (some even to the generic level[1]), and some of these belong to the oldest elements of the Japanese mammal fauna. The mammalian fauna of Central Japan falls into two groups, depending on their relationships with the mammals of other parts of East Asia. One group shows links to the north, with relatives found on the Korean Peninsula, in northern China and in regions farther north, while the other group shows affinities with species inhabiting more southerly regions of Asia.

When we consider Tsushima, the small group of islands lying in the strait between the Korean

Endo's Pipistrelle is endemic to Japan's main islands but does not occur in Hokkaidō [IiM].

Lesser Leaf-nosed Bat is endemic to the Yaeyama Islands. [BOTH IiM]

1 Two genera of shrew-moles (*Urotrichus* and *Dymecodon*) and one of dormouse (*Glirulus*) are endemic to the three main islands of Honshū, Shikoku and Kyūshū.

Peninsula and Kyūshū, it is unsurprising that they share species with both Central Japan and the continent. In fact, some continental species occur in Japan only on those islands. The islands can be seen then as a biogeographical transitional zone between Central Japan and the continent.

On the southwestern islands of Japan, lying south of the Tokara Strait and Watase's Line, but east and northeast of Taiwan, we find a particularly fascinating mammal fauna. These islands lie within, or at least straddle, the ill-defined border of a major zoogeographical region known as the Oriental Region. Although repeated connections and separations between the islands and the continent to the south have occurred, the islands themselves have remained isolated sufficiently long (either as a connected ridge, or as further isolated islands) for local endemic species to have evolved. More than 50 percent of the native non-marine mammals of this area are endemic; and some of them are endemic at the generic level, not just at the specific level. South of Okinawa there is another deep strait[2], known as the Kerama Gap, or Miyako Strait, which, like the Tokara Strait, exceeds 1,000 m in depth, and through which runs another, far less well-known zoogeographical boundary – Hachisuka's Line. Both straits are believed to have survived since the Pliocene (5·3–2·6 Mya), essentially isolating the Amami and Okinawa islands (as a unit) from the north and the south and serving to explain the high rates of local mammalian endemism.

It is the considerable species diversity and high rate of endemism within the relatively small land area of Japan (slightly larger than either New Zealand or the British Isles) that make the mammalian fauna of Japan such an interesting one.

Conservation Matters

The unique mammal fauna of Japan faces particular problems in this overcrowded nation. Unlike in many other countries, these do not include widespread hunting for sport (this is an uncommon and declining pastime targeting relatively few species), nor is there a bushmeat trade driving populations down. Habitat loss and habitat degradation, however, are the immediate critical issues. The lowlands of the Japanese islands are intensively developed for agriculture, industry and urbanization[3], as a result of which few areas of natural lowland habitat survive and disturbance is commonplace. The mountainous areas are less developed or disturbed, providing more plentiful habitat for species able to survive at higher elevations. Even there, however, habitat loss and degradation are issues to consider. Conversely, some of the rural areas of Japan experiencing steady human depopulation are seeing an increasing incidence of human–animal conflict, as species such as Japanese Macaque, Japanese Deer and Wild Boar find abandoned farmland adjacent to native forest much to their liking and fewer people living there to be able to keep them at bay.

Unfortunately, five mammalian taxa have become extinct in Japan in the last 120 years: Grey Wolf (both Japanese and Hokkaidō subspecies), Japanese Sea Lion, Eurasian (or Japanese) Otter, Okinawa Flying Fox and Bonin Pipistrelle. More recently, and compounding the problem of habitat loss, the introduction of alien species is having far-reaching implications for the native mammal fauna, and deserves considerable attention from the conservation community.

In addition to a number of feral species, such as cats, dogs, rabbits and goats, various introduced alien species are impacting the native fauna to varying degrees; these include Pallas's Squirrel and North American Raccoon. Alien species, whether introduced deliberately or accidentally, occupy important ecological niches in previously natural ecosystems. They may become predators of native species (mammals or otherwise) or they may compete with native species for valuable resources such as territory, food, cavities and den sites. Alien plant species may fundamentally alter ecosystem structures and processes or affect habitat structure, while in some cases aquatic and terrestrial organisms spread parasites or disease to plants and animals.

---

2 At 250 km wide, it is also the widest strait in the southwestern islands, making it a formidable barrier against the dispersal of non-volant species.

3 And, alarmingly, for rapidly spreading ground-hungry solar arrays.

An Overview of Japan's Birds

Almost 750 species of birds have been recorded in Japan, these comprising regularly occurring native species, introduced species, and a large number of rare migrants or accidentals.

The birds of Japan (along with those of Sakhalin, the Kuril Islands, Korea and Taiwan) have much in common with the Asian continental avifauna. The Japanese archipelago, once the mountainous rim of the Asian continent itself, was subsequently separated by the opening of the Sea of Japan basin. Certain Siberian avian species, adapted to tundra, taiga and subarctic conifer forest, reach Japan as these essentially northern birds penetrate south on high mountains in central Honshū. There are also a few Philippine and Malayan elements. While Japan inevitably shares much of its avifauna with the adjacent islands, peninsulas and the nearby continent, a proportion of its species are endemic – found nowhere else in the world.

Japan has been isolated from the Asian continent for long enough in some cases for unique forms to have evolved in that country, and the further isolation of smaller archipelagos within Japan has been sufficient for island endemics to have evolved there, too. Thus, there are endemic bird species occurring on the four main islands of Japan as a group (Hokkaidō, Honshū, Shikoku and Kyūshū), and others restricted just to the Izu Islands or the Nansei Shotō, or even to just parts of those island chains.

A scientific understanding of the differences between the birds and other animals of Japan and the Asian continent and among the various parts of Japan began during the latter half of the 19th century. An understanding of the early separation of Hokkaidō from Honshū by the Tsugaru Strait allowed the recognition that Hokkaidō was strongly influenced by the Siberian region farther north, whereas Honshū was most conspicuously influenced from the south. The Tsugaru Strait forms a crucial divide through which passes Blakiston's Line (see *p. 11*). The second most important zoogeographical division is that between the Palaearctic Region and the Oriental Region, which falls between the islands of the southern Nansei Shotō. Although generally impoverished in biological terms relative to the main islands, the Nansei

Steller's Eagle is a winter visitor to northern Japan [PP].

Shotō show many more southern characteristics than do other areas of Japan, yet are clearly influenced also from the north. The whole chain represents a transition zone between the two regions.

Japan's avifauna is, as described above, a complex mixture of north and south. This mixture, combining elements of the Palaearctic (Sino-Manchurian and Siberian) and Oriental/Indo-Malayan, even Himalayan, Regions, with the addition of a number of endemic species, means that Japan's avifauna is a complex and fascinating Asian cross-section. This mixture is emphasized further by the seasons. In winter, Japan hosts many northern species, particularly waterfowl, raptors and many woodland passerines. In summer, Japan sees birds arriving from Southeast Asia, from the Philippines and from as far as Australasia. Summer visitors and year-round residents make up a breeding avifauna of more than 250 species. The birds of Japan can be separated into six types: residents present throughout the year, summer visitors, winter visitors, passage migrants[1], wanderers[2], and accidentals[3].

Despite the great length of the Japanese archipelago, and its wide range of habitats and climates, a number of species are resident throughout a large part or even all of this range. These include Eastern Spot-billed Duck, Ural Owl, Varied Tit, Japanese White-eye, Oriental Greenfinch and Japanese Crow. Others are resident but occur only in certain parts of Japan. Some of these are endemic; two examples are Copper Pheasant and Japanese Woodpecker, both of which are absent from Hokkaidō and the Nansei Shotō.

Many birds occur in Japan only as summer visitors, and spend their winters in southeastern China, Malaysia or elsewhere in Southeast Asia, including the Philippines and Indonesia. Summer visitors include Eastern Cattle Egret and Intermediate Egret [NT][4], Grey-faced Buzzard, Northern Boobook, Siberian Blue Robin, Japanese Thrush, Blue-and-white Flycatcher, Yellow Bunting and Chestnut-cheeked Starling.

A wide range of species visits Japan only in winter. These migrate here from places that include the Kuril Islands, Kamchatka, southeastern and northeastern Siberia, Ussuriland, Sakhalin, northeastern China and even Mongolia. Winter visitors include Hooded Crane and White-naped Crane, the swans, geese and many ducks, Common Guillemot[5], Common Black-headed Gull, and passerines such as Eastern Rook, Daurian Jackdaw, Dusky Thrush and Pale Thrush, Brambling, and Elegant Bunting and Rustic Bunting.

Because Japan lies on the migratory route between more northerly breeding grounds and more southerly wintering areas, naturalists can see large numbers of a wide range of migrants passing through. In particular, large numbers of many species of shorebird and seabird visit each year, as do species such as Garganey and Grey-streaked Flycatcher.

Many resident bird species undertake short-distance seasonal movements within the country, either altitudinally, between regions, or between islands. There is a major autumnal southerly exodus of small passerines from Hokkaidō to Honshū, Shikoku and Kyūshū. Such movements are complemented by influxes of the same species from outside Japan, so that a species may be both a visitor and a seasonal wanderer. Typical wanderers include Japanese Accentor, Japanese White-eye, Meadow Bunting and Masked Bunting, Bull-headed Shrike and Oriental Greenfinch.

A large number of species from well outside their normal ranges have also been recorded as accidentals, with new species added to the Japanese list in most years. Individuals may be diverted from their normal course by storms or typhoons, or overshoot the ends of their normal routes. Species that occur irregularly or may have been observed only once or twice include Wilson's Phalarope, Franklin's

1 In this context, migrants are species that pass through Japan between breeding grounds farther north and wintering grounds farther south, but typically neither breed nor winter here.
2 Some species reach Japan occasionally, but not having followed clear migration patterns.
3 In ornithological terms, accidentals (sometimes referred to also as vagrants) are individual birds appearing far away from their normal range. They may be blown by storms, or be far off course because of inappropriate migration instincts.
4 Some occasionally spend the winter here, too.
5 Once it was an abundant breeding species on islands and islets around Hokkaidō. A very small number of pairs currently nests, but only on Teuri-tō.

Gull and Common Yellowthroat from North America, Eastern Grass Owl probably from Southeast Asia, Collared Kingfisher and White-breasted Wood Swallow from the Philippines or possibly Micronesia, Yellow-browed Bunting from continental East Asia and Fieldfare, Mistle Thrush and Wood Warbler from Eurasia.

The isolated Japanese islands are home to a number of resident endemic birds. Some are widespread throughout Japan, while others are restricted to a single island. Another group of species can be called endemic breeders, as they breed in Japan and nowhere else, but migrate out of the country for the winter. Finally, a third group (near-endemics) reflects the fact that far-east Asia is itself a centre for endemism; a number of species are found primarily in Japan, but they occur also on adjacent islands or the continental coast, such as Taiwan, the Korean Peninsula, Sakhalin, the Kuril Islands or the Sea of Okhotsk coast.

Birds endemic to Japan and adjacent territories include Japanese Murrelet, Okinawa Rail, Amami Woodcock, Green Pheasant and Copper Pheasant, Japanese Woodpecker, Pryer's Woodpecker and Owston's Woodpecker, Japanese Wagtail, Ryukyu Minivet, Japanese Accentor, Izu Robin, Ryukyu Robin and Okinawa Robin, Amami Thrush and Izu Thrush, Owston's Tit and Orii's Tit, Bonin Honeyeater and Lidth's Jay.

Endemic breeders include Short-tailed Albatross, Matsudaira's Storm Petrel, Japanese Night Heron, Latham's Snipe, Ijima's Leaf Warbler, Yellow Bunting and Chestnut-cheeked Starling. The near-endemic breeders include Temminck's Cormorant, Streaked Shearwater, Swinhoe's Storm Petrel, Black-tailed Gull, Ryukyu Scops Owl, Japanese Wood Pigeon, Japanese Pygmy Woodpecker, Brown-eared Bulbul, Japanese Robin, Brown-headed Thrush, Narcissus Flycatcher, Japanese Paradise Flycatcher, Varied Tit and Grey Bunting.

The endemic Copper Pheasant, here a female, consists of several regional subspecies [KT].

Japanese Night Heron is an endemic breeder in Japan but winters elsewhere [KuM].

Okinawa Rail is a single-island endemic breeder in Japan [MAB].

Regional Distribution of Japanese Birds

The topography of Japan, with numerous islands and major mountain ranges, presents many barriers to species' dispersal and interaction. The regional distribution of birds has for long been a subject of great interest.

The Japanese avifauna readily lends itself to regional subdivision into four major geographical areas: (1) Hokkaidō; (2) Central Japan (Honshū, Shikoku and Kyūshū); (3) the Nansei Shotō; and (4) the Izu, Ogasawara and Iwō Islands. As a result of the great distances separating some of these, a species may well be of different status in different regions; thus, both Common Redshank [VU] and Eastern Yellow

Wagtail are summer visitors to Hokkaidō, uncommon migrants in Central Japan, and winter visitors to the Nansei Shotō, while Bewick's Swan is a migrant in Hokkaidō and a winter visitor in Central Japan. As a result, general statements about the avifauna of Japan must always be made in the context of the particular region. Having said that, some species are found virtually throughout all or most of Japan, including Oriental Turtle Dove, Japanese Pygmy Woodpecker, Japanese Bush Warbler, Japanese Tit, Japanese White-eye, Japanese Crow, Eurasian Tree Sparrow and Oriental Greenfinch.

Hokkaidō

The avifauna of Hokkaidō is, with few exceptions, exclusively northern Palaearctic (the exceptions are the penetration northwards of Northern Boobook, Japanese White-eye and Brown-eared Bulbul), and bears a strong similarity to that of regions to the north and northwest, such as Sakhalin and Ussuriland.

Hokkaidō is separated only by the narrow Sōya Strait from Sakhalin, just to the north, and shares with it many species. While Sakhalin is very similar to Siberia in its avifauna, however, Hokkaidō lacks a number of true Siberian species. Even so, Hokkaidō forms the southern limit for a number of northern cold-adapted species which breed or have bred no farther south in Japan than Hokkaidō. Such species include Red-necked Grebe, Red-faced Cormorant, White-tailed Eagle, Falcated Duck, Gadwall, Northern Shoveler, Tufted Duck, Smew, Common Merganser, Hazel Grouse, Red-crowned Crane, Common Redshank, Common Guillemot and Spectacled Guillemot, Tufted Puffin, Blakiston's Fish Owl, Grey-headed Woodpecker, Lesser Spotted Woodpecker and Eurasian Three-toed Woodpecker, Sand Martin, Sakhalin Grasshopper Warbler, Marsh Tit and Pine Grosbeak.

Other birds breed almost exclusively in Hokkaidō, but have bred in or have colonized northern Honshū. These include Leach's Storm Petrel, Northern Hobby, Slaty-backed Gull, Eurasian Wryneck, Siberian Rubythroat and Long-tailed Rosefinch.

Hokkaidō is also the northern limit for a range of southern, warm adapted species such as Japanese Grosbeak, Japanese Wagtail, Japanese Accentor, Bull-headed Shrike and Blue-and-white Flycatcher, all of which occur as far north as Hokkaidō, but do not breed in Sakhalin. The southern islands of the Kuril chain, including Kunashiri, Etorofu (Iturup) and Urup, form an extension of the Hokkaidō avifauna. Hokkaidō is also on the migration route of, for example, Greater White-fronted Goose, Taiga Bean Goose [NT] and Bewick's Swan, all of which migrate through Hokkaidō to winter farther south in Central Japan.

Central Japan – Honshū, Shikoku and Kyūshū

By far the majority of species on the Japanese list have been recorded from the central region of Japan consisting of Honshū, Shikoku and Kyūshū, which although predominantly Palaearctic is greatly influenced by the proximity of the Oriental Region. In fact, while in winter it is easy to accept the predominant Palaearctic influence, it is far more difficult to do so in summer, when the climate, the vegetation and the birds suggest an almost tropical setting. The central region receives the greatest number of migrants and accidentals, but also attracts certain species that breed neither farther north nor farther south in Japan, and is home to several endemics. This group of species breeding only in Central Japan includes Swinhoe's Storm Petrel and Band-rumped Storm Petrel, Striated Heron, Eastern Cattle Egret and Little Egret, Common Kestrel, Rock Ptarmigan, Copper Pheasant and Green Pheasant, Dollarbird, Japanese Green Woodpecker, Fairy Pitta, Grey-headed Lapwing, Greater Painted Snipe, Tiger Shrike [VU], Alpine Accentor, Marsh Grassbird, Japanese Reed Bunting and Yellow Bunting, Azure-winged Magpie and also, thanks to a successful captive-breeding and reintroduction programme, Crested Ibis and Oriental Stork. Two further endemics, Japanese Wagtail and Japanese Accentor, breed in Central Japan and in Hokkaidō, but not in the Nansei Shotō. Fairy Pitta and Dollarbird, both summer visitors, and Greater Painted Snipe, a local resident, are notable since they are of tropical origin yet occur north to central and even northern Honshū on occasion.

A number of species have colonized Japan from the south with only limited success, and are relatively more abundant in the warm temperate region of the south; examples include Japanese Night Heron,

Striated Heron, Pacific Reef Egret, Grey-faced Buzzard, House Swift and Japanese Paradise Flycatcher. Other colonists have spread rapidly, extending their range even to the cooler north. One, Red-rumped Swallow, has even bred in Hokkaidō on occasion.

The high-elevation habitats provided by the mountains of central Honshū enable some northern species, such as Spotted Nutcracker, to breed farther south in Japan than they might otherwise do. Conversely, the warm current along the Pacific coast allows other species of southern origin but which have not fully adapted to cooler regions, such as Pacific Reef Egret, Intermediate Egret and Greater Painted Snipe, to breed farther north than might otherwise be expected. There is also a relict species, Rock Ptarmigan, a small isolated population of which remains at high elevations in the Japan Alps – the result of climatic changes.

The Nansei Shotō

The Nansei Shotō, the island chain in Japan where the northern limit of the Oriental Region and the southern limit of the Palaearctic Region intergrade, has elements from both regions, making the avifauna and other fauna of the Nansei Shotō of great interest and importance.

A number of endemic birds have evolved in the Nansei Shotō. Some of these range throughout the island chain, such as Ryukyu Robin (which occurs from the Danjo Gunto islands west of Kyūshū in the north to the Yaeyama Islands in the south) and Ryukyu Minivet (which not only occurs from Yaku-shima to the Yaeyama Islands, but has recently extended its range beyond the islands into Kyūshū and even Honshū). Amami Woodcock breeds on Amami Ōshima, and some winter south to Okinawa. Lidth's Jay, Owston's Woodpecker and Amami Thrush are restricted to just Amami Ōshima. Pryer's Woodpecker, Okinawa Rail and Okinawa Robin are restricted to northern Okinawa; while Ryukyu Scops Owl is virtually an endemic, occurring commonly from Amami Ōshima to Yonaguni-jima, then only on Lanyu Island off southeast Taiwan, and on the tiny Batanes and Babuyan Islands of the north Philippines.

In addition, many southern species breed in the Nansei Shotō but no farther north. These include Cinnamon Bittern, Malayan Night Heron, Purple Heron, Barred Buttonquail, Slaty-legged Crake [VU], White-breasted Waterhen, Ryukyu Green Pigeon, Pacific Swallow and Light-vented Bulbul. The islands thus form the northern limit for these species, yet at the same time they are the southern limit for other species, such as Japanese Pygmy Woodpecker.

A number of seabirds, such as Greater Crested Tern, Black-naped Tern, Roseate Tern, Sooty Tern and Brown Noddy, have important colonies among the Nansei Shotō. Moreover, in 1988, Short-tailed Albatross[1] was confirmed as breeding on the isolated Senkaku Islands. Compared with Central Japan and Hokkaidō, Nansei Shotō hosts relatively few winter and summer visitors, but instead these islands receive many migrants and accidentals and are the wintering grounds for small numbers of a wide range of shorebirds.

The Izu, Ogasawara and Iwō Islands

These isolated oceanic islands, particularly the Ogasawara Islands and Iwō Islands, have few resident species and few seasonal visitors, but they have attracted a long list of accidentals, and have a number of interesting breeding species such as: Short-tailed Albatross and Black-footed Albatross, Bonin Petrel, Matsudaira's Storm Petrel, Bulwer's Petrel, Greater Crested Tern and Black-naped Tern, Brown Noddy, Izu Thrush, Ijima's Leaf Warbler, Izu Robin, Owston's Tit and Bonin Honeyeater.

Migration

The East Asian–Australasian Flyway is one of the world's most important bird-migration routes linking points from northeastern Russia all the way to southeast Australia through the birds that travel between them. Conserving those birds means protecting not only their breeding grounds, often in the far north,

1 There is evidence to suggest that this population may be a separate, and as yet unnamed species.

but also the wintering grounds in the south and their vitally important refuelling stop-over locations along the way.

The shapes of present-day migration routes carry imprints of past coastlines prevalent during times of glaciation. At the height of ice ages, trans-ocean migration routes were shorter, whereas now when we are in an interglacial the distances that birds must cover are very much greater. On spring migrations some birds of the East Asian–Australasian Flyway in effect follow a fossilized route, while on their return in autumn they follow newly derived, more direct routes.

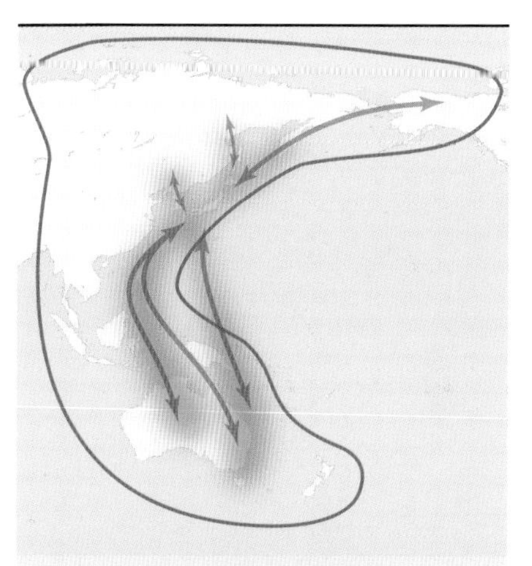

The East Asian–Australasian Flyway
The green boundary indicates the extent of the flyway and the red areas and arrows show the main routes and directions taken.

Bird-migration routes that include Japan have likely evolved in relation to long-term changes in the Asian continental coastline, and the elevation of mountain ranges now forming the spine of Japan. Initially, a major route is presumed to have extended along the then Asian continental coast, eventually becoming the Kuril Islands–Japanese Islands–Nansei Shotō–Taiwan route. Routes along the coasts of the newly formed Sea of Okhotsk depression led to an extra route down what is now Sakhalin, while farther south migration routes along the coasts of the Sea of Japan depression led to current routes from Asia to Japan via the Noto Peninsula and down the eastern side of peninsular Korea into southwestern Japan and Kyūshū. The East China Sea depression led to a further route extending along what is now the continental coast of the East China Sea, some birds crossing directly from Kyūshū to Taiwan, bypassing the Nansei Shotō.

Because of Japan's geographical position, mountainous terrain and climatic zones, the classification of birds into various categories is not an easy task. Nevertheless, some are clearly residents, winter visitors, summer visitors, passage migrants or accidentals, while some make vertical migrations, and others are resident within the country yet make local migrations. In some cases, such as Oriental Greenfinch, certain subspecies breed in Japan but move south and are replaced in winter by subspecies from farther north.

In short, 'Japanese birds' come from as far north as the arctic coasts and tundra of Siberia and even Alaska and from as far east as continental USA, and they range as far west as Myanmar, southwest to the Malay Peninsula, south to the Philippines, to New Guinea and even to Australia as far south as Tasmania and, in the case of South Polar Skua, Antarctica.

The study of migration is advanced mainly by the bird-ringing scheme organized in Japan through the Yamashina Institute for Ornithology, while direct observation complements scientific bird-ringing. For example, the migration routes of raptors, including Eastern Buzzards and various sparrowhawks, have been shown by direct observation to run across the straits from southwest Hokkaidō to northwest Honshū, and a major migration route of Chinese Sparrowhawk through the Nansei Shotō was discovered. Although the study of migration through Japan is well advanced, much remains to be learned.

South Polar Skua is the southernmost Southern Hemisphere species to visit Japan [JoH].

Bibliography and Recommended Reading

Abe, H., Ishii, N., Itoo, T., Kaneko, Y., Maeda, K., Miura, S. & Yoneda, M. 2008. *A Guide to the Mammals of Japan*. 2nd Revised Edition. Tōkai University Press; Japan.

Ando, K. 1993. Kodonospora, a new staurosporous hyphomycete genus from Japan. *Mycological Research* 97(4): 506–508.

Aoki, K., Tamaki, I., Nakao, K., Ueno, S., Kamijo, T., Setoguchi, H., Murakami, N., Kato, M. & Tsumura, Y. 2019. Approximate Bayesian computation analysis of EST-associated microsatellites indicates that the broadleaved evergreen tree *Castanopsis sieboldii* survived the Last Glacial Maximum in multiple refugia in Japan. *Heredity* 122(3): 326–340.

Ara, K., Mikami, K. & Mikami, O. 2019. Nut-dropping behavior of Carrion Crows in Hakodate City, Hokkaidō. *Japanese Journal of Ornithology* 68(1): 43–51.

Beolens, B., Watkins, M. & Grayson, M. 2011. *The Eponym Dictionary of Reptiles*. Johns Hopkins University Press; Baltimore.

Brasor, P. 2018. Japan's bears are widely vilified and little understood. *The Japan Times* 18 August 2018.

Brazil, M. A. 1987. *A Birdwatcher's Guide to Japan*. Kodansha International; Tōkyō.

Brazil, M. A. 1991. *The Birds of Japan*. Helm; London.

Brazil, M. A. 1999. Islands of diversity and divergence. *The Japan Times* 4 August 1999.

Brazil, M. A. 2002. Common Raven *Corvus corax* at play; records from Japan. *Ornithological Science* 1: 150–152.

Brazil, M. A. 2009. *Helm Field Guides: Birds of East Asia*. A&C Black; London.

Brazil, M. A. 2018. *Helm Field Guides: Birds of Japan*. Bloomsbury; London.

Brazil, M. A. 2018. *The Nature of Japan*. 2nd Edition. Japan Nature Guides; Sapporo.

Burgin, C. J., Colella, J. P. & Upham, N. S. 2018. How many species of mammals are there? *Journal of Mammalogy* 99(1): 1–14.

Chakraborty, A. & Jones, T. E. 2018. Mount Fuji: The Volcano, the Heritage, and the Mountain. Pp 167–175 in: Chakraborty, A. *et al.* (eds) 2018, *Natural Heritage of Japan. Geological, Geomorphological, and Ecological Aspects*. Springer International; Cham.

Chakraborty, A., Mokudai, K., Cooper, M., Watanabe, M. & Chakraborty, S. (eds) 2018. *Natural Heritage of Japan. Geological, Geomorphological, and Ecological Aspects*. Springer International; Cham.

Chakraborty, S. 2018. The Interface of Geology, Ecology, and Society: The Case of Aso Volcanic Landscape. Pp 117–130 in: Chakraborty, A. *et al.* (eds) 2018, *Natural Heritage of Japan. Geological, Geomorphological, and Ecological Aspects*. Springer International; Cham.

Chiba, A., Uchida, H. & Imanishi, S. 2014. Physical traits of male Japanese bush warblers (*Cettia diphone*) in summer and winter: hyperactive aspects of the vocal system and leg muscles in summer males. *Zoological Science* 31(11): 741–747.

Chiba, S. 1999. Accelerated Evolution of Land Snails *Mandarina* in the Oceanic Bonin Islands: Evidence from Mitochondrial DNA Sequences. *Evolution* 53(2): 460–471.

Chytrý, M., Horsák, M., Danihelka, J., Ermakov, N., German, D. A., Hájek, M., Hájková, P., Kočí, M., Kubešová, S., Lustyk, P., Nekola, J. C., Pavelková, V., Zdenka, R., Preislerová, Z., Resl, P. & Valachovič, M. 2018. A modern analogue of the Pleistocene steppe-tundra ecosystem in southern Siberia. *Boreas* 48(1): 36–56.

Cohen, K. M., Finney, S. C., Gibbard, P. L. & Fan, J.-X. 2013. The ICS International Chronostratigraphic Chart. *Episodes* 36: 199–204 v2019/05 update.

Cooper, M. 2018. The Lake Akan Area: A Future Geopark? Pp 153–159 in: Chakraborty, A. *et al.* (eds) 2018, *Natural Heritage of Japan. Geological, Geomorphological, and Ecological Aspects*. Springer International; Cham.

Davies, N. 2015. *Cuckoo: Cheating by Nature*. Bloomsbury Publishing; London.

de Kort, S. R. & Clayton, N. S. 2006. An evolutionary perspective on caching by corvids. *Proc. Biol. Sci.* 273(1585): 417–423.

Ecological Society of Japan (eds). 2002. *Handbook of Alien Species in Japan*. Chijinshokan; Tōkyō.

Endo, H. & Tsuchiya, K. 2006. A new species of Ryukyu spiny rat, Tokudaia (Muridae: Rodentia), from Tokunoshima Island, Kagoshima Prefecture, Japan. *Mammal Study* 31(1): 47–57.

Frazer, J. G. 1890. *The Golden Bough: A Study in Comparative Religion*. Macmillan Publishers; New York.

Fujita, K., Arakaki, T., Denda, T., Hidaka, M., Hirose, E. & Reimer, J. D. (eds) 2015. *Nature in the Ryukyu archipelago: coral reefs, biodiversity, and the natural environment*. International Research Hub Project for Climate Change and Coral Reef/Island Dynamics, Faculty of Science, University of the Ryukyus.

Fukada, K. 1964. *Nihon Hyaku-meizan (100 Famous Japanese Mountains)*. Shinchōsha; Tōkyō.

Fukui, A. 1995. The role of the brown-eared bulbul *Hypsipetes amaurotis* as a seed dispersal agent. *Researches on Population Ecology* 37(2): 211–218.

Fukui, A. 2003. Relationship between seed retention time in bird's gut and fruit characteristics. *Ornithological Science* 2(1): 41–48.

Gadow, H. 1901. *Amphibia and Reptiles. Cambridge Natural History Vol. 8*. Macmillan; London.

Goodenough, A. E., Little, N., Carpenter, W. S. & Hart, A. G. 2017. Birds of a feather flock together: Insights into starling murmuration behaviour revealed using citizen science. *PLoS One* 12(6): e0179277.

Goris, R. C. & Maeda, N. 2004. *Guide to the Amphibians and Reptiles of Japan*. Krieger Publishing; Malabar.

Gray, D. 2015. *Simple Forms: Essays on Medieval English Popular Literature*. Oxford University Press; Oxford.

Harada, M. 1995. Minamata disease: methylmercury poisoning in Japan caused by environmental pollution. *Critical Reviews in Toxicology* 25(1): 1–24.

Harato, T. & Ozaki, K. 1993. Roosting Behavior of the Okinawa Rail. *Journal of the Yamashina Institute for Ornithology* 25(1): 40–53.

Hasebe, M. 2013. Characteristics of roofs utilized by Slaty-backed Gulls *Larus schistisagus* for nesting in coastal Northwest Hokkaidō, Japan. *Bird Research* 9: S13–S18.

Hasegawa, A., Nakajima, J. & Zhao, D. 2016. Deep seismic structure. In: Moreno, T., Wallis, S., Kojima, T. & Gibbons, W. (eds) *The Geology of Japan*. Geological Society; London.

Hashimoto, M. (ed.) 1991. *The Geology of Japan*. Terra Scientific Publishing; Tokyo.

Hearn, L. 1899. *In Ghostly Japan*. Tuttle; Tōkyō.

Hearn, L. 1904. Kwaidan, Stories and Studies of Strange Things. Tuttle; Tōkyō.

Heinrich, B. & Smolker, R. 1998. Play in common ravens (Corvus corax). Pp 27–44 in: Bekoff, M. & Byers, J. A. (eds) Animal Play: Evolutionary, Comparative, and Ecological Perspectives. Cambridge University Press; Cambridge.

Higashi, S., Osawa, A. & Kanagawa, K. (eds) 1993. Biodiversity and Ecology in the Northernmost Japan. Hokkaidō University Press; Sapporo.

Honda, M., Kurita, T., Toda, M. & Ota, H. 2014. Phylogenetic Relationships, Genetic Divergence, Historical Biogeography and Conservation of an Endangered Gecko, *Goniurosaurus kuroiwae* (Squamata: Eublepharidae), from the Central Ryukyus, Japan. *Zoological Science* 31(5): 309–321.

Imaizumi, Y. 1967. A new genus and species of cat from Iriomote, Ryukyu Islands. Journal of the Mammalogical Society of Japan 3: 74–105.

Ishii, S., Gomi, F., Sasayama, H. & Takano, T. (eds) 2008. Shosetsu Nihonshi. Yamakawa Shuppansha; Tōkyō.

Japan Butterfly Conservation Society (Suda, S., Nagahata, Y., Nakamura, Y., Hasegawa, T. & Yago, M.). 2017. *Field Guide to the Butterflies of Japan*. Seibundo-Shinkosha; Tōkyō.

Kaneko, Y. 2001. Morphological discrimination of the Ryukyu spiny rat (genus *Tokudaia*) between the islands of Okinawa and Amami Oshima, in the Ryukyu Islands, southern Japan. *Mammal Study* 26(1): 17–33.

Kasuya, T. 2017. Small Cetaceans of Japan: Exploitation and Biology. CRC Press; Boca Raton.

Kathiresan, K. & Bingham, B. L. 2001. Biology of Mangroves and Mangrove Ecosystems. *Advances in Marine Biology* 40: 81–251.

Kato, M. 1998. Unique leafmining habit in the bark beetle clade: A new tribe, genus, and species of Platypodidae (Coleoptera) found in the Bonin Islands. *Annals of the Entomological Society of America* 91(1): 71–80.

Kikuchi, N. & Mokudai, K. 2018. San'in-Kaigan UNESCO Global Geopark: Geology and Conservation of the Oriental White Stork. Pp 95–101 in: Chakraborty, A. *et al.* (eds) 2018, *Natural Heritage of Japan. Geological, Geomorphological, and Ecological Aspects*. Springer International; Cham.

Kitamoto, A. 2019. All Japan Active Volcano Map and Latest Eruption Warning Level. National Institute of Informatics: http://agora.ex.nii.ac.jp/eruption/volcano/. Accessed on 21 December 2019.

Kobayashi, Y., Kariya, T., Chishima, J., Fujii, K., Wada, K., Baba, S., Itoo, T., Nakaoka, T., Kawashima, M., Saito, S., Aoki, N., Hayama, S., Osa, Y., Osada, H., Niizuma, A., Suzuki, M., Uekane, Y., Hayashi, K., Kobayashi, M., Ohtaishi, N. & Sakurai, Y. 2014. Population trends of the Kuril harbour seal *Phoca vitulina stejnegeri* from 1974 to 2010 in southeastern Hokkaidō, Japan. *Endangered Species Research* 24: 61–72.

Korea National Arboretum. 2015. *English Names for Korean Native Plants*. Korea National Arboretum; Pocheon.

Kotaka, N. & Sawashi, Y. 2004. The Road-kill of the Okinawa Rail *Gallirallus okinawae*. *Journal of the Yamashina Institute for Ornithology* 35(2): 134–143.

Kron, TM. 1985. Japan's Salmon Culture Program and Coastal Salmon Fisheries. FRED report No 50. Alaska Department of Fish & Game Division of Fisheries Rehabilitation, Enhancement and Development.

Kubota, Y., Shiono, T. and Kusumoto, B. 2015. Role of climate and geohistorical factors in driving plant richness patterns and endemicity on the east Asian continental islands. *Ecography* 38: 639-648. https://doi.org/10.1111/ecog.00981

Kuhn, R. A., Ansorge, H., Godynicki, S. & Meyer, W. 2010. Hair density in the Eurasian otter *Lutra lutra* and the Sea otter *Enhydra lutris*. *Acta Theriologica* 55(3): 211–222.

Kuroda, A. & Tetsu, S. 2017. Vegetation zonation and distribution of threatened dune plant species along shoreline–inland gradients on sandy coasts in the eastern part of the San'in region, western Japan. *Vegetation Science* 34: 23–37.

Kushida, M. & Takahashi, K. 1987. *The Spirit of Japan in Nature*. Graphicsha; Tōkyō.

Livezey, B. C. 2003. Evolution of Flightlessness in Rails (Gruiformes: Rallidae): Phylogenetic, Ecomorphological, and Ontogenetic Perspectives. *Ornithological Monographs* 53: 1–654.

Martín-Torrijos, L., Kawai, T., Makkonen, J., Jussila, J., Kokko, H. & Diéguez-Uribeondo, J. 2018. Crayfish plague in Japan: A real threat to the endemic *Cambaroides japonicus*. *PLoS ONE* 13(4): e0195353. https://doi.org/10.1371/journal.pone.0195353.

Matsui, M. 1991. Original Description of the Brown Frog from Hokkaidō, Japan (Genus *Rana*). *Japanese Journal of Herpetology* 14(2): 63–78.

Matsui, M., Misawa, Y., Nishikawa, K. & Shimada, T. 2017. A new species of lentic breeding salamander (Amphibia, Caudata) from central Japan. *Current Herpetology* 36: 116–126.

Matsuo, T. 1983. Vegetation and climate during the Last Glacial Maximum in Japan. *Quaternary Research* 19: 212–235.

Matsutani, M. 2012. First glaciers in Japan recognized. *The Japan Times* 6 April 2012.
McKay, B. D., Mays, H. L. Jr, Yao, C.T., Wan, D., Higuchi, H. & Nishiumi, I. 2014. Incorporating Color into Integrative Taxonomy: Analysis of the Varied Tit (*Sittiparus varius*) Complex in East Asia. *Systematic Biology* 63(4): 505–517.
Meteorological Agency. 2019. List of Active Volcanoes in Japan. https://www.data.jma.go.jp/svd/vois/data/tokyo/STOCK/souran_eng/intro/volcano_list.pdf. Accessed 21 December 2019.
Milewski, K. P. 2013. The Flight of the Nightingale: From Romans to Romantics. *Vanderbilt Undergraduate Research Journal* 9: 1–9.
Minagawa, M. 2000. Japan: mangrove areas and their utilization. In: Primavera, J. H., Garcia, L. M. B., Castaños, M. T. & Surtida, M. B. (eds) Mangrove-Friendly Aquaculture: Proceedings of the Workshop on Mangrove-Friendly Aquaculture organized by the SEAFDEC Aquaculture Department, January 11–15, 1999, Iloilo City, Philippines. Southeast Asian Fisheries Development Center; Iloilo, Philippines.
Minato, M. 1977. *Japan and its Nature*. Heibonsha; Tōkyō.
Minato, M., Gorai, M. & Hunahashi, M. (eds) 1965. *The geologic developments of the Japanese Islands*. Tsukiji Shokan; Tokyo.
Ministry of the Environment. 2000. *Threatened Wildlife of Japan: Red Data Book Reptilia/Amphibia* 2nd edition. Japan Wildlife Research Center; Tōkyō.
Ministry of the Environment. 2002. *Threatened Wildlife of Japan: Red Data Book Aves* 2nd edition. Japan Wildlife Research Center; Tōkyō.
Ministry of the Environment. 2008. Beautiful Nature of Four Seasons National Parks of Japan & Japan's Strategy for a Sustainable Society. Yama-Kei; Tōkyō.
Ministry of the Environment. 2016. Natural Beauty of the Four Seasons National Parks of Japan. Yama-Kei; Tōkyō.
Ministry of the Environment. 2017. Result of research on traces of River Otter on Tsushima. http://www.env.go.jp/press/104655.html. Accessed on 21 December 2019.
Miura, W. 2018. Three snow patches in central Japan certified as glaciers. *Asahi Shimbun* 7 February 2018.
Miyazaki, K., Ozaki, M., Saito, M. & Toshimitsu, S. 2016. The Kyūshū–Ryukyu Arc. In: Moreno, T., Wallis, S., Kojima, T. & Gibbons, W. (eds) *The Geology of Japan*. Geological Society; London.
Moreno, T., Wallis, S., Kojima, T. & Gibbons, W. (eds) 2016. *The Geology of Japan*. Geological Society; London.
Motokawa, M. & Kajihara, H. (eds) 2017. *Species Diversity of Animals in Japan*. Springer; New York.
Nakada, S. 2018. Volcanic Archipelago: Volcanism as a Geoheritage Characteristic of Japan. Pp 19–28 in: Chakraborty, A. *et al.* (eds) 2018, *Natural Heritage of Japan. Geological, Geomorphological, and Ecological Aspects*. Springer International; Cham.
Nakada, S., Yamamoto, T. & Maeno, F. 2016. Miocene–Holocene volcanism. In: Moreno, T., Wallis, S., Kojima, T. & Gibbons, W. (eds) *The Geology of Japan*. Geological Society; London.
Nakamura, Y. 2011. Conservation of butterflies in Japan: Status, actions and strategy. *Journal of Insect Conservation* 15(1): 5–22.
Newhall, C. G. & Self, S. 1982. The Volcanic Explosivity Index (VEI): An Estimate of Explosive Magnitude for Historical Volcanism. *Journal of Geophysical Research* 87(C2): 1231–1238.
Nihei, T. 2018. The Regional Geography of Japan. Hokkaido University Press; Sapporo.
Nihonsan Ari-rui Database Group. 2003. *Nihonsan Ari-rui Zenshu Zukan* [*Ants of Japan*]. Gakken; Tōkyō.
Nishida, Y. & Takagi, M. 2019. Male bull-headed shrikes use food caches to improve their condition-dependent song performance and pairing success. *Animal Behaviour* 152: 29–37.
Numata, M. (ed.) 1974. *The Flora and Vegetation of Japan*. Kodansha; Tōkyō.
Obara, H., Maejima, Y., Kohyama, K., Ohkura, T. & Takata, Y. 2015. Outline of the Comprehensive Soil Classification System of Japan – First Approximation. *Japan Agricultural Research Quarterly* 49(3): 217–226.
Ohdachi, S. & Aoi, T. 1987. Food habits of brown bears in Hokkaidō, Japan. *International Conference of Bears Research and Management* 7: 215–220.
Ohdachi, S. D., Ishibashi, Y., Iwasa, M. A., Fukui, D. & Saitoh, T. 2015. *The Wild Mammals of Japan*. Second Edition. Shoukadoh Book Sellers and the Mammal Society of Japan; Kyoto.
Ohwi, J. 1965. *Flora of Japan: A combined, much revised, and extended translation* (eds Meyer, F. G. & Walker, E. H.). Smithsonian Institution; Washington DC.
Ono, T. 2019. *Kita-arupusu Karamatsuzawa o hyoga ni nintei kokunai nanbanme*. 7 October 2019. Asahi Shimbun.
Ono, Y. 1990. The Northern Landbridge of Japan. *The Quaternary Research* 29(3): 183–192.
Ono, Y. 1991. Glacial and Periglacial Paleoenvironments in the Japanese Islands. *The Quaternary Research* 30(2): 203–211.
Ozaki, K. 2009. Morphological differences of sex and age in the Okinawa Rail *Gallirallus okinawae*. *Ornithological Science* 8(2): 117–124.
Ozono, A., Kawashima, I. & Futahashi, R. 2017. *Nihon no Tombo* [*Dragonflies of Japan*]. Bun-ichi Shuppan; Tōkyō.
Ratcliffe, D. 1997. *The Raven*. T. & A. D. Poyser; London.
Rozenstein, O. & Adamowski, J. 2017. Linking Spaceborne and Ground Observations of Autumn Foliage Senescence in Southern Québec, Canada. *Remote Sensing* 9(6): 630. https://doi.org/10.3390/rs9060630. Accessed 21 December 2019.
Sato, J. J. 2017. A Review of the Processes of Mammalian Faunal Assembly in Japan: Insights from Molecular Phylogenetics. In: Motokawa, M. & Kajihara, H. (eds) *Species Diversity of Animals in Japan. Diversity and Commonality in Animals*. Springer; Tōkyō.
Sato, Y. 2017. The Future of Urban Brown Bear Management in Sapporo, Hokkaidō, Japan: a Review. *Mammal Study* 42(1): 17–30.

Saito, Y., Ikehara, K. & Tamura, T. 2016. Coastal geology and oceanography. In: Moreno, T., Wallis, S., Kojima, T. & Gibbons, W. (eds) *The Geology of Japan*. Geological Society; London.

Senda, M. 1992. Japan's traditional view of nature and interpretation of landscape. *GeoJournal* 26(2): 129–134.

Shakhovskoy, I. 2018. Specific features of distribution in the World Ocean of some flying fishes of the genera *Exocoetus*, *Hirundichthys* and *Cypselurus* (Exocoetidae). *FishTaxa* 3(4): 40–80.

Shibasaki, S. 2018. Yakushima Island: Landscape History, World Heritage Designation, and Conservation Status for Local Society. Pp 73–83 in: Chakraborty, A. *et al.* (eds) 2018, *Natural Heritage of Japan. Geological, Geomorphological, and Ecological Aspects*. Springer International; Cham.

Snow, H. J. 1910. *In forbidden seas: Recollections of sea-otter hunting in the Kurils*. E. Arnold; London.

Stern, H. P. 1976. *Birds, Beasts, Blossoms, and Bugs. The Nature of Japan*. Harry N Abrams; New York.

Sugita, N., Inaba, M. & Ueda, K. 2009. Roosting Pattern and Reproductive Cycle of Bonin Flying Foxes (*Pteropus pselaphon*). *Journal of Mammalogy* 90(1): 195–202.

Sutherland, M. & Britton, D. 1980. *National Parks of Japan*. Kodansha International; Tōkyō.

Taira, A., Ohara, Y., Wallis, S. R., Ishiwatari, A. & Iryu, Y. 2016. Geological evolution of Japan: an overview. In: Moreno, T., Wallis, S., Kojima, T. & Gibbons, W. (eds) *The Geology of Japan*. Geological Society; London.

Takagi, M. 2006. Bull-headed Shrike *Mozu Lanius bucephalus*. *Bird Research News* 3(6) 2006.6.19.

Takahashi, A., Kuroki, M., Niizuma, Y., Kato, A., Saitoh, S. & Watanuki, Y. 2001. Importance of the Japanese anchovy (*Engraulis japonicus*) to breeding rhinoceros auklets (*Cerorhinca monocerata*) on Teuri Island, Sea of Japan. *Marine Biology* 139(2): 361–371.

Takanose, Y. & Kamitani, T. 2003. Fruiting of fleshy-fruited plants and abundance of frugivorous birds: Phenological correspondence in a temperate forest in central Japan. Ornithological Science 2(1): 23–32.

Tatsumi, Y., Tamura, Y., Nichols, A. R. L., Ishizuka, O., Takahashi, N. & Tani, K. 2016. Izu–Bonin Arc. In: Moreno, T., Wallis, S., Kojima, T. & Gibbons, W. (eds) *The Geology of Japan*. Geological Society; London.

Teikoku-shoin (ed.). 2009. Discovering Japan: A New Regional Geography. Teikoku-shoin; Tōkyō.

Thoresen, A. C. 1983. Diurnal Activity and Social Displays of Rhinoceros Auklets on Teuri Island, Japan. *The Condor* 85(3): 373–375.

Toda, S. 2016. Crustal earthquakes. In: Moreno, T., Wallis, S., Kojima, T. & Gibbons, W. (eds) *The Geology of Japan*. Geological Society; London.

Tokita, M., Hasegawa, Y., Yano, W. & Tsuji, H. 2018. Characterization of the Adaptive Morphology of Japanese Stream Toad (*Bufo torrenticola*) Using Geometric Morphometrics. *Zoological Science* 35(1): 99–108.

Tominaga, A., Matsui, M. & Nishikawa, K. 2019. Two new species of lotic breeding salamanders (Amphibia, Caudata, Hynobiidae) from western Japan. *Zootaxa* 4550(4): 525–544.

Tsukada, M. 1983. Vegetation and climate during the Last Glacial Maximum in Japan. *Quaternary Research* 19: 212–235.

Tsukada, M. 1985. Map of Vegetation during the Last Glacial Maximum in Japan. *Quaternary Research* 23(3): 369–381.

Uchiyama, R., Maeda, N., Numata, K. & Seki, S. 2002. *A Photographic Guide: Amphibians and Reptiles in Japan*. Heibonsha; Tōkyō.

Ueda, H. 2016. Hokkaidō. In: Moreno, T., Wallis, S., Kojima, T. & Gibbons, W. (eds) *The Geology of Japan*. Geological Society; London.

van der Geer, A., Lyras, G., de Vos, J. & Dermitzakis, M. 2011. *Evolution of Island Mammals: Adaptation and Extinction of Placental Mammals on Islands*. John Wiley & Sons; New Jersey.

Veronese, G. 2017. Hunting Japanese Azaleas. http://giulioveronese.com/hunting-japanese-azaleas/. Accessed 5 June 2019.

Wakita, K. 2018. Geology of the Japanese Islands: An Outline. Pp 9–17 in: Chakraborty, A. *et al.* (eds) 2018, *Natural Heritage of Japan. Geological, Geomorphological, and Ecological Aspects*. Springer International; Cham.

Walker, B. L. 2005. *The Lost Wolves of Japan*. University of Washington Press; Seattle.

Washitani, I. 2004. *Invasive Alien Species Problems in Japan: an Introductory Ecological Essay. Global Environmental Research* 8(1): 1–11.

Yamamoto, T., Takahashi, A., Katsumata, N., Sato, K. & Trathan, P. N. 2010. At-Sea Distribution and Behavior of Streaked Shearwaters (*Calonectris leucomelas*) During the Nonbreeding Period. *The Auk* 127(4): 871–881.

Yamashina, Y. & Mano, T. 1981. A New Species of Rail from Okinawa Island. *Journal of the Yamashina Institute for Ornithology* 13(3): 147–152.

Yashima, K. & Miyauchi, T. 1990. The Tsugaru Land Bridge Problem Related to Quaternary Coastal Tectonics, Northeast Japan. *The Quaternary Research* 29(3): 267–275.

Yoshida, M. 2018. Ogasawara Islands World Heritage Area: An Outstanding Ecological Heritage. Pp 61–72 in: Chakraborty, A. *et al.* (eds) 2018, *Natural Heritage of Japan. Geological, Geomorphological, and Ecological Aspects*. Springer International; Cham.

Yoshida, T. (ed.) 1975. *An Outline of the Geology of Japan, 3rd edition*. Geological Survey of Japan; Kawasaki.

Zhdanova, O. L., Kuzin, A. E., Skaletskaya, E. I. & Frisman, E. Y. 2017. Why the population of the northern fur seals (*Callorhinus ursinus*) of Tyuleniy Island does not recover following the harvest ban: Analysis of 56 years of observation data. *Ecological Modelling* 363: 57–67.

Acknowledgements

In one sense writing a book of this kind is a prolonged and solitary endeavour, yet on nearing completion it becomes very much a team effort. I thank my team of friends and colleagues who have given so generously of their time to aid me by reading and commenting on early drafts of the manuscript: Dr Deborah Ablin, Dr Mark Carmody, Ethel Cebra, Nigel Coldicott, Jenny & Terry Cloudman, Chris Cook, Richard Cook, Chris Harbard, Jeremy Hogarth, Dr Stefan Hotes, Kurosawa Nobumichi and Kurosawa Yūko, Angus Macindoe, Yann Muzika, Dr Tom Sharpe, Rich Pagen, Bob Quaccia, Dr Pepper Trail, John Williams and Jim Wilson. I am deeply grateful to Dr Andrew Clarke, Dr Andrew Davis and Dr Paul Green, each of whom gave substantial help during the final stages of this project. Each has done his or her utmost to help me in this challenge, to eradicate errors of fact, and to ensure that my subject matter is both readable and accessible. Any mistakes that remain are entirely my own responsibility.

I thank the team at Princeton University Press and **WILD***Guides*, Robert Kirk, Andy Swash, Rob Still and David Christie, for supporting this project and helping me to realize my dream in producing *The Natural History of an Asian Archipelago*. I especially thank Rob Still for designing and creating the many maps and illustrations that greatly enhance the text.

Although many of the photographs are from my own collection [MAB], numerous other photographers kindly made their images available for this project. So many wonderful images were offered but, with limited space available only a selection could be included. Each image is credited with the initials of the photographer [indicated after the relevant names in this list]. In particular I thank: Abashiri Tourism Association [ABTA], Akan–Mashū National Park, Aida Junichi[1] [AJ], Akahira Kaoru [AK], Aomori Tourist Association [AOTA], Arikawa Masako, Aso City Tourism Association, Aso Geopark Promotion Council [AGPC], Mayumi Brazil [MB], Terry Cloudman [TC], Collection Nationaal Museum van Wereldculturen [CNMW], Chris Cook [CC], Richard Cook [RC], Creative Commons [CCO], Mark Curley [MC], Neil Davidson [ND], Vladimir Dinets [VD], Jack Donachy [JD], Gustavianum Uppsala University Museum [GUUM], Sue Flood [SF], Fukuda Noriko, Fukuda Toshiji [FT], Fukuyama Ryobu [FR], Gifu Prefecture Tourism Federation [GPTF], Gojiraiwa Kankō [GOKA], Hagiwara Toshio [HT], Jon Hall [JH], Mark Hamilton [MH], Marijke Jensen [MJ], Herb Lingl Aerial Archives, Hiroshima Prefecture [HIPR], John Holmes [JoH], Iijima Masahiro [IiM], Imai Mitsuo [ImM], Itō Akihiro [IA], Itō Yoshihiro [IY], Kamada Sachiko [KS], Kanouchi Takuya [KT], Kantō Regional Environment Office [KREO], Katase Shinobu [KaS], Kasaoka City Horseshoe Crab Museum [KHCM], Kobayashi Masahiro, Kōbe City [KOCI], Komazawa Masaki [KoM], Kudaka Masakazu [KuM], Kurasawa Kouta, Kurasawa Miho [KM], Kyōto Prefecture, Lake Mashū Tourism Association [LMTA], Dr Matsubara Hajime [MaH], Matsuoka Ratsunari [MR], Chris McCooey [ChM], Shawn Miller [SM], Ministry of the Environment [MOEN], Ministry of the Environment Iriomote Wildlife Conservation Center [MEIW], Ministry of the Environment Nasu Ranger Station [MENR], Ministry of the Environment Ogasawara Islands Nature Information Center, Miyazaki Manabu [MM], Mizutani Hiroyuki [MiH], Yann Muzika [YM], Nakagawa Ruriko, Nakamura Sayaka [NS], Natural History Museum Botanic Garden Hokkaido University [NHHU], Naturalis Biodiversity Center Leiden [NBCL], Nihon-zaru Field Station [NZFS], Nikkō City Tourism Association [NCTA], Nishizawa Bungo [NB], Niwaki Zukan Uekipedia [NZUE], Nojiriko Naumann Elephant Museum [NNEM], Ogasawara Village Office [OGVO], Ogasawara Village Tourism Bureau [OVTB], Ohno Mutsumi [OM], Okinawa Convention & Visitors Bureau [OCVB], Okiyama Makoto Miyake-jima Tourist Association [OMTA], Osawa Toshihiro [OT], Rich Pagen, Peter Porazzo [PP], Shutterstock [SHUT], Stuart Price [SP], PublicdomainQ [PUDQ], Tokachi Tourism Federation [TOTF], Tui de Roy [TR], Sado Kankō Photo [SAKP], Saga Prefectural Tourism Federation [SPTF], Shibetsu Salmon Science Museum, Shimada Tadashi [ST], Siebold-Archiv Burg Brandenstein [SABB], Sugawara Takanori [SuT], Tabata Osamu [TO], Tagi Koji [TK], Dr Takagi Masaoki [TaM], Dr Eric Sohn Joo Tan [ET], Tateyama Caldera Sabō Museum [TCSM], Terasawa Takaki [TT], Tochigi

1 Japanese names are given in the typical Japanese way, with the family name preceding the given name.

Prefecture and Local Products Association, Dr Tominaga Atsushi [TA], Tourism Oita [TOOI], Tsuneda Mamoru [TsM], Umezawa Shun [US], Unsplash [UNSP], Wakasa Masanobu [WaM], Wada Masahiro [WM], Dr Wakana Isamu, Wakayama University Faculty of Education and Kuwabata Eibun [WUFE], John Williams [JW], Rob Williams [RW], the late John Wright [JoW], the late Yabuuchi Masayuki [YaM], Yabuuchi Ryuta, Yamamoto Sumio [YS], and Zamami Whale Watching Association [ZWWA].

In addition to my team of readers and the numerous photographers who have contributed, many others have kindly offered assistance in tracing information or images. Here I thank Wilhelm Graf Adelmann (Siebold-Archiv Burg Brandenstein), Ingeborg Eggink (National Museum van Wereldculturen), Rebecca Flodin (Curator Gustavianum, Uppsala University Museum, Art Collections), Dr Matthi Forrer (Nagasaki Deshima Restauratie Project Adviseur), Kobayashi Yukiko, Kyōto Prefectural Government Office, Karien Lahaise (Naturalis Biodiversity Center), Seki Megumi (Nojiriko Naumann Elephant Museum), Dr Suzuki Hitoshi (Professor Hokkaidō University), and Tachibana Toshiki (owner of The Trout Inn).

Red-crowned Crane is an iconic Japanese species symbolising both longevity and happiness [MAB].

Index

This index includes the animals, plants, places, people, categories and concepts mentioned in the text. It also includes the scientific names (in *italics*) of species referred to, as English names may not be familiar to some readers. The index does not include information included on the maps and Illustrations.

Regular black text is used for species and general terms, **bold text** for geographical locations, and SMALL CAPITALS for people.

Page numbers in **bold text** highlight main sections in the book.

Page numbers in *italics* relate to the location of a photograph (in some cases there is also associated text on these pages).

Page numbers in regular text indicate other key references.

C

I

J

T

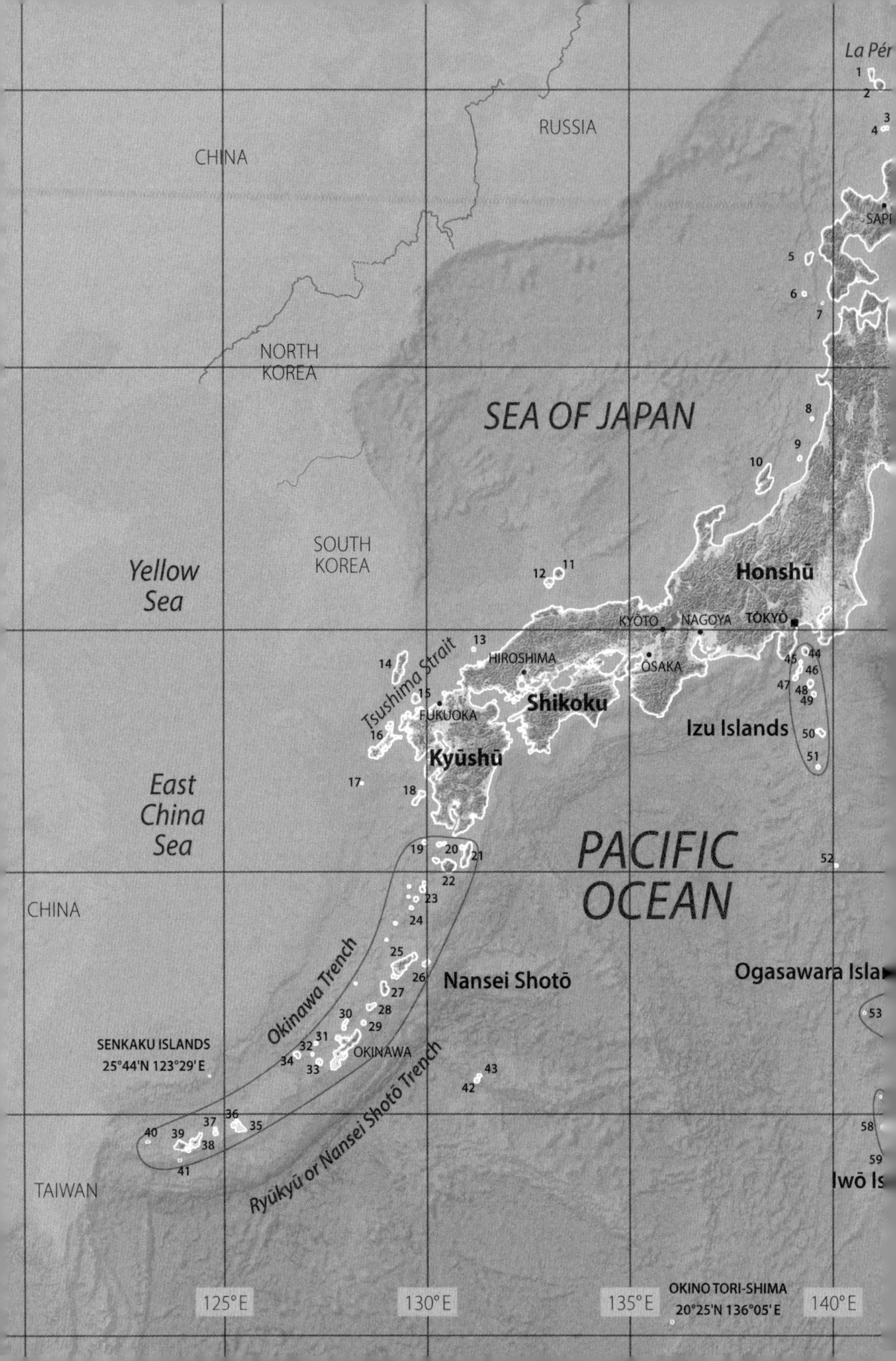

RUSSIA
CHINA
NORTH KOREA
SOUTH KOREA
SEA OF JAPAN
Yellow Sea
East China Sea
PACIFIC OCEAN
Honshū
Shikoku
Kyūshū
Tsushima Strait
KYOTO
NAGOYA
TOKYO
HIROSHIMA
OSAKA
FUKUOKA
Izu Islands
Nansei Shotō
Okinawa Trench
Ryūkyū or Nansei Shotō Trench
OKINAWA
SENKAKU ISLANDS
25°44'N 123°29'E
CHINA
TAIWAN
Ogasawara Isla
Iwō Is
OKINO TORI-SHIMA
20°25'N 136°05'E
125°E
130°E
135°E
140°E